LIE GROUPS: HISTORY, FRONTIERS AND APPLICATIONS

VOLUME XII

INVERSE SCATTERING PAPERS: 1955-1962

IRVIN KAY AND HARRY E. MOSES

MATH SCI PRESS

1982

Library of Congress Cataloging in Publication Data

Kay, Irvin W.
Inverse scattering papers, 1955-1962.

(Lie groups ; 12)
Includes bibliographical references.
1. Potential scattering--Addresses, essays, lectures.
I. Moses, Harry E. II. Title. III. Series.
QC794.6.S3K38 530.1'5 82-152
ISBN 0-915692-32-5 AACR2

MATH SCI PRESS

53 JORDAN ROAD
BROOKLINE, MASSACHUSETTS 02146

Printed in the United States of America

Dedicated to Peggy Kay and Barbara Moses

TABLE OF CONTENTS

INTRODUCTION

The present book consists of reproductions of our papers on inverse scattering and spectral theory written in the period 1955 to 1963. We have several reasons for putting out this book.

First, the series of papers that we published in Nuovo Cimento represented an attempt to investigate systematically the underlying principles of a general formalism for dealing with inverse spectral theory and scattering in the context of quantum mechanics and other areas. The fortuitous editorial policy of Nuovo Cimento that permitted the publication of papers written in much greater detail and at much greater length than is customary in modern journals was of inestimable importance to our pursuit of this goal. Moreover, it permitted us to adopt an expository style that provided the series with tutorial value. An heuristic approach allowed us to present the essential ideas in a form that should make them more accessible to readers not acquainted with the subject of inverse scattering than a completely rigorous treatment would have permitted.

Second, because of the generality of our approach, many of the ideas presented in those papers have not yet been fully exploited and therefore may, even at this date, provide a source of research topics.

Third, the formalism developed in our Nuovo Cimento series and the other papers in this volume has proven to be a useful tool for constructing explicit potentials which lead to exact solutions of the eigenvalue problem and illustrate operators with unusual spectral properties. For example, by this means one can obtain non-local potentials for which the spectral theory can be examined with ease.

Finally, the inverse problem for the one-dimensional Schrödinger equation and the particular case of the reflectionless potentials discussed in the present volume have provided the basis of the use of the inverse spectral transform method for obtaining soliton and other solutions of certain non-linear problems.

The remaining paragraphs of this introduction will outline the early history of the inverse problem insofar as it influenced the work presented in this book and will discuss briefly some of the more important results. A fuller discussion of the history of the inverse problem in a broader context is given by Newton in his introduction to Reference 1.

Our interest began shortly after Gel'fand and Levitan (Ref. 2) gave their startling prescription for recovering the potential from the spectral measure function for the radial Schrödinger equation. Levinson (Ref. 3) had given a simple heuristic derivation of the Gel'fand and Levitan results and applied them to the problem of obtaining the scattering potential for the radial Schrödinger equation from the scattering phases. Jost and Kohn (Ref. 4) also used the Gel'fand-Levitan equation to obtain the scattering potential from the scattering phases. In both References 3 and 4 the relationship of the scattering phase to the spectral measure function was obtained from the analytic properties of the Jost function derived by Jost and Kohn in Reference 5, where they had previously given a different, non-linear method, for solving inverse scattering problem for the radial equation.

We were not aware of the work of Marchenko (Ref. 6 and genrealized with Agronovich in Ref. 7 and earlier papers quoted there). Indeed, their work had not yet appeared when we started ours. The relation of their work to the Gel'fand-Levitan equation and to our approach is explained by Faddeev in Reference 8 and Kay in Reference 9.

Our approach to the inverse problem was motivated initially by Kay's interest in obtaining the electron density of the ionosphere from appropriate ionosonde data. The equation for the electromagnetic field in the ionosphere in certain simple cases is identical to the Schrödinger equation in one dimension where the number density of electrons is proportional to the potential. In notes of K.O. Friedrichs for a course at New York University on spectral representation of linear operators, the spectral measure function for the one-dimensional Schrödinger equation was seen to involve the reflection coefficient in a simple way. Kay (Paper 2 in the present volume) then used that relationship to obtain the Gel'fand-Levitan equation for the one-dimensional scattering problem by noting the analogue of the completeness relation with that used in the Gel'fand-Levitan equation for the radial equation. In Paper 2 it was shown how the potential could be recovered when the reflection coefficient was a rational function of the wave number k. An alternative version of this latter result is given in Paper 11 of the present book. In Paper 2 Kay also showed that an alternative derivation of the one-dimensional Gel'fand-Levitan equation could be obtained by using causality arguments, which are also reproduced in Paper 11.

A very surprising aspect was that the only data needed in the one-dimensional Gel'fand-Levitan equation was the reflection coefficient from one side, the point eigenvalues, and the normalizations of the eigenfunctions corresponding to the point spectrum. It followed that this data must determine the potential uniquely.

Moses discovered a different approach for the one-dimensional inverse problem, based on a generalization of the non-linear method of Reference 5. He gave an iteration method which to the first two iterations produced the same result for the potential as that obtained by using the first two iterations of a Neumann expansion solution of the one-dimensional Gal'fand-Levitan equation. This paper appears as Paper 8 of the present book. It was in this paper that the inverse problem for the general three-dimensional Schrödinger equation appears to have been first considered. The approach was non-linear and a generalization of the method of Reference 5. The potential was obtained formally in terms of an iteration procedure. It was found that only a very restricted portion of the amplitude of the scattered spherical wave was needed for reconstructing the potential. Recently Prosser (Ref. 10) showed that the iterative procedure converges to a unique potential if the scattering is sufficiently weak. He also gave analogous results in terms of other scattering data and for the acoustical equation as well.

We then decided to see whether a fairly general abstract formalism could be set up to obtain equations of the Gel'fand-Levitan type. In particular one wanted to obtain the radial and one-dimensional Gel'fand-Levitan equations as special cases and, if possible, the three-dimensional equation as well. It was recognized that spectral theory would play a central role. In particular, the multiplicity in the continuous spectrum and the special nature of the discrete spectrum would have to be taken into account. The papers reprinted in the Nuovo Cimento series result from these studies.

It was shown that the Gel'fand-Levitan equation is equivalent to a completeness relation for the eigenfunctions of the perturbed problem taken together with the decomposition of the operator (now known as an intertwining operator), which maps the perturbed Hamiltonian into the unperturbed Hamiltonian, into the sum of the identity and a triangular or Volterra operator. Under quite general conditions a Gel'fand-Levitan equation was obtained. It was shown that the one-dimensional inverse problem could be reconstructed and that the radial equation inverse problem was also contained in the general formalism. It should be mentioned that Faddeev, using our concepts (Ref. 8) provided rigorous details, especially for the radial equation.

We have already stated that the three-dimensional Gel'fand-Levitan equation came out of the general formalism. The results appear in Papers 7 and 12 of the present volume. Because of the difficulties of defining the concept of "triangularization" in higher dimensions, it generally led to non-local potentials and required more than the minimal scattering data, which the non-linear inverse method indicated was needed for a local potential. Solving the three-dimensional inverse problem for local potentials in the sense of obtaining explicit closed-form examples has yet to be accomplished. The subject is under active study; see, for example, References 10, 11, 12, and 13.

One of the results of our study of the inverse problem was to construct potentials with unusual properties. The best known of them are the reflectionless potentials, i.e., potentials for which the refelction coefficient is identically zero and which support a finite number of point eigenvalues. These reflectionless potentials, together with the solution of the one-dimensional inverse problem, which appear in the present book, formed the basis for solving a class of non-linear equations (Ref. 14). In particular, certain solutions of great stability, called solitons, have been of considerable interest. The applications to non-linear equations has led to an explosion of research into other inverse problems and other non-linear equations associated with them.

We shall end this introduction by summarizing what, in our opinion, are still useful ideas contained in the papers reprinted in this volume.

1. The triangularity condition on the Gel'fand-Levitan kernel and its relation to causality, the local (or non-local) properties of the potential, and the boundary conditions on the eigenfunctions.
2. The connection between the Gel'fand-Levitan equation and the relation and linear filter theory, referred to as "the generalized Wiener-Hopt equation" in the papers.
3. The Gel'fand-Levitan equation as an equation for the intertwining operator in terms of the spectral decomposition of the perturbed and unperturbed Hamiltonians.
4. The use of "formal eigenfunctions" to complete the space of the unperturbed Hamiltonian and thus allow the perturbed Hamiltonian to have a spectrum different from that of the unperturbed problem in Hilbert space.
5. The relation of the inverse problem for acoustics and transmission lines to the inverse problem for the Schrödinger equation.
6. Solution of the one-dimensional inverse problem. Techniques for obtaining new potentials for which exact, closed form, solutions of the one-dimensional Schrödinger equation exist.
7. Potentials that can be obtained from reflection coefficients that are rational functions of the wave number. The calculation of all such potentials by means that can be justified independently, i.e., without reference to the Gel'fand-Levitan theory.
8. Reflectionless potentials and other "pathological" potentials.
9. Minimal scattering data needed to reconstruct the scattering potential in three dimensions and a possible iterative method to obtain the potential.
10. The utility of characterizing the Jost function and the spectral measure function as operators.

We believe that this volume offers a useful framework for the general theory of inverse problems. Furthermore, there are many specific inverse problems, particularly those involving one space dimension, that are covered precisely by the abstract methods that we have formulated. For example, the methodology can be used to modify Schrödinger operators having a pure point spectrum so as to change, add or delete, a finite number of eigenvalues.

Coupled first order equations which are analogues of the Dirac equation in one dimension can be treated by the same means, and one can include some effects of complex potentials.

We also believe that there are many applications not yet explored even within the one-dimensional context. Thus we offer the present volume as an introduction to an active field that offers considerable opportunity for further research.

We would like to thank the publishers of Nuovo Cimento, Physical Review, Journal of Applied Physics, and Communications on Pure and Applied Mathematics for permission to reprint articles that originally appeared in their journals.

I. Kay
Institute for Defense Analysis

H.E. Moses
University of Lowell
Center for Atmospheric Research

References

1. K. Chadan and P.C. Sabatier, *Inverse Problems in Quantum Scattering*, Springer, 1977.

2. I.M. Gel'fand and B.M. Levitan, *Amer. Math. Soc. Trans.* 1, 253 (1951).

3. N. Levinson, *Phys. Rev.* 89, 755 (1953).

4. R. Jost and W. Kohn, *Det. Kg. Danske Vidensk. Selskab, Mat.-fys. Medd.* 27, n.9 (1953).

5. R. Jost and W. Kohn, *Phys. Rev.* 87, 977 (1952).

6. V.A. Marchenko, *Dokl. Akad. Nauk. SSSR* 104, 695 (1955).

7. Z.S. Agranovich and V.A. Marchenko, *The Inverse Problem of Scattering Theory*, Gordon and Breach, New York (1963).

8. L.D. Faddeev, *J. Math. Phys.* 4, 72 (trans. by B. Seckler) (1963).

9. I. Kay, *Mathematics of Profile Inversion*, NASA Technical Memorandum TM X-62, 150, L. Colin (ed.), 6-2 (1972).

10. R.T. Prosser, *J. Math. Phys.* 21, 2648 (1980).

11. L.D. Faddeev, *J. Sov. Math.* 5, 334 (1976).

12. R.G. Newton, *J. Math. Phys.* 21, 1698 (1980). Also earlier papers quoted in this article.

13. H.E. Moses, *J. Math. Phys.* 21, 83 (1980).

14. C.S. Gardner, J.M. Green, M.D. Kruskal, and R.M. Miura, *Phys. Rev. Lett.* 19, 1095 (1967).

IL NUOVO CIMENTO VOL. I, N. 1 1° Gennaio 1955

The Scattering Operator and the Adiabatic Theorem (*).

H. E. MOSES

New York University, Institute of Mathematical Sciences, New York, N.Y.

(ricevuto il 16 Settembre 1954)

Summary. — It is shown that the outgoing and incoming eigenfunctions which are used in scattering theory can be obtained by specifying a relationship between the initial or final states and the state at time $t=0$. Thus appeal to the adiabatic theorem is unnecessary.

1. – Introduction.

In the older treatments of scattering problems the eigenfunctions of the total Hamiltonian corresponding to the continuous spectrum were specified in a time-independent way by requiring « outgoing wave » conditions in terms of the $\boldsymbol{x}$-representation. This is the procedure used, for example, by DIRAC (1) and Mott and MASSEY (2). More recent writers, in trying to free themselves from imposing conditions in terms of the $\boldsymbol{x}$-representation, have considered the initial value problem associated with the time-dependent Schrödinger equation. The solution is prescribed to be an eigenfunction of the unperturbed Hamiltonian in the infinite past and then the outgoing eigenfunctions are defined as being the solution of the time-dependent Schrödinger equation when the time t equals zero. To obtain sensible results most of these writers resort to the device of assuming that the perturbation is switched off adiabatically

(*) The research reported in this article was done at the Institute of Mathematical Sciences, New York University and has been made possible through support and sponsorship extended by Geophysics Research Directorate, Air Force Cambridge Research Center under Contract No. AF 19(122)-463.

(1) P. A. M. DIRAC: *The Principles of Quantum Mechanics*, Third Edition (London, 1947), Chapter VIII, § 49.

(2) N. F. MOTT and H. S. W. MASSEY: *The Theory of Atomic Collisions*, Second Edition (London, 1949).

as the time t approaches $-\infty$. This is the procedure used, for example, by LIPPMANN and SCHWINGER (3) and GELL-MANN and GOLDBERGER (4).

However, FRIEDRICHS (5) has shown that if one works with an appropriate class of solutions of the time-dependent equation, namely those which are *wave packets* of the eigenfunctions corresponding to the continuous spectrum of the total Hamiltonian, the adiabatic switch-off of the perturbation is unnecessary. This fact has been called «the irrelevance of the adiabatic hypothesis» by SNYDER (6) who has considered the problem from a different point of view.

In the present paper we use Friedrich's approach to derive expressions for the «outgoing and incoming» eigenfunctions and the scattering operator arising from the time-dependent Schrödinger equation. The use of adiabatic switch-off procedures will be found unnecessary and the mathematical procedures will appear more natural then heretofore.

For the sake of completeness we shall include a discussion of the adiabatic approach in order to compare it with the procedure used herein.

2. – Definition of the Scattering Operator.

The introduction of the scattering operator is perhaps best motivated in the following fashion. Let us consider the Schrödinger equation which must be satisfied by the state vector $\Phi(t)$.

$$-\frac{1}{i}\frac{\partial}{\partial t}\Phi(t) = H\Phi(t)\,, \qquad (\hbar = 1) \tag{2.01}$$

where H is the Hamiltonian of the system. The equation for the state $\Phi(t)$ is of first order in time. Hence $\Phi(t)$ is uniquely determined, when it is prescribed for one value of time. One could, for example, take $\Phi(t)$ as being given at time $t = t_0$. Then $\Phi(t)$ is determined uniquely in terms of $\Phi(t_0)$.

In the scattering operator formalism one is interested in the case where the time t_0 is made to approach $-\infty$ and where t approaches $+\infty$. That is, one prescribes that state at a time infinitely distant in the past and seeks the state at a time infinitely distant in the future. The physical situation corresponding to such a calculation is one in which the length of time between the preparation of a state and the measurement of the state after an «exper-

(3) B. LIPPMANN and J. SCHWINGER: *Phys. Rev.*, **79**, 469 (1950).
(4) M. GELL-MANN and M. L. GOLDBERGER: *Phys. Rev.*, **91**, 398 (1953).
(5) K. O. FRIEDRICHS: *Comm. on Pure and. App Math.*, **1**, 361 (1948).
(6) H. S. SNYDER: *Phys. Rev.*, **83**, 1154 (1951).

iment » is very long compared to the natural periods occuring in the system. In practice the physical situation is generally of this type. The advantage, as far as theory goes, of considering such situations is that the various expressions which one obtains for the state become considerably simpler.

In scattering experiments one thinks of the system as being separated into two parts corresponding to the following splitting up of the Hamiltonian:

$$H = H_0 + \varepsilon V \, . \tag{2.02}$$

Here H_0 is a time-dependent « unperturbed » Hamiltonian, ε is a smallness parameter, and εV is the « perturbation ».

In scattering problems this separation corresponds to the requirement that physically interesting solutions $\Phi(t)$ of (2.01) shall also be solutions, in some sense, of the unperturbed Schrödinger equation

$$-\frac{1}{i}\frac{\partial}{\partial t}\Phi(t) = H_0\Phi(t) \tag{2.03}$$

as $t \to \pm\infty$. This requirement means that as $t \to \pm\infty$, the influence of the perturbation εV is to disappear. Thus we shall prescribe solutions $\Phi(t)$ of the perturbed equation (2.01) to behave as specified solutions of the unperturbed equation (2.03) as $t \to -\infty$ and find the solutions $\Phi(t)$ as $t \to +\infty$ under the restriction that $\Phi(t)$ shall also behave as solutions of equation (2.03) $t \to +\infty$.

This way of looking at things appears as intuitively reasonable if one considers the problem of a particle which is scattered by a potential which vanishes outside some finite region in space. When $t \to -\infty$ one may think of the particle as being far from the scattering region so that the state vector obeys the unperturbed Schrödinger equation (2.03). When t assumes finite values the particle will be in the region where the potential exerts an influence so that the state vector satisfies the perturbed Schrödinger equation (2.01). As $t \to +\infty$, the particle passes into a region where the potential again has no effect and the state again is a solution of the unperturbed Schrödinger equation (2.03). The scattering operator is defined in such a manner as to give the dependence of the solution for $t = +\infty$ upon the solution at $t = -\infty$.

Influenced by such considerations we shall now define the scattering operator. Consider states $\Phi(t)$ which are solutions of (2.01) such that following limits exist:

$$\lim_{t\to\pm\infty} \exp[iH_0 t]\Phi(t) = \Phi_\pm \, . \tag{2.04}$$

Then the scattering operator S is defined by the relation

$$\Phi_+ = S\Phi_- \, . \tag{2.05}$$

If the limits (2.04) exist, we have obtained $\Phi(t)$ in terms of $\Phi(t_0)$ as $t \to +\infty$ and $t_0 \to -\infty$ as desired, for we can write

$$(2.06) \quad \begin{cases} \Phi(t) \simeq \exp[-iH_0 t]\Phi_+ , \\ \quad\quad \simeq \exp[-iH_0 t]S\Phi_- , \\ \quad\quad \simeq \exp[-iH_0 t]S \exp[iH_0 t_0]\Phi(t_0) . \end{cases}$$

In practice one prescribes Φ_- (which is equivalent to prescribing $\Phi(t_0)$ for $t_0 \to -\infty$) and by means of the scattering operator obtains Φ_+ (which is equivalent to obtaining $\Phi(t)$ for $t \to +\infty$). It will not be necessary to use (2.06) explicitly.

In the present paper we shall consider the case where the perturbation is constant in time. In addition, we restrict ourselves to Hamiltonians H and H_0 such that the spectrum of H_0 is purely continuous and is bounded from below, and the spectrum of H consists of a continuous part which coincides with the spectrum of H_0 and possible point eigenvalues which lie below the continuous spectrum. Though these restrictions on H and H_0 appear to be severe, many of the results can be generalized immediately to the case where H and H_0 suffer the less severe restrictions that their spectra coincide in a range where they are both continuous. It should be pointed out that the above restrictions on the spectra of H and H_0 are implicit in the usual treatment of scattering operators. However, we make a point of indicating such restrictions explicitly.

3. – The Spectral Representation of the Unperturbed Hamiltonian H_0.

We should like to set up a representation in terms of the unperturbed Hamiltonian H_0. This representation is convenient because in this representation the expression $\exp[iH_0 t]\Phi(t)$ of equation (2.04) takes a particularly simple form. Moreover, it is customary to prescribe Φ_- and to ask for Φ_+ in this representation.

Since, in general, H_0 will have a degenerate spectrum, it is necessary to assume that there exist other operators which commute with H_0 and which, together with H_0, form a complete set of commuting variables. We shall designate the quantum number associated with H_0 by E and the quantum numbers associated with the other operators collectively by α. The quantum number E lies in a continuum $E_a \leqslant E \leqslant E_b$. In many problems E_b may be taken as $+\infty$. E_a will always be finite in what follows.

To give a more physical meaning we may consider the scattering of a partial by a potential in which case α would consist of the polar angles θ, φ

which give the direction of the momentum and which could also include discrete variables such as spin.

Let us designate the (improper) eigenvectors or eigenstates of H_0 by $\omega_0(E, \alpha)$. Then the eigenvectors $\omega_0(E, \alpha)$ satisfy the equation

$$H_0 \omega_0(E, \alpha) = E \omega_0(E, \alpha) \tag{3.01}$$

and also satisfy the orthogonality relation

$$\big(\omega_0(E, \alpha), \omega_0(E', \alpha')\big) = \delta(E - E')\, \delta(\alpha, \alpha'), \tag{3.02}$$

where the left-hand side of (3.02) denotes the usual Hermitian inner product of $\omega_0(E, \alpha)$ and $\omega_0(E', \alpha')$. The function $\delta(E - E')$ is the Dirac δ-function, and $\delta(\alpha, \alpha')$ is a generalized δ-function defined by

$$\left\{ \begin{aligned} \int_R r(\alpha)\, \delta(\alpha, \alpha')\, \mathrm{d}\alpha &= r(\alpha') && \text{if the region of integration } R \text{ includes } \alpha' \\ &= 0 && \text{otherwise}. \end{aligned} \right. \tag{3.03}$$

Integrations over α are to be interpreted as summations over the discrete variables included in α.

Let us expand any state vector Φ as follows

$$\Phi = \int_{E_a}^{E_b} \int f(E, \alpha) \omega_0(E, \alpha)\, \mathrm{d}\alpha\, \mathrm{d}E. \tag{3.04}$$

From (3.02) the coefficients $f(E, \alpha)$ of the expansion are given by

$$f(E, \alpha) = \big(\omega_0(E, \alpha), \Phi\big). \tag{3.05}$$

The function $f(E, \alpha)$ is the representative of Φ in the H_0 representation and we write

$$\Phi \underset{E}{\longleftrightarrow} f(E, \alpha) \tag{3.06}$$

to indicate the one-to-one correspondence between the state Φ and its H_0 representative.

We note that

$$H_0 \Phi = \int_{E_a}^{E_b} \int E f(E, \alpha) \omega_0(E, \alpha)\, \mathrm{d}\alpha\, \mathrm{d}E, \tag{3.07}$$

so that we may write

$$H_0\Phi \underset{E}{\longleftrightarrow} Ef(E, \alpha) . \tag{3.08}$$

If to every state Φ we can uniquely assign a function $f(E, \alpha)$ such that the function $Ef(E, \alpha)$ is assigned to the state $H_0\Phi$, we shall call such an assignment a spectral representation of the operator H_0.

We note that for any function of H_0, $L(H_0)$ say, we have

$$L(H_0)\Phi \underset{E}{\longleftrightarrow} L(E)f(E, \alpha) . \tag{3.09}$$

In fact, in a really rigorous treatment, (3.09) is essentially used to *define* $L(H_0)$. In particular we have

$$\exp[iH_0t]\Phi \underset{E}{\longleftrightarrow} \exp[iEt]f(E, \alpha) . \tag{3.10}$$

The inner product of two states $\Phi^{(1)}$, Φ can be written as

$$(\Phi^{(1)}, \Phi) = \int_{E_a}^{E_b}\int \bar{f}^{(1)}(E, \alpha)f(E, \alpha)\, d\alpha\, dE, \tag{3.11}$$

where $f^{(1)}$ and f are the representatives of $\Phi^{(1)}$ and Φ respectively. Equation (3.11) follows from the orthogonality relations (3.02). Throughout this report, unless otherwise noted, we shall restrict ourselves to proper states. That is, we shall əsn as representatives $f(E, \alpha)$ only functions which are quadratically integrable unless we specifically state otherwise. The use of such proper states simplifies much of our work. Since these states are those which are observed in nature, the motivation of much of what follows is made easier, and we are able to avoid some of the artificialities which one is forced to resort to if one works with improper states. An eigenstate of H_0 corresponding to the eigenvalues E', α' has the following H_0-representer

$$\omega_0(E', \alpha') \underset{E}{\longleftrightarrow} (\omega_0(E, \alpha), \omega_0(E', \alpha')) = \delta(E - E')\delta(\alpha, \alpha') . \tag{3.12}$$

We shall consider this function of E and α as being a limiting form of a representer $f(E, \alpha)$ which has a sharp peak at E', α'.

We shall designate any operator K in the H_0 by K^E. By definition, if $\Phi \underset{E}{\longleftrightarrow} f(E, \alpha)$, then

$$K\Phi \underset{E}{\longleftrightarrow} K^E f(E, \alpha) . \tag{3.13}$$

Note that from (3.08) we have

$$H_0^E = E . \tag{3.08a}$$

Usually it will be convenient to express K^E as an integral operator in the following way:

$$K^E f(E, \alpha) = \int_{E_a}^{E_b}\!\!\int K^E(E, \alpha \,|\, E', \alpha') f(E', \alpha')\, d\alpha'\, dE' . \tag{3.14}$$

The kernel $K^E(K, \alpha \,|\, E', \alpha')$ of the integral operator is just the usual « matrix » of K in terms of the H_0 representation, i.e.,

$$K^E(E, \alpha \,|\, E', \alpha') = (\omega_0(E, \alpha),\, K\omega_0(E', \alpha')) . \tag{3.15}$$

4. – The Spectral Representation of the Total Hamiltonian H.

Since the solution $\Phi(t)$ of the Schrödinger equation (2.01) is most simply expressed in terms of the spectral representation of the total Hamiltonian H, we shall introduce this representation. Let us designate the quantum number associated with H by F. As mentioned earlier, we assume that the spectrum of H consists of a continuous part which coincides with the spectrum of H_0 and point eigenvalues which lie below the continuous spectrum. Hence F lies in a continuum when $E_a \leqslant F \leqslant E_b$ and consists of points F_i when $F < E_1$. Furthermore, since we have assumed that the continuous spectrum has the same degeneracy as the spectrum of H_0, we shall have to introduce auxiliary quantum numbers β which are analogous to α and which have the same range. For the bound states or point eigenvalues the degeneracy may have any character whatsoever. We shall introduce auxiliary quantum numbers also denoted by β to describe the degeneracy of the spectrum of the bound states, though for these states β may have an entirely different character than when F lies in the continuum.

We can now introduce eigenvectors or eigenstates of H corresponding to the eigenvalue F and denote these eigenstates by $\omega(F, \beta)$ where F may either lie in the continuum $E_a \leqslant F \leqslant E_b$ or be one of the point eigenvalues F_i. We have

$$H\omega(F, \beta) = F\omega(F, \beta) . \tag{4.01}$$

The eigenfunctions also satisfy the orthogonality relations

$$(\omega(F, \beta),\, \omega(F', \beta')) = \delta(F - F')\delta(\beta, \beta') . \tag{4.02}$$

The function $\delta(F - F')$ is to be interpreted as a Dirac δ-function when both F and F' lie in the continuous spectrum, a Kronecker δ when both F,

F' lie in the point spectrum, and zero otherwise.

An arbitrary state Φ can be expanded in terms of the eigenfunctions $\omega(F, \beta)$ as follows

$$\Phi = \iint g(F, \beta)\omega(F, \beta)\, d\beta\, dF, \tag{4.03}$$

where the integration over the values of F and β corresponding to point eigenvalue are to be replaced by summation. We shall use this integration convention in order to minimize, in a formal fashion, the difference in character between the eigenstates of the continuous spectrum and those of the discrete spectrum.

From (4.02) and (4.03) we have

$$g(F, \beta) = (\omega(F, \beta), \Phi) \tag{4.04}$$

and we write

$$\Phi \underset{F}{\longleftrightarrow} g(F, \beta)\,. \tag{4.05}$$

We call $g(F, \beta)$ the representative of Φ in the H representation.

We note that

$$H\Phi \underset{F}{\longleftrightarrow} Fg(F, \beta) \tag{4.06}$$

and in general that a function of the operator H, $L(H)$ say, has the property

$$L(H)\Phi \underset{F}{\longleftrightarrow} L(F)f(F, \beta)\,. \tag{4.07}$$

In particular equation (4.07) gives us

$$\exp[-iHt]\Phi \underset{F}{\longleftrightarrow} \exp[-iFt]g(F, \beta)\,. \tag{4.08}$$

In terms of the spectrum representation of H, the inner product of two states $\Phi^{(1)}$, Φ whose representatives are $g^{(1)}(F, \beta)$ and $g(F, \beta)$ respectively is given by

$$(\Phi^{(1)}, \Phi) = \iint \bar{g}^{(1)}(F, \beta)g(F, \beta)\, d\beta\, dF, \tag{4.09}$$

where, in accordance with our previously mentioned convention, integrations over the quantum variables are to be understood as summations over those values which are point eigenvalues.

As in the case of the H_0 representation, we say that if to every state a unique assignment can be made to a function $g(F, \beta)$ such that $H\Phi$ is assigned to $Fg(F, \beta)$, then we say that the set of functions $g(F, \beta)$ constitutes a spectral representation of the operator H.

We can also introduce operators in the H representation. The operator K in the H representation will be denoted by K^F which is defined by

$$K\Phi \underset{F}{\leftrightarrow} K^F g(F, \beta) , \tag{4.10}$$

where

$$\Phi \underset{F}{\leftrightarrow} g(F, \beta) .$$

5. – The Transformation from the H to the H_0 Representation.

It is desired to find the transformation which relates the H representer of an arbitrary state Φ to the H_0 representer of that state. The reason for wanting this transformation is that we can solve for the solution $\Phi(t)$ of the Schrödinger equation (2.01) in terms of the H representation in a simple way. We can then transform to the H_0 representation and thereby obtain explicitly the limits which occur in equation (2.04).

To be more specific let us take, as representatives of an arbitrary state

$$\Phi \underset{E}{\leftrightarrow} f(E, \alpha)$$

$$\Phi \underset{F}{\leftrightarrow} g(F, \beta)$$

and express the desired transformation as an integral transformation with the kernel $u(E, \alpha | F, \beta)$

$$f(E, \alpha) = \iint u(E, \alpha | F, \beta) g(F, \beta) \, d\beta \, dF . \tag{5.01}$$

Then if we write the representatives of the solution $\Phi(t)$ of the Schrödinger equation (2.01) as

$$\Phi(t) \underset{F}{\leftrightarrow} g(F, \beta; t) \tag{5.02}$$

$$\Phi(t) \underset{E}{\leftrightarrow} f(E, \alpha; t) , \tag{5.03}$$

we have from (5.01)

$$f(E, \alpha; t) = \iint u(E, \alpha | F, \beta) g(F, \beta; t) \, d\beta \, dF . \tag{5.04}$$

But the solution of (2.01) may be written as

$$\Phi(t) = \exp[-iHt]\Phi(0), \tag{5.05}$$

or in terms of the H representation (see (4.08))

$$g(F, \beta; t) = \exp[-iFt] g(F, \beta; 0), \tag{5.06}$$

so that from (5.04) we have

$$(5.07) \qquad f(E,\alpha;t) = \iint u(E,\alpha|F,\beta)\exp[-iFt]g(F,\beta;0)\,d\beta\,dF\,.$$

Now in the H_0 representation $\exp[iH_0t]\Phi(t)$ is represented by

$$(5.08) \qquad \exp[iH_0t]\Phi(t) \underset{E}{\longleftrightarrow} \exp[iEt]f(E,\alpha;t) =$$
$$= \iint u(E,\alpha|F,\beta)\exp[-i(F-E)t]g(F,\beta;0)\,d\beta\,dF\,.$$

Thus if we know the kernel $u(E,\alpha|F,\beta)$, we can evaluate explicitly the limits $\lim_{t\to\pm\infty}\exp[iH_0t]\Phi(t)$ which occur in (2.04) in terms of the H_0 representation. If these limits exist, we can find $\Phi_\pm$ and the scattering operator in the H_0 representation. We shall therefore examine the properties of the kernel $u(E,\alpha|F,\beta)$.

It should be noted that $u(E,\alpha|F,\beta)$ is a symbolic function rather than a proper function. For example, if the perturbation is identically zero, the functions $f(E,\alpha)$ and $g(F,\beta)$ of equation (5.01) must be identical and $u(E,\alpha|F,\beta)$ must be a δ-function. MØLLER [7] calls a function equivalent to $u(E,\alpha|F,\beta)$ « the wave matrix ». FRIEDRICHS [5] interprets (5.01) as being the result of an operation in Hilbert space rather than a transformation between two representations of Hilbert space.

Perhaps a more convenient way of looking at $u(E,\alpha|F,\beta)$ is to recognize it as an eigenfunction of H, corresponding to the eigenvalues F, β as given in the H_0 representation, i.e.,

$$(5.09) \qquad \omega(F,\beta) \underset{E}{\longleftrightarrow} u(E,\alpha|F,\beta)\,,$$

or equivalently

$$(5.10) \qquad u(E,\alpha|F,\beta) = \big(\omega_0(E,\alpha),\,\omega(F,\beta)\big)\,.$$

Equation (5.10) and hence equivalently (5.09) follows from the equality

$$\iint_{E_a}^{E_b} f(E,\alpha)\omega_0(E,\alpha)\,d\alpha\,dE = \iint g(F,\beta)\omega(F,\beta)\,d\beta\,dF$$

where

$$\Phi \underset{E}{\longleftrightarrow} f(E,\alpha)$$
$$\Phi \underset{F}{\longleftrightarrow} g(F,\beta)\,,$$

(7) C. MØLLER, *General Properties of the Characteristic Matrix im the Theory of Elementary Particles* in *Det. Kgl. Danske Videnskabernes Selskab*, (København, 1945).

the state Φ being arbitrary. The use of the orthogonality relation (3.02) for $\omega_0(E, \alpha)$ yields (5.01) where $u(E, \alpha|F, \beta)$ is given by (5.10).

6. – Characterization of the Transformation.

We want to find an equation for the kernel $u(E, \alpha|F, \beta)$ of the transformation. Since $u(E, \alpha|F, \beta)$ is an eigenfunction of H in the H_0 representation corresponding to the eigenvalue F, we must have

$$H^E u(E, \alpha|F, \beta) = Fu(E, \alpha|F, \beta)\,, \tag{6.01}$$

where H^E is the operator H as given in the H_0 representation. The operator H^E operates on the variables E, α of $u(E, \alpha|F, \beta)$. Now,

$$\begin{aligned} H^E &= H_0^E + \varepsilon V^E \\ &= E + \varepsilon V^E\,. \end{aligned} \tag{6.02}$$

Hence (6.01) becomes

$$(F - E)u(E, \alpha|F, \beta) = \varepsilon V^E u(E, \alpha|F, \beta)\,. \tag{6.03}$$

Equation (6.03) must be satisfied by all eigenfunctions of H. However, the eigenfunctions are not completely characterized by this equation. The eigenfunctions corresponding to eigenvalues F lying in the continuum $E_a \leqslant F \leqslant E_b$ must have further conditions imposed upon them, as we shall now see.

However, let us first consider equation (6.03) when F is one of the point eigenvalues F_i. The factor $(F_i - E)$ on the left-hand side of (6.03) is nowhere zero, since all values of E lie above F_i. Hence we can divide both sides of (6.03) by $F_i - E$ for all values of E. Hence we obtain for the eigen-functions belonging to point eigenvalues of H the following equation

$$u(E, \alpha|F_i, \beta) = \frac{\varepsilon}{F_i - E} V^E u(E, \alpha|F_i, \beta)\,. \tag{6.04}$$

It should be noted that $u(E, \alpha|F, \beta)$ is a proper, i.e., quadratically integrable, function of the variables E, α.

When F lies in the continuum spectrum, it has the same range of values as E. Hence the factor $F - E$ can be zero. Now generally when we wish to divide by a number which can have the value zero we have the possibility of getting symbolic functions. That is, if

$$x k(x) = r(x)\,, \tag{6.05}$$

when x can take the value zero, we can conclude that

$$k(x) = c\delta(x) + \frac{P}{x} r(x), \tag{6.06}$$

where c is an arbitrary constant and P/x means that for those integrals where $1/x$ appears as one of the factors of the integrand, the principal part of the integral is to be used (see, e.g., DIRAC ([8])). That $k(x)$ as given (6.06) satisfies equation (6.05) can be verified readily.

From the above considerations we obtain from (6.03) the following expression for $u(E, \alpha | F, \beta)$ when F lies in the continuum spectrum

$$u(E, \alpha | F, \beta) = \lambda(E, \alpha | E, \beta)\, \delta(E - F) + \varepsilon \frac{P}{F - E} V^E u(E, \alpha | F, \beta), \tag{6.07}$$

or

$$u(E, \alpha | F, \beta) = \lambda(E, \alpha | E, \beta)\, \delta(E - F) + \tag{6.07a}$$

$$+ \varepsilon \frac{P}{F - E} \int_{E_a}^{E_b} \int V^E(E, \alpha | E', \alpha')\, u(E', \alpha' | F, \beta)\, d\alpha'\, dE',$$

where $\lambda(E, \alpha | E, \beta)$ is an arbitrary function of its arguments. In order to determine λ we must now impose additional conditions on the eigenfunctions of H belonging to the continuous spectrum.

7. – Determination of the Eigenfunctions of the Continuous Spectrum.

We shall now determine the function $\lambda(E, \alpha | E, \beta)$ of equation (6.07). This function is usually obtained by following Dirac's method ([9]) in which the eigenfunctions $u(E, \alpha | E, \beta)$ are transformed to the x representation and boundary conditions are imposed on the eigenfunctions with respect to the variable x. We shall follow a different procedure. In this procedure we specify a relationship between either Φ_- or Φ_+ of equation (2.04) and $\Phi(0)$.

Let us first note that for solutions $\Phi(t) = \exp[-iHt]\Phi(0)$ of equation (2.01), the limits $\lim_{t\to\pm\infty} \exp[iH_0 t]\Phi(t)$ occuring in (2.04) exist only if $\Phi(0)$ — or equivalently $\Phi(t)$ — is orthogonal to the bound eigenstates of H. To show this let us write equation (5.08) which gives $\exp[iH_0 t]\Phi(t)$ in the H_0 repres-

([8]) P. A. M. DIRAC: loc. cit., Chapter III, § 15.
([9]) P. A. M. DIRAC: loc. cit., Chapter VIII, § 50.

entation as

$$\text{(7.01)}\quad \exp[iH_0t]\Phi(t) \underset{E}{\longleftrightarrow} \exp[iEt]f(E,\alpha;t) =$$

$$= \int\int_{E_a}^{E_b} u(E,\alpha|F,\beta)\exp[-i(F-E)t]g(F,\beta;0)\,d\beta\,dF +$$

$$+\sum_i \int u(E,\alpha|F_i,\beta)E\exp[-i(F_i-E)t]g(F_i,\beta;0)\,d\beta\,.$$

In (7.01) we have broken up the formal integration over F into a true integration over the continuous spectrum and a summation over the discrete spectrum. We shall later show that as $t \to \pm\infty$ in (7.01) the integral approaches a limit for many functions $g(F,\beta;0)$ defined for $E_a \leqslant F \leqslant E_b$. However, the sum over the point eigenvalues, as can be seen, merely oscillates and does not approach a limit as $t \to \pm\infty$.

Hence, in order to have limits $\Phi_\pm$, we must take $g(F,\beta;0) = 0$ whenever F belongs to the discrete spectrum. Since $g(F_i,\beta;0) = (\omega(F_i,\beta), \Phi(0))$ we have our results that $\Phi(0)$ must be orthogonal to the eigenfunctions of the discrete spectrum.

Let us therefore now restrict ourselves to state $\Phi(0)$ such that its H representer $g(F,\beta;0)$ is identically zero for values of F corresponding to the discrete spectrum. We shall now characterize the states Φ_- or Φ_+ in terms of $\Phi(0)$.

It is clear that the H_0 representer of $\Phi_\pm$ will be linearly related to the H representer of $\Phi(0)$. Furthermore it is clear that

$$\text{(7.02)}\qquad (\Phi_+,\Phi_+) = (\Phi_-,\Phi_-) = (\Phi(0),\Phi(0))$$

in order that probability may be preserved. Let us express $\Phi(0)$ in the H representation and $\Phi_\pm$ in the H_0 representation in the following way

$$\Phi(0) \underset{F}{\longleftrightarrow} g(F,\beta;0) \qquad (E_a \leqslant F \leqslant E_b)$$

$$\Phi_\pm \underset{E}{\longleftrightarrow} f_\pm(E,\alpha)\,.$$

Then one may write the relation between $f_\pm$ and g as an integral transformation

$$\text{(7.03)}\qquad f_\pm(E,\alpha) = \int\int_{E_a}^{E_b} K_\pm(E,\alpha|F,\beta)\,g(F,\beta;0)\,d\beta\,dF\,,$$

where $K_\pm$ satisfy the orthogonality relations

$$\text{(7.04)}\quad \begin{cases} \displaystyle\iint_{E_a}^{E_b} \bar{K}_\pm(E, \alpha | F, \beta) K_\pm(E, \alpha | F', \beta') \, \mathrm{d}\alpha \, \mathrm{d}E = \delta(F - F') \, \delta(\beta, \beta') , \\ \displaystyle\iint_{E_a}^{E_b} \bar{K}_\pm(E, \alpha | F, \beta) K_\pm(E', \alpha' | F, \beta) \, \mathrm{d}\beta \, \mathrm{d}F = \delta(E - E') \, \delta(\alpha, \alpha') , \end{cases}$$

where in the above F and F' are restricted to the continuous spectrum of H. Each different way of choosing K_+ or K_- will lead to a different choice of eigenfunctions belonging to the continuous spectrum. We can prescribe *either* K_+ *or* K_-; we cannot choose them both.

Let us first consider the simplest choice of K_-, namely

$$\text{(7.05)}\quad K_-(E, \alpha | F, \beta) = \delta(E - F) \delta(\alpha, \beta)$$

or from (7.03)

$$\text{(7.06)}\quad f_-(E, \alpha) = g(E, \alpha; 0) .$$

We may re-express condition (7.05) in the following way: *We shall prescribe the solution $\Phi(t)$ of the perturbed Schrödinger equation* (2.01) *to be in an asymptotic sense also a solution of the unperturbed Schrödinger equation* (2.03).

$$\text{(7.07)}\quad \Phi(t) \underset{E}{\longleftrightarrow} \exp[- iEt] f_-(E, \alpha) .$$

We shall then have as a solution $\Phi(t)$ of the perturbed equation (2.01) *at finite times*

$$\text{(7.08)}\quad \Phi(t) \underset{F}{\longleftrightarrow} g(F, \beta; t) = \exp[- iFt] f_-(F, \beta) .$$

It will not be possible to obtain **bound states** *of H from such initial conditions.*

In the problem where a particle is scattered by a potential, the function $f_-(E, \alpha)$ would be the assigned probability amplitude that the particle initially had the kinetic energy E and direction designated by α before collision. Even in the case of an attractive potential which is capable of having a bound state it is not possible for the particle to be captured.

We shall now show how (7.06) characterizes our eigenfunctions.

It is convenient to introduce the matrix $T(E, \alpha | F, \beta)$ defined by

$$\text{(7.09)}\quad T(E, \alpha | F, \beta) = \iint_{E_a}^{E_b} V^E(E, \alpha | E', \alpha') u(E', \alpha' | F, \beta) \, \mathrm{d}\alpha' \mathrm{d}E' .$$

The integral equation (6.07a) is then written

$$u(E,\alpha|F,\beta)=\lambda(E,\alpha|E,\beta)\,\delta(E-F)+\varepsilon\frac{P}{F-E}T(E,\alpha|F,\beta)\,. \tag{7.10}$$

Equation (7.01) is then

$$\exp[iEt]f(E,\alpha;t)=\int\int_{E_a}^{E_b}u(E,\alpha|F,\beta)g(F,\beta;0)\exp[-i(F-E)t]\,\mathrm{d}\beta\,\mathrm{d}F= \tag{7.11}$$

$$=\int\lambda(E,\alpha|E,\beta)g(E,\beta;0)\,\mathrm{d}\beta+$$

$$+\varepsilon\int\int_{E_a}^{E_b}\frac{T(E,\alpha|F,\beta)}{F-E}\,g(F,\beta;0)\,\exp[-i(F-E)t]\,\mathrm{d}\beta\,\mathrm{d}F\,.$$

In equation (7.11) and in the integrals which follow we drop the symbol P but understand that the principal values of the integrals containing the factor $1/(F-E)$ in the integrands are to be taken.

We shall choose λ by using our above-mentioned requirement that

$$\lim_{t\to-\infty}\exp[iEt]f(E,\alpha;t)\equiv f_-(E,\alpha)=g(E,\alpha;0)\,. \tag{7.12}$$

Therefore, let us consider the second term of equation (7.11)

$$\lim_{t\to-\infty}\int\int_{E_a}^{E_b}\frac{T(E,\alpha|F,\beta)}{F-E}\,g(F,\beta;0)\exp[-i(F-E)t]\,\mathrm{d}\beta\,\mathrm{d}F\,.$$

To evaluate this limit we shall use the Riemann-Lebesgue theorem for Fourier Transforms which says that

$$\lim_{\alpha\to\infty}\int_A^B f(x)\exp[i\alpha x]\,\mathrm{d}x=0\,, \tag{7.13}$$

if $f(x)$ is bounded and integrable in (A,B), or the integral $\int_A^B f(x)x$ is absolutely convergent [10]. We write

$$\text{(7.14)} \qquad \lim_{t\to-\infty} \iint\limits_{E_a}^{E_b} \frac{T(E,\alpha|F,\beta)}{F-E} g(F,\beta;0)\exp[-i(F-E)t]\,\mathrm{d}\beta\,\mathrm{d}F =$$

$$= \lim_{t\to-\infty} \iint\limits_{E_a}^{E_b} \frac{T(E,\alpha|F,\beta)g(F,\beta;0)-T(E,\alpha|E,\beta)g(E,\beta;0)}{F-E}\cdot$$

$$\cdot\exp[-i(F-E)t]\,\mathrm{d}F\,\mathrm{d}\beta + \lim_{t\to-\infty}\int T(E,\alpha|E,\beta)g(E,\beta;0)\,\mathrm{d}\beta\cdot$$

$$\cdot\int\limits_{E_a}^{E_b} \frac{\exp[-i(F-E)t]}{F-E}\,\mathrm{d}F\,.$$

The expression

$$\frac{T(E,\alpha|F,\beta)g(F,\beta;0)-T(E,\alpha|E,\beta)g(E,\beta;0)}{F-E},$$

as a function of F is absolutely integrable for a large class of functions $g(F,\beta;0)$, if $T(E,\alpha|F,\beta)\,g(F,\beta;0)$ considered as a function of F has a derivative at $F=E$. Hence the first term on the right hand side of equation (7.14) vanishes for appropriate functions $g(F,\beta;0)$.

We shall now consider the second term of (7.14). We find by a suitable change of variable

$$\text{(7.15)} \qquad \lim_{t\to-\infty}\int\limits_{E_a}^{E_b} \frac{\exp[-i(F-E)t]}{F-E}\,\mathrm{d}F = \lim_{t\to-\infty}\int\limits_{(E_a-E)t}^{(E_b-E)t} \frac{\exp[-i\zeta]}{\zeta}\,\mathrm{d}\zeta\,.$$

Let us first take the case where $E_a<E<E_b$, that is where E lies *within* the limits of the spectrum of H_0. Equation (7.15) becomes

$$\text{(7.16)} \qquad \lim_{t\to-\infty}\int\limits_{E_a}^{E_b} \frac{\exp[-i(F-E)t]}{F-E}\,\mathrm{d}F = \int\limits_{+\infty}^{-\infty} \frac{\exp[-i\zeta]}{\zeta}\,\mathrm{d}\zeta = 2i\int\limits_{0}^{\infty}\frac{\sin\zeta}{\zeta}\,\mathrm{d}\zeta = i\pi\,.$$

Hence we have from (7.11), (7.14), (7.16) and also our initial condition (7.12)

$$\text{(7.17)} \qquad \lim_{t\to-\infty}\exp[iEt]f(E,\alpha;t) = \int\lambda(E,\alpha|E,\beta)g(E,\beta;0)\,\mathrm{d}\beta +$$

$$+\,\varepsilon i\pi\int T(E,\alpha|E,\beta)g(E,\beta;0)\,\mathrm{d}\beta = g(E,\alpha;0)\,.$$

([10]) H. S. CARSLAW: *Theory of Fourier Series and Integrals*, Third Edition (New York, 1945).

Thus from the definition of $\delta(\alpha, \alpha')$ (eq. (3.03))

$$\lambda(E, \alpha | E, \beta) + \varepsilon i\pi T(E, \alpha | E, \beta) = \delta(\alpha, \beta) , \tag{7.17a}$$

or

$$\lambda(E, \alpha | E, \beta) = \delta(\alpha, \beta) - i\varepsilon\pi T(E, \alpha | E, \beta) . \tag{7.18}$$

Our integral equation (6.07a) for $u(E, \alpha | F, \beta)$ is thus

$$u_-(E, \alpha | F, \beta) = \delta(E - F)\delta(\alpha, \beta) + + \varepsilon\left[- i\pi\delta(F - E) + \frac{P}{F - E}\right] T_-(E, \alpha | F, \beta) , \tag{7.19}$$

or

$$u_-(E, \alpha | F, \beta) = \delta(E - F)\delta(\alpha, \beta) + \varepsilon\left[- i\pi\delta(F - E) + \frac{P}{F - E}\right] \cdot \cdot \int_{E_a}^{E_b}\int V^E(E, \alpha | E', \alpha')u_-(E', \alpha' | F, \beta)\, d\alpha'\, d'E . \tag{7.19a}$$

In (7.19) and (7.19*a*) we have introduced the negative subscript to indicate that the eigenfunctions of the continuous spectrum have been characterized by initial conditions given at $t = -\infty$.

Having shown that the limit $\lim_{t\to-\infty} \exp[iEt]f(E, \alpha; t) = f_-(E, \alpha)$ exists when E lies within the end points of the spectrum of H_0, let us now consider the case where E equals one of the end points, $E = E_a$, say. In the development carried out above equation (7.15) is replaced by

$$\lim_{t\to-\infty}\int_{E_a}^{E_b} \frac{\exp[-i(F, E)t]}{F - E_a}\, dF = \int_0^{-\infty} \frac{\exp[-i\zeta]}{\zeta}\, d\zeta . \tag{7.15a}$$

The integral diverges. Therefore we have to put some requirement on $T(E, \alpha | F, \beta)$ or $g(F, \beta; 0)$ in order to obtain $\lim_{t\to-\infty} \exp[iE_a t]f(E_a, \alpha; t)$. Now the second term on the right hand side of equation (7.14) will cause no trouble for $E = E_a$, if either $T(E_a, \alpha | E_a, \beta) = 0$ or $g(E_a, \beta; 0) = 0$.

Using the definition of $T(E, \alpha | F, \beta)$ given by (7.09) the first condition implies a condition on $V^E(E_a, \alpha | E', \alpha')$, namely

$$V^E(E_a, \alpha | E', \alpha') = 0 . \tag{7.20}$$

The second condition limits the class of states Φ_- or $\Phi(0)$ which can be used.

Similarly when $E = E_b$ we have two possibilities, namely we can either take

$$(7.21) \qquad V^E(E_b, \alpha | E', \alpha') = 0 ,$$

or take $g(E_b, \beta; 0) = 0$.

Then with either the above conditions on V or on $g(F, \beta; 0) = f_-(F, \beta)$ one obtains the limit of $\exp[-iEt]f(E, \alpha; t)$ as t goes to $-\infty$ for all values of E.

We shall then consider $u_-(E, \alpha | F, \beta)$ as given by (7.19) as being an appropriate set of eigenfunctions for the continuous spectrum, provided they are applied either to suitable states $g(F, \beta; 0)$ or the perturbation satisfies conditions (7.20) and (7.21).

The conditions on $g(F, \beta; 0)$ or on $V^E(E, \alpha | E', \alpha')$ will apply to all other processes in which time limits are considered and we shall not repeat these conditions.

Now we shall look at the eigenfunctions which we obtain when conditions at $t = +\infty$ are prescribed. In analogy to the derivation of (7.19) and (7.19*a*), we take $f_+(E, \alpha)$ to be $g(E, \alpha; 0)$ and find the following equations for the eigenfunctions, which we denote by $u_+(E, \alpha | F, \beta)$

$$(7.22) \quad u_+(E, \alpha | F, \beta) = \delta(E-F)\delta(\alpha, \beta) + \varepsilon\left[i\pi\delta(F-E) + \frac{P}{F-E}\right] T_+(E, \alpha | F, \beta) ,$$

or

$$(7.22a) \qquad u_+(E, \alpha | F, \beta) = \delta(E-F)\delta(\alpha, \beta) + \varepsilon\left[i\pi\delta(F-E) + \frac{P}{F-E}\right] \cdot$$

$$\cdot \int\int_{E_a}^{E_b} V^E(E, \alpha | E', \alpha') u_+(E', \alpha' | F, \beta)\, d\alpha'\, dE' .$$

For convenience let us introduce a notation for the functions $[\pm i\pi\delta(F-E) + P/(F-E)]$ which occur in (7.19) and (7.22). Accordingly we set

$$(7.23) \qquad \gamma_\pm(x) = \pm i\pi\delta(x) + \frac{P}{x} .$$

As is well known the functions $\gamma_\pm$ can be considered as being the limits of proper functions

$$(7.24) \qquad \gamma_\pm(x) = \lim_{\zeta \to 0} \frac{1}{x \mp i\zeta} = \left[\pm i\pi\delta(x) + \frac{P}{x}\right] ,$$

ζ being positive.

To summarize: The eigenfunctions of H corresponding to the discrete spectrum satisfy the integral equation

$$u(E, \alpha | F_i, \beta) = \frac{\varepsilon}{F_i - E} \int_{E_a}^{E_b} \int V^E(E, \alpha | E', \alpha') u(E', \alpha' | F_i, \beta) \, d\alpha' \, dE' , \tag{7.25}$$

while for the eigenfunctions corresponding to the continuous spectrum we can choose among others either of the two sets of eigenfunctions $u_-(E, \alpha | F, \beta)$ or $u_+(E, \alpha | F, \beta)$. The eigenfunctions $u_\pm$ satisfy the following integral equation

$$u_\pm(E, \alpha | F, \beta) = \delta(E - F)\delta(\alpha, \beta) + \varepsilon \gamma_\pm(F - E) T_\pm(E, \alpha | F, \beta) , \tag{7.26}$$

where

$$T_\pm(E, \alpha | F, \beta) = V^E u_\pm(E, \alpha | F, \beta) = \int_{E_a}^{E_b} \int V^E(E, \alpha | E', \alpha') u_\pm(E', \alpha' | F, \beta) \, dE' \, d\alpha' . \tag{7.27}$$

The eigenfunctions $u_-(E, \alpha | F, \beta)$ are used to construct solutions of the Schrödinger equation when initial conditions at $t = -\infty$ are given. When Φ_- is prescribed in the H_0 representation in the following way

$$\Phi_- \underset{E}{\leftrightarrow} f_-(E, \alpha) ,$$

then the solution $\Phi(t)$ in the H representation is given by

$$\Phi(t) \underset{F}{\leftrightarrow} g(F, \beta; t) = \exp[-iFt] f_-(F, \beta) , \tag{7.28}$$

or in the H_0 representation by

$$\Phi(t) \underset{E}{\leftrightarrow} f(E, \alpha; t) = \int_{E_a}^{E_b} \int u_-(E, \alpha | F, \beta) \exp[-iFt] f_-(F, \beta) \, d\beta \, dF . \tag{7.29}$$

The eigenfunctions $u_+(E, \alpha | F, \beta)$ are used to construct solutions of the Schrödinger equation when final conditions at $t = +\infty$ are prescribed. When Φ_+ is prescribed in the following way

$$\Phi_+ \underset{E}{\leftrightarrow} f_+(E, \alpha) ,$$

the solution $\Phi(t)$ is given by

$$\Phi(t) \underset{F}{\leftrightarrow} g(F, \beta; t) = \exp[-iFt] f_+(F, \beta) , \tag{7.30}$$

or

$$(7.31)\qquad \Phi(t) \underset{E}{\leftrightarrow} f(E, \alpha; t) = \int\int_{E_a}^{E_b} u_+(E, \alpha | F, \beta) \exp[-iFt] f_+(F, \beta)\, d\beta\, dF\,.$$

It might be mentioned that instead of using the functions $\gamma_\pm$ MØLLER and others prefer to use the functions $\delta_\pm$ defined in the following way

$$(7.32)\qquad \delta_\pm(x) = \mp \frac{i}{2\pi} \gamma_\pm(x)\,.$$

However, we find it more convenient to use the functions $\gamma_\pm$, which in effect are also used in reference (3) and (5). Incidently, the functions $u_\mp$ are the eigenstates $\Psi^\pm$ reference (3).

Let us now consider the case where K_+ or K_- is a more general operator than that of (7.05). Then from equations (7.03) and (7.12) we have as the initial conditions which determine the eigenfunctions u_-

$$(7.33)\qquad \lim_{t\to-\infty} \exp[iEt] f(E, \alpha; t) = f_-(E, \alpha) = \int\int_{E_a}^{E_b} K_-(E, \alpha | F, \beta) g(F, \beta; 0)\, d\beta\, dF.$$

Equation (7.17a), the equation which determines $\lambda(E, \alpha | E, \beta)$ is replaced by

$$(7.34)\qquad [\lambda(E, \alpha | E, \beta) + \varepsilon i\pi T(E, \alpha | E, \beta)]\, \delta(E - F) = K_-(E, \alpha | F, \beta)\,.$$

Equation (7.34) implies a restriction on the function K_- which can be used or, equivalently, a restriction on the relation which may be specified between Φ_- and $\Phi(0)$. From (7.34) and the unitarity relation (7.04) we see that $K_-(E, \alpha | F, \beta)$ must have the form

$$(7.35)\qquad K_-(E, \alpha | F, \beta) = \delta(E - F) \exp[i\eta(E)] \mu(\alpha, \beta)\,,$$

where we may choose the functions $\eta(E)$ and $\mu(\alpha, \beta)$ in any way we wish, provided $\eta(E)$ is a real function of its argument and $\mu(\alpha, \beta)$ satisfies the orthogonality relations

$$(7.36)\qquad \begin{cases} \int \bar{\mu}(\alpha, \beta) \mu(\alpha', \beta)\, d\beta = \delta(\alpha, \alpha')\,, \\ \int \bar{\mu}(\alpha, \beta) \mu(\alpha, \beta')\, d\alpha = \delta(\beta, \beta')\,. \end{cases}$$

With such a choice for K_- we find from (7.34)

$$\lambda(E, \alpha | E, \beta) = \exp[i\eta(E)]\mu(\alpha, \beta) - \varepsilon i\pi T(E, \alpha | E, \beta) . \tag{7.37}$$

Denoting the eigenfunctions u for which λ is chosen in this way by $u_{K_-}(E, \alpha | F, \beta)$ we obtain from (6.07*a*) the integral equation for these eigenfunctions

$$u_{K_-}(E, \alpha | F, \beta) = \exp[i\eta(E)]\mu(\alpha, \beta)\delta(E - F) + \gamma_-(F - E) \cdot \\ \cdot \int V^E(E, \alpha | E', \alpha') u_{K_-}(E', \alpha' | F, \beta) \, dE' \, d\alpha' . \tag{7.38}$$

We thus obtain a set of eigenfunctions of H which is not usually considered explicitly in the literature. If we denote the eigenfunctions whose H_0 representatives are $u_-(E, \alpha | F, \beta)$ and $u_{K_-}(E, \alpha | F, \beta)$ by $\omega_-(F, \beta)$ and $\omega_{K_-}(F, \beta)$ respectively, it can be shown that ω_{K_-} can be obtained by the following relation from $\omega_-(F, \beta)$

$$\omega_{K_-}(F, \beta) = \exp[i\eta(F)] \int \mu(\beta', \beta)\omega_-(F, \beta') \, d\beta' . \tag{7.39}$$

Hence all eigenfunctions of the continuous spectrum of H, obtained by specifying a relationship between $\Phi(0)$ and Φ_- can be expressed in terms of ω_-. There are many interesting consequences of (7.39) but we shall not pursue them further here.

8. – The Scattering Operator.

We are now in a position to express the scattering operator and its inverse in terms of the eigenfunctions u_- and u_+ respectively. Let us then use the eigenfunctions u_- and construct the solution of the Schrödinger equation $\Phi(t)$ in the H_0 representation, having prescribed the state Φ_-

$$\Phi(t) \overset{}{\underset{E}{\leftrightarrow}} f(E, \alpha; t) = \int_{E_a}^{E_b}\!\!\int u_-(E, \alpha | F, \beta) \exp[-iFt] f_-(F, \beta) \, d\beta \, dF , \tag{8.01}$$

where

$$\Phi_- \underset{E}{\leftrightarrow} f_-(E, \alpha) .$$

Then we can evaluate the limit

$$\lim_{t \to +\infty} \exp[iEt] f(E, \alpha; t) = f_+(E, \alpha) = \\ = \lim_{t \to +\infty} \int_{E_a}^{E_b}\!\!\int u_-(E, \alpha | F, \beta) \exp[-i(F - E)t] f_-(F, \beta) \, d\beta \, dF .$$

by replacing $u_-(E, \alpha | F, \beta)$ by the right-hand side of (7.27) and treating the integral which contains the factor $P/(F-E)$ in a manner similar to that of the previous section. We find that

$$f_+(E, \alpha) = \int_{E_a}^{E_b}\!\!\int S^E(E, \alpha | E', \alpha') f_-(E', \alpha') \, d\alpha' \, dE, \tag{8.02}$$

where

$$S^E(E, \alpha | E', \alpha') = \delta(E-E')\delta(\alpha, \alpha') - 2\pi\varepsilon i \delta(E-E') T_-(E, \alpha | E', \alpha') . \tag{8.3}$$

Since $f_-(E, \alpha)$ and $f_+(E, \alpha)$ are the H_0 representatives of Φ_- and Φ_+ respectively, it is clear from the definition of the scattering operator (2.05) that $S^E(E, \alpha | E', \alpha')$ is the matrix of the scattering operator in the H_0 representation.

It might also be noted that $S^E(E, \alpha | F, \beta)$ is just the kernel $K_+(E, \alpha | F, \beta)$ of the transformation (7.03) when $K_-(E, \alpha | F, \beta)$ is given by (7.05). We should also note that it is not necessary that K_- be given by (7.05) in order to get an expression for the scattering operator. If we had chosen K_- to be the kernel of another unitary transformation, we should have obtained another representation of the scattering operator.

In a similar way, we construct the solution $\Phi(t)$ with the eigenfunctions u_+, and write

$$\Phi(t) \underset{E}{\leftrightarrow} f(E, \alpha; t) = \int_{E_a}^{E_b}\!\!\int u_+(E, \alpha | F, \beta) \exp[-iFt] f_+(F, \beta) \, d\beta, dF, \tag{8.04}$$

where $f_+(E, \alpha)$ is a given function such that

$$\Phi_+ \underset{E}{\leftrightarrow} f_+(E, \alpha) .$$

We can show

$$\lim_{t\to-\infty} \exp[iEt] f(E, \alpha; t) = f_-(E, \alpha) = \int_{E_a}^{E_b}\!\!\int \hat{S}^E(E, \alpha | E', \alpha') f_+(E', \alpha') \, d\alpha' \, dE', \tag{8.05}$$

where

$$\hat{S}^E(E, \alpha | E', \alpha') = \delta(E-E')\delta(\alpha, \alpha') + 2i\pi\varepsilon\delta(E-E') T_+(E, \alpha | E, \alpha') . \tag{8.06}$$

It is clear that $\hat{S}^E(E, \alpha | E', \alpha')$ is the matrix of the inverse scattering operator S^{-1} in the H_0 representation.

We can give $S^E(E, \alpha | E', \alpha')$ a more familiar interpretation in the following way. In taking the limits as $t \to \pm\infty$ we have assumed that Φ_- or, equivalently, its H_0 representative $f_-(E, \alpha)$ in normalizable. However, having obtained expressions for the limits in time for such proper states, we can then take Φ_- to be an eigenstate of H_0, this being a state which is not normalizable and hence not really a physical state. We wan consider such an eigenstate as having been constructed from a proper state by a suitable limiting process. Let us then prescribe Φ_- to be an eigenstate of H_0 corresponding to the eigenvalues E', α'. Then $f_-(E, \alpha) = \delta(E - E')\delta(\alpha, \alpha')$ and $f_+(E, \alpha) = S^E(E, \alpha|E', \alpha')$. Hence the matrix of the scattering operator can be interpreted as the H_0 representer of the state after scattering has taken place. The first term of $S^E(E, \alpha|E', \alpha')$, namely $\delta(E - E')\,\delta(\alpha, \alpha')$, which is also equal to $f_-(E, \alpha)$, is called the «incident» or «incoming» wave. The second term is called the «scattered wave».

9. – Adiabatic Switch-on of the Perturbation and the Adiabatic Theorem.

It has been shown that for a large class of solutions

$$\Phi(t) = \exp[-iHt]\Phi(0) \tag{9.01}$$

of the Schrödinger equation (2.01), one can evaluate the limits

$$\lim_{t\to\pm\infty} \exp[iH_0 t]\Phi(t) = \Phi_\pm\,. \tag{9.02}$$

Such solutions satisfy the unperturbed Schrödinger equation asymptotically as $t \to \pm\infty$ as pointed out in Section **2**. The solutions thus behave as though the perturbation had been «switched off» as $t \to \pm\infty$. In particular it was shown by means of the eigenfunctions u_- that if the initial condition on the solution of the Schrödinger equation (2.01) were that the H_0 representative $f(E, \alpha; t)$ should behave like $\exp[-iEt]f_-(E, \alpha)$ at $t \to -\infty$, then at finite times the H representative $g(F, \beta; t)$ was given by

$$g(F, \beta; t) = \exp[-iFt]f_-(F, \beta)$$

this condition being, in fact the condition used to specify u_- completely.

We shall now consider the problem in which the perturbation is «really» switched off at a very slow rate, i.e., adiabatically, as $t \to -\infty$. We shall show that if we prescribed as the initial condition that the solution of the Schrödinger equation shall behave as above as $t \to -\infty$, then at finite times t the H_0 and H representative, $f(E, \alpha; t)$ and $g(F, \beta; t)$, respectively, are the same as those obtained using the eigenfunctions u_- and assuming the perturbation constant in time.

SNYDER [6] appears to have been the first to have discussed the generally assumed equivalence of the adiabatic switch-off procedure and the procedure of using the spectral representation of H. FRIEDRICHS [11] has also discussed the equivalence of the adiabatic switch-off procedure to the procedure of using the spectral representation but in a more rigorous fashion.

In their discussion of the use of adiabatic switch off, both SNYDER and FRIEDRICHS have restricted themselves to the case where the total Hamiltonian at all times has a continuous spectrum only, which coincides with the spectrum of H_0. In our treatment we shall follow more closely the spirit of the work of LIPPMANN and SCHWINGER and FUBINI [12] in which no explicit assumption is made regarding the total Hamiltonian. However, in our discussion we shall indicate conditions needed to obtain meaningful results.

Accordingly we introduce the Hamiltonian $H_\zeta(t)$ defined by

(9.04) $$H_\zeta(t) = H_0 + \varepsilon \exp[\zeta t]V \qquad (\zeta > 0).$$

We note that

(9.05) $$\lim_{t\to-\infty} H_\zeta(t) = H_0$$

and

(9.06) $$\lim_{\zeta\to 0} H_\zeta(t) = H\,.$$

Let us denote by $f_\zeta(E, \alpha; t)$ the H_0 representative of $\Phi_\zeta(t)$ which is a solution of the Schrödinger equation

(9.07) $$H_\zeta(t)\,\Phi_\zeta(t) = i\frac{\partial}{\partial t}\Phi_\zeta(t)\,.$$

The representative $f_\zeta(E, \alpha; t)$ thus satisfies the equation

(9.08) $$i\frac{\partial f_\zeta(E, \alpha; t)}{\partial t} = Ef_\zeta(E, \alpha; t) + \\ + \varepsilon \exp[\zeta t]\int_{E_a}^{E_b}\int V^E(E, \alpha|E', \alpha')f_\zeta(E', \alpha'; t)\,d\alpha'\,dE'\,.$$

We take as initial conditions

(9.09) $$\lim_{t\to-\infty} \exp[iEt]f_\zeta(E, \alpha; t) = f_-(E, \alpha)$$

[11] K. O. FRIEDRICHS: *Report on the Adiabatic Theorem*, New York University Report (in preparation).

[12] S. FUBINI: *Nuovo Cimento*, **9**, 846 (1952).

where $f_-(E, \alpha)$ is prescribed. Let us write $f_\zeta(E, \alpha; t)$ in terms of $\exp[-iEt]f_-(E,\alpha)$, which is the solution of (9.08) if no perturbation were present, in terms of an integral operator as follows:

$$f_\zeta(E, \alpha; t) = \int_{E_a}^{E_b}\int w_{\zeta,t}(E, \alpha | E', \alpha') \exp[-iE't] f_-(E', \alpha') d\alpha' dE'. \tag{9.10}$$

It is our objective to show that in many cases $w_{\zeta,t}$ becomes the eigenfunction u_- in the adiabatic limit. That is,

$$\lim_{\zeta\to 0} w_{\zeta,t}(E, \alpha | E', \alpha') = u_-(E, \alpha | E', \alpha') . \tag{9.11}$$

We shall then have, as follows from (7.28) and (7.29), the result that $f_0(E, \alpha; t)$ and $\exp[-iFt]f_-(F, \beta)$ are the H_0 and H representatives, respectively, of the solution of $\Phi(t)$ of the Schrödinger equation

$$H\Phi(t) = i \frac{\partial \Phi(t)}{\partial t}, \tag{9.12}$$

with the initial condition

$$\lim_{t\to-\infty} \exp[iH_0 t]\Phi(t) = \Phi_- \underset{E}{\longleftrightarrow} f_-(E, \alpha) .$$

This statement is essentially the adiabatic theorem for the continuous spectrum which can be worded as follows: When a perturbation is switched on infinitely slowly, a superposition of eigenstates of the unperturbed Hamiltonian in the infinite past goes over into the *identical* superposition of corresponding eigenstates of the total Hamiltonian at finite times.

In our discussion it will be convenient to introduce another integral operator with the kernel $v_{\zeta,t}(E, \alpha | E', \alpha')$ defined by

$$v_{\zeta,t}(E, \alpha | E', \alpha') = w_{\zeta,t}(E, \alpha | E', \alpha') \exp[i(E - E')t] . \tag{9.13}$$

Hence

$$\exp[iEt] f_\zeta(E, \alpha; t) = \int_{E_a}^{E_b}\int v_{\zeta,t}(E, \alpha | E', \alpha') f_-(E', \alpha') d\alpha' dE'. \tag{9.14}$$

The initial condition (9.09) can now be interpreted as an initial condition on $v_{\zeta,t}(E, \alpha | E', \alpha')$, namely,

$$\lim_{t\to-\infty} v_{\zeta,t}(E, \alpha | E', \alpha') = \delta(E - E')\delta(\alpha, \alpha') . \tag{9.15}$$

The simplicity of the initial condition on $v_{\zeta,t}$ motivates its introduction in the present context. It should be pointed out that the kernel or matrix $v_{\zeta,t}(E, \alpha | E', \alpha')$ is just the operator U_+ of LIPPMANN and SCHWINGER as expressed in the H_0 representation. This operator also appear in FUBINI's paper.

Substitution of (9.14) into the differential equation (9.08) for $f_\zeta(E, \alpha; t)$ and the use of the fact that $f_-(E, \alpha)$ is largely arbitrary leads us to the following equation for $v_{\zeta,t}$:

$$(9.16) \qquad \frac{\partial}{\partial t} v_{\zeta,t}(E, \alpha | E', \alpha') = -i\varepsilon \int_{E_a}^{E_b}\!\!\int V^E(E, \alpha | E'', \alpha'') \exp[[i(E-E'')+\zeta]t] \cdot v_{\zeta,t}(E'', \alpha'' | E', \alpha') \mathrm{d}\alpha'' \mathrm{d}E'',$$

or on integration with respect to t and on using the initial condition (9.15) we find

$$(9.17) \qquad v_{\zeta,t}(E, \alpha | E', \alpha') = \delta(E-E')\delta(\alpha, \alpha') - i\varepsilon \int_{-\infty}^{t} \mathrm{d}t' \int_{E_a}^{E_b}\!\!\int V^E(E, \alpha | E'', \alpha'') \cdot \exp[[i(E-E'')+\zeta]t'] v_{\zeta,t'}(E'', \alpha'' | E', \alpha') \mathrm{d}\alpha'' \mathrm{d}E''.$$

Having used our initial condition, we express $v_{\zeta,t}$ in terms of $w_{\zeta,t}$((9.13)) and obtain an equation for $w_{\zeta,t}$

$$(9.18) \qquad w_{\zeta,t}(E, \alpha | E', \alpha') = \delta(E-E')\delta(\alpha, \alpha') - i\varepsilon \exp[-i(E-E')t] \cdot \int_{-\infty}^{t} \mathrm{d}t' \int_{E_a}^{E_b}\!\!\int V^E(E, \alpha | E'', \alpha'') \exp[[i(E-E')+\zeta]t'] \cdot w_{\zeta,t'}(E'', \alpha'' | E', \alpha') \mathrm{d}\alpha'' \mathrm{d}E''.$$

In equation (9.18) let us now integrate the time integral by parts. We obtain

$$(9.19) \qquad w_{\zeta,t}(E, \alpha | E', \alpha') = \delta(E-E')\delta(\alpha, \alpha') - \frac{i\varepsilon \exp[\zeta t]}{i(E-E')+\zeta} \cdot \int_{E_a}^{E_b}\!\!\int V^E(E, \alpha | E'', \alpha'') w_{\zeta,t}(E'', \alpha'' | E', \alpha') \mathrm{d}\alpha'' \mathrm{d}E'' + $$
$$+ \frac{i\varepsilon \exp[-i(E-E')t]}{i(E-E')+\zeta} \int_{-\infty}^{t} \mathrm{d}t' \int_{E_a}^{E_b}\!\!\int V^E(E, \alpha | E'', \alpha'') \exp[[i(E-E')+\zeta]t'] \cdot \frac{\partial}{\partial t'} w_{\zeta,t'}(E'', \alpha'' | E', \alpha') \mathrm{d}\alpha'' \mathrm{d}E''.$$

Taking the limit as ζ approaches zero, we find

$$(9.20)\qquad w_{0,t}(E,\alpha|E',\alpha') = \delta(E-E')\delta(\alpha,\alpha') + \gamma_-(E'-E)\int\int_{E_a}^{E_b} V^E(E,\alpha|E'',\alpha'')\cdot$$

$$\cdot w_{0,t}(E'',\alpha''|E',\alpha')\,d\alpha''\,dE'' + i\varepsilon\exp[-i(E-E')t]\lim_{\zeta\to 0}\frac{1}{i(E-E')+\zeta}\cdot$$

$$\cdot\int_{-\infty}^{t} dt'\int\int_{E_a}^{E_b} V^E(E,\alpha|E'',\alpha'')\exp[[i(E-E')+\zeta]t']\frac{\partial}{\partial t'}w_{\zeta,t'}(E'',\alpha''|E',\alpha')\,d\alpha''\,dE''\,.$$

We see that if

$$(9.21)\qquad \lim_{\zeta\to 0}\frac{1}{i(E-E')+\zeta}\int_{-\infty}^{t} dt'\int\int_{E_a}^{E_b} V^E(E,\alpha|E'',\alpha'')\exp[[i(E-E')+\zeta]t']\cdot$$

$$\cdot\frac{\partial}{\partial t'}w_{\zeta,t'}(E'',\alpha''|E',\alpha')\,d\alpha''\,dE'' = 0\,,$$

then $w_{0,t}$ satisfies the same integral equation as the eigenfunction u_-. Since the solution of this integral equation can be shown to be unique, we have the result that (9.21) is a necessary and sufficient condition for $w_{0,t}$ to be identical to the eigenfunction u_-. That is, relation (9.21) is a necessary and sufficient condition for the equivalence of the adiabatic switch-off procedure and the use of the spectral representation of H in the time-dependent formalism in scattering problem. In our treatment (and that of references ([3]) and ([12])) no explicit conditions have been put on the total Hamiltonian H. However, condition (9.21) is essentially a condition on the perturbation and hence also on H.

A sufficient condition for the validity of the adiabatic theorem can be seen to be

$$(9.22)\qquad \lim_{\zeta\to 0}\frac{\partial}{\partial t}w_{\zeta,t}(E,\alpha|E',\alpha') = 0\,,$$

since in this case (9.21) is satisfied. We shall now give arguments to indicate that (9.22) and hence the adiabatic theorem is valid in many cases, even when H has point eigenvalues, a case not included in Snyder's or Friedrich's treatments.

From (9.19) it can be shown that $\partial w_{\zeta,t}/\partial t$ satisfies the following equation.

$$(9.23)\quad \frac{\partial}{\partial t} w_{\zeta,t}(E,\alpha|E',\alpha') = \frac{\varepsilon}{(E'-E)+i\zeta}\Big\{\zeta \exp[\zeta t]\int_{E_a}^{E_b}\!\!\int V^E(E,\alpha|E'',\alpha'')\cdot$$

$$\cdot w_{\zeta,t}(E'',\alpha''|E',\alpha')\mathrm{d}\alpha''\mathrm{d}E'' + i(E-E')\exp[-i(E-E')t]\cdot$$

$$\cdot\int_{-\infty}^{t}\mathrm{d}t'\int_{E_a}^{E_b}\!\!\int V^E(E,\alpha|E'',\alpha'')\cdot$$

$$\cdot\frac{\partial}{\partial t'} w_{\zeta,t'}(E'',\alpha''|E',\alpha')\exp[[i(E-E')+\zeta]t']\,\mathrm{d}\alpha''\mathrm{d}E''\Big\}.$$

We note that that we can write

$$w_{\zeta,t}(E,\alpha|E',\alpha') = \int_{-\infty}^{t}dt'\,\frac{\partial}{\partial t'} w_{\zeta,t'}(E,\alpha|E',\alpha') + w_{\zeta,-\infty}(E,\alpha|E',\alpha').$$

However, from (9.19) it is clear that

$$w_{\zeta,-\infty}(E,\alpha|E',\alpha') = \delta(E-E')\,\delta(\alpha,\alpha'),$$

so that (9.23) becomes

$$(9.24)\quad \frac{\partial}{\partial t} w_{\zeta,t}(E,\alpha|E',\alpha') = \frac{\varepsilon\exp[\zeta t]}{(E'-E)+i\zeta}\Big\{\zeta V^E(E,\alpha|E',\alpha') +$$

$$+\int_{-\infty}^{t}\mathrm{d}t'\int_{E_a}^{E_b}\!\!\int V^E(E,\alpha|E'',\alpha'')\{\zeta + i(E-E')\exp[i(E-E')+\zeta](t'-t)\}\cdot$$

$$\cdot\frac{\partial}{\partial t'} w_{\zeta,t'}(E'',\alpha''|E',\alpha')\,\mathrm{d}\alpha''\mathrm{d}E''\Big\}.$$

We can solve equation (9.24) by iteration. We find that

$$(9.25)\quad \frac{\partial}{\partial t} w_{\zeta,t}(E,\alpha|E',\alpha') = \zeta\Big[\varepsilon\exp[\zeta t](E'-E+i\zeta)^{-1}V^E(E,\alpha|E',\alpha') +$$

$$+2\varepsilon^2\exp[2\zeta t](E'-E+2i\zeta)^{-1}\int_{E_a}^{E_b}\!\!\int V^E(E,\alpha|E'',\alpha'')(E'-E''+i\zeta)^{-1}\cdot$$

$$\cdot V^E(E'',\alpha''|E',\alpha')\,\mathrm{d}\alpha''\mathrm{d}E'' +$$

$$+3\varepsilon^3\exp[3\zeta t](E'-E+3i\zeta)^{-1}\int_{E_a}^{E_b}\!\!\int\int_{E_a}^{E_b}\!\!\int V^E(E,\alpha|E'',\alpha'')(E'''-E''+i\zeta)^{-1}\cdot$$

$$\cdot V^E(E'',\alpha''|E''',\alpha''')(E'-E'''+2i\zeta)^{-1}V^E(E''',\alpha'''|E',\alpha')\cdot\mathrm{d}\alpha''\mathrm{d}E''\mathrm{d}\alpha'''\mathrm{d}E''' + \ldots\Big].$$

If the series with the square brackets converges as ζ approaches zero, then it is clear that (9.22) is satisfied. Now when $\zeta \to 0$ the series in brackets is within a factor ε the derivative with respect to ε of the series for u_- obtained by iteration from (7.19*a*). Hence if the series for u_- converges, i.e., if u_- is an analytic function in ε, the series in brackets will converge, since the derivative of an analytic function is also analytic and hence is representable as a power series with the same radius of convergence. We thus come to the conclusion that a sufficient condition for the validity of the adiabatic theorem is the convergence of the power series in ε for u_-. Since there are examples in which the series for u_- converges even when H has point eigenvalues (see e.g., ref. ([5])) we have indicated that there are cases even when the spectrum of H differs from that of H_0 where the adiabatic theorem holds.

We could have obtained the above results by solving equation (9.18) directly by iteration and then allowing ζ to approach zero. We would then have found that the series expression for $w_{0,t}$ is identical to that for u_-. This procedure is, in fact, used by SNYDER in his discussion of the «irrelevance of the adiabatic theorem». However, it appears useful to indicate explicit conditions, namely, (9.21) or (9.22) under which the adiabatic theorem will hold.

It might be noted that the convergence of the power series for u_- is a sufficient condition only, so that it is possible that the adiabatic theorem will hold even when such a series expansion is not valid. This appears to be the main point of SNYDER's treatment in which he assumes $H_\zeta(t)$ has the same spectrum of H_0 for all values of ζ and t.

RIASSUNTO (*)

Si dimostra che le autofunzioni di entrata e di uscita che si impiegano nella teoria dello scattering si possono ottenere specificando una relazione fra gli stati iniziali o finali e lo stato al tempo $t=0$. Così diviene superfluo il far ricorso al teorema adiabatico.

(*) *Traduzione a cura della Redazione.*

RESEARCH REPORT No. EM-74

NEW YORK UNIVERSITY
Institute of Mathematical Sciences
Division of Electromagnetic Research

THE INVERSE SCATTERING PROBLEM

IRVIN KAY

1955

THE INVERSE SCATTERING PROBLEM

by Irvin Kay

Abstract

The differential equation $u''(k,x) + [k^2 - V(x)]u(k,x) = 0$ is considered in the interval $-\infty < x < \infty$. From a knowledge of the reflection coefficient $b(k)$ associated with a wave e^{ikx} incident from the left it is shown how one can calculate the function $V(x)$ of the differential equation (when $V(x) = 0$ for x less than some constant). The method used is due to I.M. Gelfand and B.M. Levitan.

Sufficient conditions such that a function $b(k)$ can be the reflection coefficient corresponding to a differential equation of this form are given.

An example, in which $b(k) = -(k^2+1)^{-1}$, is worked out in detail. It is also shown how one can obtain $V(x)$ explicitly for a large class of reflection coefficients.

The differential equation $u''(k,x) + k^2n^2(x)u(k,x) = 0$ in the interval $-\infty < x < \infty$ is also considered, the problem being to determine the function $n(x)$ from a knowledge of the reflection coefficient, assuming that $n(x) =$ constant for x less than some given value.

1. INTRODUCTION

It is a customary procedure in mathematical physics to make predictions about particular physical situations on the basis of a general theory and then to compare those predictions with experimental data. This procedure is, of course, necessary to check the validity of a new theory. But when a physical theory has evoked sufficient confidence because of its continuing success over many years, one need no longer look upon such a procedure as a test of the theory. Rather, the theory becomes a tool by means of which one can describe in detail the various special cases arising when particular conditions are added to the general theoretical assumptions. Experimental data for a special case are then useful to measure or make determinate those physical quantities of the general theory which otherwise are arbitrary. In the application of a trusted theory to special cases it still seems to be customary, however, to proceed in the same way as in the verification of a new theory. One makes certain assumptions about the relevant physical situation, one predicts the consequences of these assumptions on the basis of the general theory, and finally, one makes an empirical check of the assumption by means of experimental data. The assumptions can be corrected according to the experimental results and the procedure repeated until the desired accuracy is reached. In the case of an old and tested theory, however, it would appear more logical to adopt a different procedure, if possible. A gain in time, accuracy and understanding might be expected if one were able to reconstruct directly from the experimental data the physical conditions corresponding to a special case of the general theory.

The problem of making such a reconstruction is the direct inverse of the problem ordinarily undertaken. This reconstruction is closely related to the problem of synthesis wherein the task is to construct special physcial conditions which, according to the general theory, will produce experimental data given beforehand.

An example of the "inverse problem" is the attempt to determine the ionization density of the ionosphere from the time delay of a pulse radio wave which has been transmitted from the earth and reflected back to the earth by the ionosphere. The synthesis problem is a common one for electrical engineers who are often required to design networks which produce a prescribed response to a prescribed input voltage over a given frequency band.

Some work has been done in recent years on inverse problems suggested by quantum mechanics and associated with the differential equation

$$\frac{d^2}{dx^2} u(k,x) + [k^2 - V(x)] u(k,x) = 0 .$$

For the quantum mechanical application the boundary condition $u(k,0) = 0$ is assumed and the variable x lies in the interval $0 \leq x < \infty$. The given experimental data provide a knowledge of the phase function $\phi(k)$ which appears in the asymptotic form of $u(k,x)$ for large positive x:

$$u(k,x) \sim A(k) \sin[kx + \phi(k)] , \qquad x \to +\infty .$$

The problem is: Given $\phi(k)$; determine $V(x)$. The uniqueness of the solution when the operator

$$-\frac{d^2}{dx^2} + V(x)$$

has no bound states (proper eigenfunctions) was proved by N. Levinson [1] and V.A. Marcenko [2]. A similar inverse problem for the same differential equation arises when a finite interval is chosen and homogeneous boundary conditions are applied at both ends: Given the two sets of eigenvalues of the operator

$$\frac{d^2}{dx^2} + V(x)$$

corresponding to two different pairs of boundary conditions, determine $V(x)$. The uniqueness of $V(x)$ in this situation was proved by N. Levinson [3] and

G. Borg [4] and V.A. Marcenko [2]. I.M. Gelfand and B.M. Levitan [5] have shown how to construct the quantity $V(x)$ for the interval $0 \leq x < \infty$ from the spectral measure function associated with the operator

$$- \frac{d^2}{dx^2} + V(x)$$

and boundary conditions. Another solution by an entirely different method, capable of including the case in which bound states exist, was given by W. Kohn and R. Jost [6]. M. Krein [7] has solved the inverse problem in a different way for the finite interval.

A very clear English summary of the paper by I.M. Gelfand and B.M. Levitan [5], with an additional note on the error due to a small perturbation in the spectral measure function, has been published by N. Levinson [8].

It will be the purpose of this paper to treat the same ordinary differential equation in the interval $-\infty < x < \infty$. The information assumed given will be the reflection coefficient $b(k)$ when the solution $u(k,x)$ of the differential equation has the asymptotic forms

$$u(k,x) \rightarrow e^{ikx} + b(k)e^{-ikr} , \qquad x \rightarrow -\infty ,$$

$$u(k,x) \rightarrow a(k)e^{ikx} , \qquad x \rightarrow +\infty .$$

The function $V(x)$ will be obtained from $b(k)$ by an adaptation of the method of I.M. Gelfand and B.M. Levitan [5], and an explicit example will be worked out. Moreover, sufficient conditions will be derived for a function $b(k)$ to be the reflection coefficient of such a differential equation (this is necessary for the synthesis problem), and a method for solving a large class of inverse problems explicitly will be given. Finally, the differential equation

$$\frac{d^2}{dx^2} u(k,x) + k^2 n^2(x) u(k,x) = 0 ,$$

for which the problem is to find the function $n(x)$, will be considered.

The solutions obtained in this article appear to have a number of applications. The synthesis problems for a non-uniform transmission line (in which either the inductance or the capacitance is constant) and, under certain conditions, for a waveguide of variable cross-section can be treated by the method described here. A synthesis problem in which a variable stratified dielectric is to be constructed so that the reflection coefficient of an electromagnetic wave will vary in a prescribed way with frequency can also be treated. Another synthesis problem involves a plane wave incident on a plane interface of a variable dielectric slab; the reflected wave amplitude variation as a function of the angle of incidence is prescribed, and the variation of the dielectric properties of the slab can be found so as to produce the prescribed reflected wave amplitude variation. The problem, already mentioned, of determining the ionization density of a stratified ionosphere is another possible application.

In this investigation the preliminary information used concerning the differential equation

$$\frac{d^2}{dx^2} u(k,x) + [k^2 - V(x)]\, u(k,x) = 0$$

and the general facts about the spectral theory and about the scattering matrix associated with the differential equation in the interval $-\infty < x < \infty$ can be found in the Appendix 17 of K. Friedrich's notes on the spectral theory of linear operators in Hilbert space [9].

2. THE BASIC DIFFERENTIAL EQUATION AND SOME OF ITS PROPERTIES

If a continuously differentiable wave function $u(k,x)$ satisfies the differential equation

$$\frac{d^2 u(k,x)}{dx^2} + [k^2 - V(x)]\, u(k,x) = 0 , \tag{2.1}$$

where the function $V(x)$ is real, bounded, piecewise continuous and belongs to L_1 in the interval $-\infty < x < \infty$, then it is well known that the asymptotic behavior of $u(k,x)$ can be expressed in the form

$$\begin{aligned} u(k,x) &\sim \gamma_+(k)e^{ikx} + \delta_+(k)e^{-ikx} , & x \to +\infty \\ u(k,x) &\sim \gamma_-(k)e^{ikx} + \delta_-(k)e^{-ikx} , & x \to -\infty . \end{aligned} \tag{2.2}$$

The quantities $\gamma_\pm, \delta_\pm$ are, as indicated, functions of k alone and independent of x. Moreover, a linear relation exists between the components of the vector

$$\begin{pmatrix} \gamma_+ \\ \delta_+ \end{pmatrix}$$

and those of the vector

$$\begin{pmatrix} \gamma_- \\ \delta_- \end{pmatrix}$$

This linear relation can be expressed in terms of a two by two square matrix, the so-called scattering matrix

$$S: \begin{pmatrix} \gamma_- \\ \delta_- \end{pmatrix} = S \begin{pmatrix} \gamma_+ \\ \delta_+ \end{pmatrix} .$$

The four components of S are functions of k alone which are the same for all solutions u(k,x) of (2.1) and are fixed by the function V(x) appearing in (2.1). If we write S in the form

$$S = \begin{pmatrix} \alpha(k) & \alpha'(k) \\ \beta(k) & \beta'(k) \end{pmatrix} , \qquad (2.3)$$

the following relations between the components of S hold for real K:

$$\begin{aligned} \alpha(-k) &= \overline{\alpha(k)} = \alpha'(k) \\ \beta(-k) &= \overline{\beta(k)} = \beta'(k) \qquad (2.4) \\ |\alpha|^2 - |\beta|^2 &= 1 . \end{aligned}$$

From the last equation in (2.4) it follows that

$$|\alpha|^2 > |\beta|^2 ; \qquad |\alpha|^2 \geq 1 , \qquad (2.5)$$

for finite values of $|\alpha|$ and $|\beta|$ when k is real. It is also known that $\alpha(k)$ and $\beta(k)$ are analytic functions of k in the complex plane.

It is possible to find a solution of (2.1) which has the following asymptotic behavior:

$$\begin{aligned} u(k,x) &= e^{ikx} + b(k)e^{-ikx} , \qquad x \to -\infty \\ &= a(k)e^{ikx} , \qquad x \to +\infty . \end{aligned} \qquad (2.6)$$

From the definition of the scattering matrix we have

$$\begin{aligned} b(k) &= \frac{\beta(k)}{\alpha(k)} , \\ a(k) &= \frac{1}{\alpha(k)} , \end{aligned} \qquad (2.7)$$

and from (2.4) it follows that

$$|a|^2 + |b|^2 = 1 \qquad (2.8)$$

when k is real. In accordance with the asymptotic behavior of u(k,x) given by (2.6) the quantity b(k) is called the reflection coefficient and the quantity a(k) the transmission coefficient.

In general, for the interval $-\infty < x < +\infty$, the operator $-(d^2/dx^2) + V(x)$ of Equation (2.1) has a spectral representation with a spectrum composed of a discrete part and a continuous part. The discrete spectrum has for its corresponding eigenfunctions those solutions of (2.1) which belong to L_2. For values of k in the discrete spectrum the eigenfunctions must then die out exponentially at $+\infty$ and $-\infty$; therefore, from (2.2) it is apparent that the discrete eigenvalues occur only for values of k which have non-zero imaginary parts. Since the eigenvalues of (2.1) are given by k^2 which is real only when k is pure imaginary or real and since the operator $-(d^2/dx^2) + V(x)$ is Hermitian, it follows that the values of k corresponding to the eigne-values must lie on the imaginary axis. Since when $u(k,x)$ satisfies (2.1) so does $u(-k,x)$, we can without loss of generality restrict the eigenvalues to lie on the positive imaginary axis.

Further, the eigenvalues occur for values of k for which the function $b(k)$ has poles on the positive imaginary axis. It is known that these poles are simple and that $b(k)$ is otherwise regular in the upper half-plane and on the real axis, except perhaps at zero. If we add the condition that $xV(x)$ belong to L_1, then $b(k)$ has only a finite number of poles in the upper half-plane; i.e., there are only a finite number of eigenvalues.

The completeness of the spectral resolution of the operator $-(d^2/dx^2) + V(x)$ of (2.1) can be expressed succinctly by the relation

$$\sum_{\nu} u(k_\nu,x)\, u^*(k_\nu,y) + \int_{-\infty}^{\infty} u(k,x)\, u^*(k,y)\, dk = \delta(x-y) \quad , \tag{2.9}$$

where $\delta(x-y)$ is the Dirac delta-function which has the property that

$$\int_{-\infty}^{\infty} f(x)\, \delta(x)\, dx = f(0)$$

for arbitrary $f(x)$. The function $u^*(k,x)$ is a particular solution of (2.1) corresponding to the given solution $u(k,x)$; the sum in (2.9) is taken over the values of k corresponding to the eigenvalues. When $u(k,x)$ and $u^*(k,x)$ are analytic functions of k the relation on the left side of (2.9) can be written as a single integral in the complex plane,

$$\int_C u(k,x)\, u^*(k,y)\, ds \quad ,$$

in which the contour of integration, C, is a large semi-circle in the upper half-plane. The relationship between the proper eigenfunctions and the improper eigenfunctions [$u(k,x)$, k in the continuous spectrum] is made clear by this contour integral representation.

3. FORMULATION OF THE MAIN PROBLEM AND DERIVATION OF THE FUNDAMENTAL INTEGRAL EQUATION

The problem we are going to consider is: Given the reflection coefficient $b(k)$ as a function of k for an equation of the form (2.1) with an unknown real $V(x)$, to find $V(x)$ as a function of x. The method we shall use for this investigation is due to Gelfand and Levitan [5] and has been summarized by M. Levinson [8] in English.

The quantity $V(x)$ in (2.1) can be looked upon as a perturbation operator which, when added to the operator $-(d^2/dx^2)$ of the equation

$$\frac{d^2}{dx^2} u_0(k,x) + k^2 u_0(k,x) = 0 , \tag{3.1}$$

produces the operator $-(d^2/dx^2) + V(x)$ of (2.1). We shall refer to functions $u_0(k,x)$ satisfying (3.1) as solutions of the unperturbed equation, in contrast to the functions $u(k,x)$ which are solutions of the perturbed equation (2.1). From this point on, for simplicity, we shall assume that the function $V(x)$ vanishes for $x < -\delta/2$, where δ is some real positive constant.

The first step in the Gelfand-Levitan method is to recognize a fundamental property of the relationship between a solution $u_0(k,x)$ of the unperturbed equation and a solution $u(k,x)$ of the perturbed equation. There exists a linear transformation $T(x,y)$ which transforms a solution $u_0(k,x)$ of the unperturbed equation into a solution $u(k,x)$ of the perturbed equation, namely,

$$T(x,y) = \int_{-\infty}^{\infty} u(k,x)\, u_0^*(k,y)\, dk$$

where $u_0^*(k,y)$ is the conjugate function of $u_0(k,y)$ in the completeness relation

$$\int_{-\infty}^{\infty} u_0(k,x)\, u_0^*(k,y)\, dk = \delta(x-y) .$$

The fundamental observation to be made is this: If $u(k,x) = u_0(k,x)$ for $x \leq -\delta$ [we recall that $V(x) = 0$ for $x < -\delta/2$], then $T(x,y)$ is *triangular* in the sense that $T(x,y) = 0$ for $y < x$. It is convenient to write $T(x,y) = \delta(x-y) + K(x,y)$. Then $K(x,y)$ is also triangular, that is, $K(x,y) = 0$ for $y < x$. We now state the following:

<u>Theorem</u>. If $u(k,x) = u_0(k,x)$ for $x \leq -\delta/2$, then

$$u(k,x) = u_0(k,x) + \int_{-(x+\delta)}^{x} K(x,y)\, u_0(k,y)\, dy ,$$

and $K(x,y)$ is uniformly bounded, continuous, and has piecewise continuous first and second derivatives.

To prove the theorem we apply the operator $(d^2/dx^2) - V(x)$ to (3.2) on both sides. Using (2.1) and (3.1) we obtain

$$
\begin{aligned}
-k^2u(k,x) &= -k^2u_0(k,x) - V(x)u_0(k,x) + u_0(k,x)\frac{dK(x,x)}{dx} \\
&+ K(x,x)\frac{du_0(k,x)}{dx} + u_0(k,-x-\delta)\frac{dK(x,-x-\delta)}{dx} \\
&+ K(x,-x-\delta)\frac{du_0(k,-x-\delta)}{dx} + K_x(x,x)u_0(k,x) \\
&+ K_x(x,-x-\delta)u_0(k,-x-\delta) + \int_{-(x+\delta)}^{x} K_{xx}(x,y)u_0(k,y)\,dy \\
&- V(x)\int_{-(x+\delta)}^{x} K(x,y)u_0(k,y)\,dy \quad .
\end{aligned}
$$

But from (3.2) and (3.1), integrating by parts, we have

$$
\begin{aligned}
-k^2u(k,u) &= -k^2u_0(k,x) + \int_{-(x+\delta)}^{x} K(x,y)\frac{d^2}{dy^2}u_0(k,y)\,dy \\
&= -k^2u_0(k,x) + K(x,x)\frac{du_0(k,x)}{dx} + K(x,-x-\delta)\frac{du_0(k,-x-\delta)}{dx} \\
&- K_y(x,x)u_0(k,x) + K_y(x,-x-\delta)u_0(k,-x-\delta) \\
&+ \int_{-(x+\delta)}^{x} K_{yy}(x,y)u_0(k,y)\,dy \quad .
\end{aligned}
$$

Then combining these two results and combining terms appropriately we obtain

$$
\begin{aligned}
0 &= \int_{-(x+\delta)}^{x} \left\{K_{xx}(x,y) - K_{yy}(x,y) - V(x)K(x,y)\right\} u_0(k,y)\,dy \\
&+ \left\{2\frac{dK(x,x)}{dx} - V(x)\right\} u_0(k,x) + 2\frac{dK(x,-x-\delta)}{dx}u_0(k,x) \quad .
\end{aligned}
$$

This equation will be satisfied if

$$
K_{xx}(x,y) - K_{yy}(x,y) - V(x)K(x,y) = 0
$$

and

$$\frac{dK(x,x)}{dx} = \frac{1}{2} V(x) \quad , \tag{3.3}$$

and, if,

$$\frac{dK(x, -x-\delta)}{dx} = 0 \quad ,$$

that is, if $K(x, -x-\delta) = \text{constant}$.

On the other hand we know from the theory of hyperbolic partial differential equations that there exists a $K(x,y)$ satisfying (3.3) (this is a characteristic boundary-value problem). It is easy to see that having determined $K(x,y)$ from (3.3) we could retrace our steps and arrive at (3.2). Thus, the existence of $K(x,y)$ is proved (obviously this function satisfies the conditions of the theorem). It should be observed that, according to (3.3), $K(x,y)$ is independent of the choice of particular solutions $u(k,x)$ of (2.1) and $u_0(k,x)$ of (3.1) except that (3.2) implies: $u(k,x) = u_0(kx)$ when $x \le -\delta/2$. Thus $K(k,x)$ transforms any solution of (3.1) into a solution of (2.1) which is the same function for $x \le -\delta/2$.

By a completely analogous argument we can demonstrate the existence of a triangular operator kernel $K_0(x,y)$ such that

$$u_0(k,x) = u(k,x) + \int_{-(x+\delta)}^{x} K_0(x,y)u(k,y)\, dy \quad , \tag{3.4}$$

when $u(k,x) = u_0(k,x)$ for $x \le -\delta/2$.

In order to make use of contour integration in the complex plane we shall work with two solutions $v(k,x)$ and $\hat{v}(k,x)$ of (2.1) which are analytic functions of k for $\text{Im } K > 0$ and continuous functions of k for $\text{Im } k \ge 0$. They can be defined by their asymptotic behavior:

$$v(k,x) \begin{cases} \sim \alpha(k)e^{ikx} + \beta(k)e^{-ikx} \quad , & x \to -\infty \\ \sim e^{ikx} \quad , & x \to +\infty \end{cases}$$

$$\hat{v}(k,x) \begin{cases} \sim e^{-ikx} \quad , & x \to -\infty \\ \sim -\overline{\beta(k)}\, e^{ikx} + \alpha(k)e^{-ikx} \quad , & x \to +\infty \end{cases} \tag{3.5}$$

From the spectral representation theory of the operator $-(d^2/dx^2) + V(x)$ we have the relation

$$\frac{1}{2\pi} \int_C \frac{v(k,x)\hat{v}(k,y)}{\alpha(k)}\, dk = \delta(x-y) \quad ,$$

where the contour C is a large semi-circle in the upper half of the complex k-plane and the integration direction is clockwise. The completeness relation (3.6) could also be written

$$\frac{1}{2\pi} \int_{-\infty}^{\infty} \frac{v(k,x)\hat{v}(k,y)}{\alpha(k)} \, dx - \sum_{\nu} a_{\nu} v(k_{\nu},x) \, \hat{v}(k_{\nu},y) = \delta(x-y) , \tag{3.7}$$

where the sum runs over the poles of $1/\alpha(k)$ in the upper half-plane, and the coefficients a_{ν} are i times the residues $1/\alpha'(k)$ of $1/\alpha(k)$ at these poles.

Before we can proceed we must establish the following identity:

$$\frac{1}{2\pi} \int_{-\infty}^{\infty} \frac{v(k,x)}{\alpha(k)} \, e^{-iky} \, dk - \sum_{\nu} a_{\nu} v(k_{\nu},x) e^{-ik_{\nu}y} = 0 , \quad x > y . \tag{3.8}$$

To this end we consider the kernel $\hat{K}(x,y)$ in the relation

$$e^{-iky} = \hat{v}(k,y) + \int_{-(y+\delta)}^{y} \hat{K}(y,z)\hat{v}(k,z) \, dz . \tag{3.9}$$

Assume $x > y$. Multiply (3.9) by $v(k,x)/2\pi\alpha(k)$ and integrate with respect to k from $-\infty$ to $+\infty$; then add to this the result of multiplying (3.9) with $k = k_{\nu}$ by $a_{\nu} u(k_{\nu},x)$ and summing over ν. The left side of the result is identical with the left side of (3.8). By (3.7) the right side of the result is

$$\delta(x-y) + \int_{-(y+\delta)}^{y} \hat{K}(y,z)\delta(x-z) \, dz .$$

Thus, for $x > y$ the right side of the result vanishes, and (3.8) follows.

We are now prepared to obtain an integral equation for $K(x,y)$, from which by (3.3) we can determine $V(x)$. Consider the relation (3.2) in the particular form

$$v(k,x) = \alpha(k)e^{ikx} + \beta(k)e^{-ikx} + \int_{-(x+\delta)}^{x} K(x,z) \left[\alpha(k)e^{ikz} + \beta(k)e^{-ikz}\right] dz. \tag{3.10}$$

We multiply (3.10) by $e^{-iky}/2\pi\alpha(k)$ and integrate over the contour C of (3.6). From (3.8) it is apparent that the left side of the resulting equation vanishes for $x > y$. We have then an equation of the form

$$0 = B(x+y) + K(x,y) + \int_{-(x+\delta)}^{x} K(x,z)B(y+z)\,dz \quad , \quad x > y \tag{3.11}$$

where

$$B(\tau) = \frac{1}{2\pi}\int_{-\infty}^{\infty} b(k)e^{-ik\tau}\,dk - i\sum_{\nu} b_{\nu}e^{-ik_{\nu}\tau} \quad . \tag{3.12}$$

In (3.12) we have written b_{ν} for the residues of $b(k)$ at the poles k_{ν} of $b(k)$ in the upper half-plane, and we have used the fact that

$$\frac{1}{2\pi}\int_{-\infty}^{\infty} e^{ik\tau}\,dk = \delta(\tau)$$

as well as the definition

$$b(k) = \frac{\beta(k)}{\alpha(k)}$$

to arrive at (3.11).

For every fixed value of x Equation (3.11) is a Fredholm integral equation of the second kind for $K(x,y)$ when y ranges over the interval $-\infty < y < x$. The kernel $B(y+z)$ is given by (3.12) in terms of the reflection coefficient $b(k)$.

If (3.11) has a unique solution and we solve for $K(x,y)$, the solution must be the function $K(x,y)$ in (3.10). The necessary step remaining, therefore, is to prove the uniqueness of the solution of (3.11). We can do this by showing that the operator kernel $\delta(x-y) + B(x+y)$ is positive definite.

First we establish the identity

$$\begin{aligned}\delta(x-y) + \frac{1}{2\pi}\int_{-\infty}^{\infty} b(k)e^{-ik(x+y)}\,dk &= \frac{1}{2\pi}\int_{0}^{\infty} |a(k)|^2 e^{ik(x-y)}\,dk \\ &+ \frac{1}{2\pi}\int_{-\infty}^{0}\left[e^{ikx} + b(k)e^{-ikx}\right] \\ &\cdot \left[\overline{e^{iky} + b(k)e^{-iky}}\right]dk \quad .\end{aligned} \tag{3.13}$$

Here $a(k)$ is the transmission coefficient. Using (2.8) we can write (3.13) in the form

$$\delta(x-y) + \frac{1}{2\pi}\int_{-\infty}^{\infty} b(k)e^{-ik(x+y)}\,dk$$

$$= \frac{1}{2\pi}\int_{0}^{\infty} e^{ik(x-y)}\,dk - \frac{1}{2\pi}\int_{0}^{\infty} |b(k)|^2 e^{ik(x-y)}\,dk$$

$$+ \frac{1}{2\pi}\int_{-\infty}^{0} e^{ik(x-y)}\,dk + \frac{1}{2\pi}\int_{-\infty}^{0} |b(k)|^2 e^{-ik(x-y)}\,dk$$

$$+ \frac{1}{2\pi}\int_{-\infty}^{0} b(k)\,e^{-ik(x+y)} + \frac{1}{2\pi}\int_{-\infty}^{0} \overline{b(k)}\,e^{ik(x+y)}\,dk\ .$$

The sum of the first and third terms on the right side is $\delta(x-y)$, and the second and fourth terms on the right side cancel. We have finally

$$\frac{1}{2\pi}\int_{-\infty}^{\infty} b(k)\,e^{-ik(x+y)}\,dk = \frac{1}{2\pi}\int_{-\infty}^{0} b(k)\,e^{-ik(x+y)}\,dk$$
$$+ \frac{1}{2\pi}\int_{-\infty}^{0} \overline{b(k)}\,e^{ik(x+y)}\,dk\ . \tag{3.14}$$

Changing variables, by (2.4) and (2.7), from k to $-k$ in the second terms on the right, we see that (3.14) is, indeed, an identity. Since our steps are obviously reversible, it follows that (3.13) is an identity.

The term

$$-i\sum_{\nu} b_{\nu}\,e^{-ik_{\nu}(x+y)}$$

can be shown to be real and positive for all real x and y. The exponential factor is obviously real and positive, since the k_{ν} are pure imaginary quantities. Now

$$b_{\nu} = \frac{\beta(k_{\nu})}{\alpha'(k_{\nu})}\ ,$$

by the definition of b_{ν} as a residue of a pole of $b(k)$. From the known relations (see [9], Appendix II)

$$\alpha'(k_{\nu}) = -i\int_{-\infty}^{\infty} \hat{v}(k_{\nu},x)v(k_{\nu},x)\,dx\ ,$$

$$v(k_{\nu},x) = \beta(k_{\nu})\hat{v}(k_{\nu},x)\ ,$$

and from the fact that $\hat{v}(k_{\nu},x)$ is real for real x, it is seen without difficulty that $-i\,b_{\nu}$ is a real positive number.

Now if there were two solutions, $K_1(x,y)$ and $K_2(x,y)$, of (3.11), their difference $K_1 - K_2 = W(x,y)$ would be a solution of the corresponding homogeneous integral equation

$$\int_{-\infty}^{x} [\delta(y,z) + B(y+z)]\, W(x,z)\, dz = 0 . \tag{3.15}$$

According to (3.12) and (3.13) this can be written

$$\int_{-(x+\delta)}^{x} \left\{ \frac{1}{2\pi} \int_{0}^{\infty} |a(k)|^2 e^{ik(y-z)} dk + \frac{1}{2\pi} \int_{-\infty}^{0} \left[e^{iky} + b(k) e^{-iky}\right] \right.$$

$$\left. \cdot \left[\overline{e^{ikz} + b(k) e^{-ikz}}\right] dk - i \sum_{\nu} b_{\nu} e^{-ik_{\nu}(y+z)} \right\} W(x,z)\, dz = 0 .$$

On multiplying through by $\overline{W(x,y)}$ and integrating with respect to y from $-(x+\delta)$ to x, and then interchanging integral signs, we obtain

$$\frac{1}{2\pi} \int_{0}^{\infty} |a(k)|^2 \left| \int_{-(x+\delta)}^{x} W(x,y) e^{iky} dy \right|^2 dk$$

$$+ \frac{1}{2\pi} \int_{-\infty}^{0} \left| \int_{-(x+\delta)}^{x} W(x,y) \left[e^{iky} + b(k) e^{-iky}\right) dy \right|^2 dk \tag{3.16}$$

$$+ \sum_{\nu} (-ib_{\nu}) \left| \int_{-(x+\delta)}^{x} W(x,y)\, e^{ik_{\nu}y} dy \right|^2 = 0 .$$

Since each term on the left side of (3.16) is positive, each term must be separately zero. Thus, for example

$$\int_{-(x+\delta)}^{x} W(x,y)\, e^{\pm iky} dy = 0 ,$$

whenever $|a(k)| \neq 0$. It follows that $W(x,y) = 0$ almost everywhere when $-(x+\delta) \leq y \leq x$.

4. SUFFICIENT CONDITIONS FOR $b(k)$ TO BE A REFLECTION COEFFICIENT (THE SYNTHESIS PROBLEM)

We have seen how to obtain an integral equation for the function $K(x,y)$, from which we can determine $V(x)$, if the reflection coefficient associated with $V(x)$ is given. In the problem of synthesis, however, we are not only required to construct $V(x)$, but generally we must first fix upon an appropriate function $b(k)$ for the reflection coefficient, one that satisfies certain physical requirements. The question arises: If a function $b(k)$ is given, then by the process described in Section 3 can we always arrive at a function $V(x)$ corresponding to which $b(k)$ is the reflection coefficient? Undoubtedly, certain conditions must be imposed on $b(k)$ before this will be possible. We shall require, accordingly:

I. $b(k) = \overline{b(-k)}$ for real k;

II. $b(k)$ is analytic in k with no singularities in the upper half-plane except possibly a finite number of simple poles on the positive imaginary axis, and no singularities on the real axis except, perhaps, at zero;

III. The residues of $b(k)$ at its poles on the positive imaginary axis are pure imaginary numbers with positive imaginary parts;

IV. $|b(k)| \leq 1$ for all real k;

V. $b(k) = e^{-ik\delta}\, O(1/k)$ in the upper half-plane;

VI. The Fourier transform $B(\tau)$ of $b(k)$ exists and is continuous in $-\infty < \tau < \infty$;

VII. $B(\tau)$ has piecewise continuous first and second derivatives in any finite interval.

These conditions are sufficient for $b(k)$ to be the reflection coefficient corresponding to some quantity $V(x)$ in (2.1), as we shall see. Condition V guarantees that $B(\tau) = 0$ for $\tau < -\delta$, and therefore, by (3.11), that $K(x,y) = 0$ for $x+y < -\delta$. This implies that

$$V(x) = 2\frac{dK(x,x)}{dx}$$

$$= 0$$

(see (3.3)) for $x < -\delta/2$. Conditions VI and VII along with (3.11) imply that $K(x,y)$ is continuous and has piecewise continuous first and second partial derivatives.

We shall now prove that $b(k)$ satisfying conditions I to VII is actually the reflection coefficient corresponding to some differential equation of the form (2.1) with a $V(x)$ obtained from $b(k)$ by solving (3.11) for $K(x,y)$ and setting

$$V(x) = 2 \frac{dK(x,x)}{dx} .$$

The proof is divided into two parts. In the first part we show that $K(x,y)$ satisfies (3.3). We start with (3.11), which, because of V (observe that, since $x < y$, $-(y+\delta) > -(x+\delta)$), takes the form

$$0 = B(x+y) + K(x,y) + \int_{-(y+\delta)}^{x} K(x,z)B(y+z)\,dz ; \qquad x > y,\quad x > -(y+\delta) . \tag{4.1}$$

Differentiating (4.1) twice with respect to x gives

$$\begin{aligned} 0 = {} & B''(x+y) + K_{xx}(x,y) + B'(x+y)K(x,x) + B(x+y)\frac{d}{dx}K(x,x) \\ & + B(x+y)K_x(x,x) + \int_{-(y+\delta)}^{x} K_{xx}(x,z)B(y+z)\,dz . \end{aligned} \tag{4.2}$$

Differentiating (4.1) twice with respect to y and integrating twice by parts gives, when we use the fact that $B(\tau)$ is continuous,

$$\begin{aligned} 0 = {} & B''(x+y) + K_{yy}(x,y) + B'(x+y)K(x,x) - B(x+y)K_y(x,x) \\ & + \int_{-(y+\delta)}^{x} K_{yy}(x,z)B(y+z)\,dz . \end{aligned} \tag{4.3}$$

Now we subtract (4.3) from (4.2) and add to this result the product of $-2(dK(x,x))/dx$ with (4.1). We then have

$$\begin{aligned} 0 = {} & \left\{ K_{xx}(x,y) - K_{yy}(x,y) - 2\frac{dK(x,x)}{dx} K(x,y) \right\} \\ & + \int_{-(y+\delta)}^{x} \left\{ K_{xx}(x,z) - K_{yy}(x,z) - 2\frac{dK(x,x)}{dx} K(x,z) \right\} B(y+z)\,dz . \end{aligned} \tag{4.4}$$

It follows from the uniqueness theorem already proved that

$$K_{xx}(x,y) - K_{yy}(x,y) - 2\frac{dK(x,x)}{dx} K(x,y) = 0 ; \tag{4.5}$$

i.e., $K(x,y)$ satisfies (3.3). Further, a solution $u(k,x)$ of (2.1) is given by (3.2) when $u_0(k,x)$ satisfies (3.1). The function $u(k,x)$ given by (3.2) is evidently continuous, and because $K(x,y)$ is continuous and partial derivatives of $K(x,y)$ are piecewise continuous, $u(k,x)$ has a continuous derivative.

For the second part of the proof we must show that the solution of (2.1) given by (3.3) when $u_0(k,x) = e^{ikx} + b(x)e^{-ikx}$ is the one which has the behavior $u(k,x) \sim a(k)e^{ikx}$ for $x \to +\infty$. It is difficult to ascertain directly the behavior as $x \to +\infty$ of a solution $u(k,x)$ of (2.1) when $u(k,x)$ is given in the form (3.2), but its behavior as $x \to -\infty$ is just $u_0(k,x)$. We shall therefore investigate the character of our solution $u(k,x)$, which approaches $e^{ikx} + b(k)e^{-ikx}$ as $x \to -\infty$, by comparing it with that (unique) solution $\hat{u}(k,x)$ of (2.1) which has the behavior e^{-ikx} as $x \to -\infty$. We know: If and only if

$$\frac{1}{2\pi} \int_C u(k,x)\hat{u}(k,y)\, dk = \delta(x-y) \quad , \tag{4.6}$$

$u(k,x)$ has the behavior

$$u(k,x) \sim \begin{cases} e^{ikx} + b(k)e^{-ikx} \quad , & x \to -\infty \\ a(k)e^{ikx} \quad , & x \to +\infty \quad , \end{cases} \tag{4.7}$$

for the general spectral theory of (2.1) shows that (4.6) is a necessary condition and it follows immediately on multiplying (4.6) by the conjugate function (the one which has the behavior (4.7) of $\hat{u}(k,y)$) and integrating with respect to y that (4.6) is also a sufficient condition. We write then

$$\begin{aligned} u(k,x) &= e^{ikx} + b(k)e^{-ikx} + \int_{-(x+\delta)}^{x} K(x,z)\left[e^{ikz} + b(k)e^{-ikz}\right] dz \\ \hat{u}(k,y) &= e^{-iky} + \int_{-(x+\delta)}^{x} K(x,z)\, e^{-ikz}\, dz \quad . \end{aligned} \tag{4.8}$$

If we take the product of these two expressions and perform the integration indicated in (4.6), making use of (3.11), we find that (4.6) is satisfied. Thus the solution $u(k,x)$ of (2.1) has the required asymptotic behavior and the quantity $b(k)$ is, indeed, the reflection coefficient corresponding to

$$V(x) = 2\frac{dK(x,x)}{dx} \quad .$$

5. EXAMPLE

Let

$$b(k) = -\frac{1}{k^2+1}$$

be the reflection coefficient for some as yet unknown function $V(x)$ in (2.1). The sufficient conditions I-VII on $b(k)$ (see Section 4) are easily verified for this case. Also the function $B(\tau)$ is easily calculated by a straightforward application of the theory of residues:

$$B(\tau) = -\frac{1}{2\pi}\int_{-\infty}^{\infty} \frac{e^{-ik}}{k^2+1}\,dk + \frac{1}{2}e^{\tau}$$

$$= \begin{cases} \sinh\tau , & \tau \geq 0 \\ 0 , & \tau \leq 0 \end{cases} .$$

Then Equation (3.11) becomes in this case

$$0 = \sinh(x+y) + K(x,y) + \int_{-y}^{x} K(x,z)\sinh(y+z)\,dz , \quad x > |y| . \tag{5.2}$$

We can obtain a differential equation for $K(x,y)$ by differentiating (5.2) twice with respect to y. A single derivative gives

$$\theta = \cosh(x+y) + K_y(x,y) + \int_{-y}^{x} K(x,z)\cosh(y+z)\,dz ,$$

and a second derivative gives

$$0 = \sinh(x+y) + K_{yy}(x,y) + K(x,-y) + \int_{-y}^{x} K(x,z)\sinh(y+z)\,dz . \tag{5.4}$$

The integral term of (5.4) is the same as that of (5.2). We can therefore eliminate this term, obtaining

$$K_{yy}(x,y) + K(x,-y) - K(x,y) = 0 . \tag{5.5}$$

It is evident from (5.5) that $K_{yy}(x,y)$ is an odd function of y. Thus

$$K(x,y) = -K(x,-y) + a(x)y + 2b(x) , \tag{5.6}$$

where $a(x)$ and $b(x)$ are arbitrary functions of x alone. Then (5.5) becomes, after we substitute for $K(x,-y)$ from (5.6)

$$K_{yy}(x,y) = 2K(x,y) - a(x)y - b(x) \quad . \tag{5.7}$$

The general solution of (5.7) is

$$K(x,y) = \frac{a(x)y + 2b(x)}{2} + \frac{A(x)}{2} e^{\sqrt{2}\,y} + \frac{B(x)}{2} e^{-\sqrt{2}\,y} \quad , \tag{5.8}$$

where $A(x)$ and $B(x)$ are new arbitrary functions of x alone. After substituting the value of $K(x,y)$ obtained from (5.8) into (5.5), we see that $a(x) = 0$ and $A(x) = B(x)$. Thus, by (5.8)

$$K(x,y) = b(x) + A(x) \sinh \sqrt{2}\, y \quad . \tag{5.9}$$

To determine the functions $b(x)$ and $A(x)$ we substitute (5.9) into (5.2). Then we have the result

$$\begin{aligned} A(x) &= -\frac{\operatorname{sech} \sqrt{2}\, x}{\sqrt{2}} \quad , \\ b(x) &= -\frac{\tanh \sqrt{2}\, x}{\sqrt{2}} \quad . \end{aligned} \tag{5.10}$$

From (5.9) we now have

$$K(x,y) = \begin{cases} -\dfrac{1}{\sqrt{2}} \dfrac{\sinh\sqrt{2}\, x + \sinh \sqrt{2}\, y}{\cosh \sqrt{2}\, y} \quad , & y \geq -x \\ 0 \quad , & y \leq -x \quad . \end{cases} \tag{5.11}$$

By (3.3) we finally have

$$V(x) = \begin{cases} -4 \operatorname{sech}^2 \sqrt{2}\, x \quad , & x > 0 \\ 0 \quad , & x < 0 \end{cases} \tag{5.12}$$

For this $v(x)$, the solution $u(k,x)$ of (2.1) having the asymptotic behavior given by (2.6) and thus exhibiting the reflection coefficient

$$b(k) = -\frac{1}{k^2 + 1}$$

can be written immediately through the use of (3.2) and (5.11):

$$u(k,x) = e^{ikx} - \frac{1}{k^2+1} e^{-ikx} - \frac{\operatorname{sech}\sqrt{2}\,x}{\sqrt{2}} \tag{5.13}$$

$$\cdot \int_{-x}^{x} \left[\sinh\sqrt{2}\,x + \sinh\sqrt{2}\,y\right] \left[e^{iky} - \frac{1}{k^2+1} e^{-iky}\right] dy$$

After performing the indicated integration and simplifying the right side of (5.13) we obtain

$$u(k,x) = \begin{cases} \left[\dfrac{k^2 + \sqrt{2}\,ik \tanh\sqrt{2}\,x}{k^2+1}\right] e^{ikx} , & x \geq 0 \\ e^{ikx} - \dfrac{1}{k^2+1} e^{-ikx} , & x \leq 0 \end{cases} \tag{5.14}$$

Since, as $x < +\infty$,

$$u(k,x) \sim \frac{k^2 + \sqrt{2}\,ik}{k^2+1} e^{ikx} ,$$

the transmission coefficient is given by

$$a(k) = \frac{k^2 + \sqrt{2}\,ik}{k^2+1} . \tag{5.15}$$

Obviously $a(k)$ and $b(k)$ satisfy the required conditions

$$a(-k) = \overline{a(k)} ,$$

$$b(-b) = \overline{b(k)} ,$$

$$|a(k)|^2 + |b(k)|^2 = 1 \qquad \text{for real } k .$$

We might note in passing that if $V(x) = -4 \operatorname{sech}^2 \sqrt{2}\,x$ for the entire interval $-\infty < x < \infty$, it is "transparent"; that is, there exists a solution of (2.1) (we can obtain this solution from (5.14)) which has the behavior $u \sim e^{ikx}$ as $x \to -\infty$ and the behavior

$$u \sim \frac{k + \sqrt{2}\,i}{k - \sqrt{2}\,i} e^{ikx}$$

as $x \to +\infty$, so that no reflected wave exists for an incident wave from the left. The amplitude of the incident wave remains one as the wave approaches $+\infty$, but the phase changes by an amount depending on the frequency.

6. SOLUTION OF A GENERAL CLASS OF PROBLEMS

The method employed in Section 5 of solving integral Equation (5.2) suggests that a whole class of problems associated with reflection coefficients of a certain form might be solved explicitly.

Suppose we have a reflection coefficient $b(k)$ satisfying conditions I to VII of Section 4 and suppose also that $b(k)$ is a meromorphic function with a finite number of poles in either half-plane. The function $B(\tau)$ then has the form

$$B(\tau) = \begin{cases} \sum_{n=1}^{N} b_n e^{-ik_n \tau} , & \tau \geq -\delta \\ 0 , & \tau \leq -\delta \end{cases} \tag{6.1}$$

Accordingly the integral Equation (3.11) for $K(x,y)$ becomes

$$0 = \sum_{n=1}^{N} b_n e^{-ik_n (x+y)} + K(x,y) + \sum_{n=1}^{N} b_n \int_{-(y+\delta)}^{x} K(x,z) e^{-ik_n (y+z)} dz \tag{6.2}$$

We can obtain a new equation from (6.2) by differentiating with respect to y; each time we differentiate we obtain another equation. Let us differentiate N times, forming $N+1$ equations altogether, including (6.2) itself. The new equations have the form

$$\begin{aligned} 0 = & \sum_{n=1}^{N} (-ik_n)^{\nu} b_n e^{-ik_n (x+y)} + K^{(\nu)}(x,y) \\ & + \sum_{n=1}^{N} (-ik_n)^{\nu} b_n e^{-ik_n y} \int_{-(y+\delta)}^{x} K(x,z) e^{-ik_n z} dz \\ & + \sum_{n=1}^{N} b_n e^{ik_n \delta} \sum_{\mu=1}^{\nu-1} (-ik_n)^{\mu} K^{(\nu-\mu-1)}(x, -(y+\delta)) , \quad \nu = 1,\ldots,N \end{aligned} \tag{6.3}$$

where $K^{(\nu)}(x,y)$ is written for the ν-th y-derivative of $K(x,y)$ and where we have used the fact in (6.1) that

$$B(-\delta) = \sum_{n=1}^{N} b_n e^{ik_n \delta} = 0 .$$

We can eliminate the N quantities

$$\int_{-(y+\delta)}^{x} K(x,z)\, e^{-ik_n z}\, dz$$

from Equations (6.3) and (6.2) obtaining a single N-th order differential equation in $K(x,y)$ and $K(x,-(y+\delta))$ of the form:

$$0 = P(x,y) + \sum_{n=0}^{N} Q_n(y) K^{(n)}(x,y) + \sum_{n=0}^{N-2} R_n(y) K^{(n)}(x,-(y+\delta)) \tag{6.4}$$

By substituting $-y-\delta$ for y in (6.4) we obtain another differential equation. By differentiating (6.4) and then performing this substitution again we obtain a new equation. If we differentiate $N-2$ times and perform the substitution each time, we arrive at a system of equations involving $K(x,y)$ and its derivatives and $K(x,-(y+\delta))$ and its derivatives. It is now possible to eliminate $K(x,-(y+\delta))$ and its derivatives and finally obtain a single differential equation of order $2N-2$ for $K(x,y)$. In the example in Section 5, $N = 2$, and consequently we obtain a second-order differential equation for $K(x,y)$ in this case.

In most cases we can solve the integral Equation (3.11) by a more direct method, however. Let us assume that $K(x,y)$ has the form:

$$K(x,y) = \sum_{\nu=-N}^{N} f_\nu(x)\, e^{\alpha_\nu y}, \tag{6.5}$$

where the sum extends over all integers between $-N$ and N *except zero*, and where α_ν and $f_\nu(x)$ are quantities to be determined. If we substitute (6.5) into (3.11) we have

$$0 = \sum_{n=1}^{N} b_n e^{-ik_n(x+y)} + \sum_{\nu=-N}^{N} f_\nu(x)\, e^{\alpha_\nu y}$$

$$+ \sum_{n=1}^{N} \sum_{\nu=-N}^{N} b_n e^{-ik_n y} f_\nu(x) \int_{-(y+\delta)}^{x} e^{(\alpha_\nu - ik_n)z}\, dz$$

$$= \sum_{n=1}^{N} b_n e^{-ik_n(x+y)} + \sum_{\nu=-N}^{N} f_\nu(x)\, e^{\alpha_\nu y}$$

$$+ \sum_{n=1}^{N} \sum_{\nu=-N}^{N} b_n e^{-ik_n y} f_\nu(x) \left[\frac{e^{(\alpha_\nu - ik_n)x} - e^{-(y+\delta)(\alpha_\nu - ik_n)}}{\alpha_\nu - ik_n} \right]$$

$$= \sum_{n=1}^{N} b_n e^{-ik_n(x+y)} + \sum_{n=1}^{N} b_n e^{-ik_n(x+y)} \sum_{\nu=-N}^{N} \frac{e^{\alpha_\nu x} f_\nu(x)}{\alpha_\nu - ik_n}$$

$$+ \sum_{\nu=-N}^{N} f_\nu(x) e^{\alpha_\nu y} - \sum_{\nu=-N}^{N} e^{-\alpha_\nu y - \alpha_\nu \delta} f_\nu(x) \sum_{n=1}^{N} \frac{b_n e^{ik_n \delta}}{\alpha_\nu - ik_n} .$$

This form suggests that we assume $\alpha_\nu = -\alpha_{-\nu}$; then

$$0 = 1 + \sum_{\nu=-N}^{N} \frac{f_\nu(x) e^{\alpha_\nu x}}{\alpha_\nu - ik_n} , \qquad n = 1,\dots,N \tag{6.6}$$

and

$$0 = f_\nu(x) + f_{-\nu}(x) e^{\alpha_\nu \delta} \sum_{n=1}^{N} \frac{b_n e^{ik_n \delta}}{\alpha_\nu + ik_n} , \qquad \nu = 1,\dots,N \tag{6.7}$$

A similar set of equations goes with (6.7) and can be deduced from it by substituting $-\nu$ for ν. In order to get a non-trivial solution $f_\nu(x)$ of the resulting homogeneous system, the coefficient determinant must be set equal to zero:

$$\left[\sum_{n=1}^{N} \frac{b_n e^{ik_n \delta}}{ik_n - \alpha_\nu} \right] \left[\sum_{n=1}^{N} \frac{b_n e^{ik_n \delta}}{ik_n + \alpha_\nu} \right] - 1 = 0 . \tag{6.8}$$

Equation (6.8) reduces to an N-th degree algebraic equation for α_ν; thus in general there are N values which α_ν can have. It is evident from (6.8) that if α_ν is a solution then $\alpha_{-\nu} = -\alpha_\nu$ is also a solution. When α_ν is known, Equations (6.7) can be used to give $f_{-\nu}(x)$ in terms of $f_\nu(x)$. After substituting this result in (6.6) we have

$$-1 = \sum_{\nu=1}^{N} \left[\frac{e^{\alpha_\nu x}}{\alpha_\nu - ik_n} + \frac{e^{-\alpha_\nu (x+\delta)}}{\alpha_\nu + ik_n} \sum_{\mu=1}^{N} \frac{b_\mu e^{ik_\mu \delta}}{ik_\mu - \alpha_\nu} \right] f_\nu(x) , \tag{6.9}$$

$$n = 1,\dots,N$$

Equations (6.9) can be solved for $f_\nu(x)$ provided that the coefficient determinant does not vanish. The coefficient determinant will surely vanish if any two values of α_ν are the same as ν goes from one to N. Thus we need the condition generally that (6.8) have no multiple roots, except perhaps $\alpha_\nu = 0$, which can be a double root.

If it turns out that Equations (6.9) have no solutions, we must assume a different form for $K(x,y)$:

$$K(x,y) = \sum_{\nu=-N}^{N} f_\nu(x) P_\nu(y)\, e^{\alpha_\nu y} \quad , \tag{6.10}$$

where the $P_\nu(y)$ are polynomials in y. A similar procedure to the one just described should lead to a determination of the α_ν, $f_\nu(x)$ and the $P_\nu(y)$ and thus of $K(x,y)$.

7. THE SECOND DIFFERENTIAL EQUATION

Often the differential equation of interest in scattering problems is not (2.1) but rather an equation of the form

$$\frac{d^2u(k,x)}{dx^2} + k^2n^2(x)u(k,x) = 0 \quad , \tag{7.1}$$

where $n(x)$ is real, continuous and has a continuous derivative in $-\infty < x < \infty$. The solution of (7.1) can also be interpreted as a wave if $n(x)$ approaches a constant as $x \to \pm\infty$, and a reflection coefficient $b(k)$ can again be defined. The problem of determining $n^2(x)$ from the reflection coefficient can be dealt with by a consideration of an equation of the form (2.1) corresponding in a certain way to (7.1).

Suppose that we have a differential equation of the form (2.1):

$$w''(z) + [k^2 - V(z)]w(z) = 0 \quad , \tag{7.2}$$

where the primes indicate differentiation with respect to z. Let us make the substitution $x = x(z)$ in (7.2). We have

$$\ddot{w} + \frac{x''}{x'^2}\dot{w} + \frac{(k^2 - V)}{x'^2} w = 0 \tag{7.3}$$

where again the primes indicate differentiation with respect to z and the dots with respect to x. Now let us make the substitution

$$w = \frac{u(k,x)}{\sqrt{x'}}$$

in (7.3):

$$\ddot{u} + \left[\frac{k^2}{x'^2} + \left(\frac{3x''^2}{4x'^4} - \frac{x'''}{2x'^3} - \frac{V}{x'^2} \right) \right] u = 0 \quad . \tag{7.4}$$

If the second term in the bracketed factor of u in (7.4) is zero, (7.4) will have the form (7.1) with $n(x) = 1/x'$. Moreover, if $x(z)$ is a monotonic function of z and independent of k, the reflection coefficient is the same for equations (7.2) and (7.4). Therefore, we can find $n(x) = 1/x'$ if we can find $x(z)$. We assume that $V(z)$ is constructed by the method of the previous sections, and now it can be used to obtain $x(z)$.

We set the second term in the brackets in (7.4) equal to zero, and this implies

$$3x''^2 - 2x'x''' - 4x'^2 V(z) = 0$$

or

$$-4 \frac{x'}{x''} V - 2 \frac{x'''}{x''} + 3 \frac{x''}{x'} = 0$$

or

$$2 \frac{d}{dz} (\log x'') - 3 \frac{d}{dz} (\log x') = \frac{-4V}{\frac{d}{dz} (\log x')} \quad . \tag{7.5}$$

Now we define

$$\omega = \frac{x''}{x'} = \frac{d}{dz} (\log x') \quad ,$$

from which we have

$$\frac{d}{dz} (\log x'') = \omega + \frac{\omega'}{\omega} \quad . \tag{7.6}$$

Substituting in (7.5) gives us

$$\omega' - \frac{1}{2} \omega^2 = -2V \quad . \tag{7.7}$$

Equation (7.7) is a Ricatti equation and can be written as a linear second-order differential equation by means of the transformations

$$\omega = -2W \quad ;$$

$$W = \frac{U'}{U} \quad .$$

We have finally

$$U''(z) - V(z)U(z) = 0 \quad , \tag{7.8}$$

where

$$x' = \frac{C}{U^2(z)}$$

and C is an arbitrary constant; thus

$$x = C \int^z \frac{dz}{U^2(z)} \quad . \tag{7.9}$$

After we have found $V(z)$ by the methods of the previous sections, we can use (3.2) with k set equal to zero to obtain a solution $U(z)$ of (7.8). Then by means of (7.8) and (7.9) we get

$$\begin{aligned} n(x) &= \frac{1}{x'(z(x))} \\ &= U^2(z(x)) \quad . \end{aligned} \tag{7.10}$$

The case where $n(x) = \text{constant}$ in (7.1) corresponds to the case where $V(z) = 0$ in (7.2). According to (7.9), x would then be proportional to z or to $1/(z+A)$, where A is an arbitrary constant. The second case, however, leads in general to a singular $n(x)$.

In each problem it must be decided which of the solutions of (7.8) to use, according to the physical situation. Obviously, $n(x)$ is not determined uniquely by the reflection coefficient as is $V(z)$.

References

1. N. Levinson, On the uniqueness of the potential in a Schrödinger equation for a given asymptotic phase, *Danske Vid. Selsk. Mat.-Fys. Medd.* 25, No. 9 (1949).

2. V.A. Marcenko, Concerning the theory of a differential operator of second order, *Doklady Akad. Nauk. SSSR (N.S.)* 72, 457-460 (1950) (Russian).

3. N. Levinson, The inverse Sturm-Liouville problem, *Mat. Tidsskr. B. 1949*, pp. 25-30 (1949).

4. Gören Borg, Eine Umkehrung der Sturm-Liouvilleschen Eigenwertaufgabe. Bestimmung der Differentialgleichung durch die Eigenwerte, *Acta Math.* 78, 1-96 (1946).

5. I.M. Gelfand and B.M. Levitan, On the determination of a differential equation by its spectral function, *Doklady Akad. Nauk. SSSR (N.S.)* 77, 557-560 (1951) (Russian); *Izvestiya Akad. Nauk. SSSR Ser. Mat.* 15, 309-360 (1951) (Russian).

6. W. Kohn and R. Jost, Construction of a potential from a phase shift, *Phys. Rev.* 87, No. 6, 977-992 (1952).

7. M.C. Krein, Solution of the inverse Sturm-Liouville problem, *Doklady Akad. Nauk. SSSR (N.S.)* 76, 21-24 (1951) (Russian).

8. N. Levinson, Certain explicit relationships between phase shifts and scattering potential, *Phys. Rev.* 89, No. 4, 755-757 (1953).

9. K. Friedrichs, Spectral representation of linear operators (Notes), New York University (1951).

IL NUOVO CIMENTO VOL. II, N. 5 1° Novembre 1955

The Determination of the Scattering Potential from the Spectral Measure Function (*).

I. Continuous Spectrum.

I. KAY and H. E. MOSES

Institute of Mathematical Sciences, New York University - New York

(ricevuto il 2 Luglio 1955)

Summary. — It is shown how a generalized form of the Gelfand-Levitan procedure for obtaining scattering potentials from spectral measure functions may be obtained from the theory of operators in a vector space. Part I treats the case where the perturbed and unperturbed Hamiltonians have the same spectrum which is taken to be purely continuous. The vector space in which one works is Hilbert space.

1. – Introduction.

For special cases it has been shown by previous workers how the scattering potential can be obtained from the spectral measure function using the equation of GELFAND and LEVITAN (1). By showing, in addition, how the spectral measure function can be obtained from the scattering phases JOST and KOHN (2) and LEVINSON (3) have been able to construct the scattering

(*) The research reported in this article was done at The Institute of Mathematical Sciences, New York University, and has been made possible through support and sponsorship extended by Geophysics Research Directorate, Air Force Cambridge Research Center under Contract No. AF 19(122)-463.

(1) I. M. GELFAND and B. M. LEVITAN: *Izvest. Akad. Nauk SSSR*, Math. Series, **15**, 309 (1951).

(2) R. JOST and W. KOHN: *Det. Kgl. Danske Vidensk. Selskab, Mat.-fys. Medd.*, **27**, n. 9 (1953).

(3) N. LEVINSON: *Phys. Rev.*, **89**, 755 (1953).

potential from phases. KAY [4], CORINALDESI [5], and NEWTON and JOST [6] have adapted the method of references [1] and [2] to the one-dimensional wave equation ($-\infty < x < +\infty$), the radial Gordon-Klein-Schrödinger equation, and systems of coupled equations, respectively. In the treatments discussed above the unperturbed Hamiltonian H_0 has been given explicitly as the kinetic energy operator, and it has been assumed that the perturbation is diagonal in the x-representation.

In all of these papers one uses the analytical properties oft he eigenfunctions of the perturbed Hamiltonian H and the special nature of the unperturbed Hamiltonian H_0. In contrast, the approach in the present paper is more abstract and makes use of the « algebraic » relationships between various operators. The advantage of the present viewpoint is that the structure of the inverse problem is seen to be largely independent of the special properties of H and H_0.

We shall now describe the « direct » and « inverse » problems of spectral theory.

In the *direct* problem of spectral theory one is *given*:

a) the Hamiltonian H (which for the while we shall assume has a purely continuous spectrum);

b) the boundary conditions on the eigenfunctions of the continuous spectrum.

One *seeks*:

a) the eigenfunctions of the continuous spectrum;

b) the measure functions (or equivalently weight functions) associated with the eigenfunctions which must be used in the completeness relationship.

The *inverse* problem has a different set of given data. One is *given*:

a) the boundary condition on the eigenfunctions of H;

b) the measure function associated with these eigenfunctions.

One *seeks*:

a) the eigenfunctions of H;

b) the operator H.

The inverse problem is important because in some cases the spectral measure

---

(4) I. KAY: EM-74, *Inst. of Math. Sci., Div. of Electromagnetic Res.*, New York University.

(5) E. CORINALDESI: *Nuovo Cimento*, 11, 468 (1954).

(6) R. G. NEWTON and R. JOST: *Nuovo Cimento*, 1, 590 (1955).

function associated with the eigenfunctions satisfying certain boundary conditions can be obtained simply from the scattering operator. Hence, after obtaining the spectral measure function from the scattering operator, we can obtain the scattering potential $V = H - H_0$ from the measure function. We thereby shall have a method of obtaining the scattering potential from the scattering operator.

The present study is divided into several parts. In Part I we consider the case where H has a purely continuous spectrum. In Part II we shall generalize the results to take into account the case where H has a discrete spectrum in addition to the continuous spectrum. In Part III we show how the one-dimensional Schrödinger equation ($-\infty < x < \infty$), already discussed in reference ([4]) by means of a special procedure, appears as a case of the more general procedure discussed herein.

We shall sketch the three basic theorems which are proved.

Let there be given a Hamiltonian H_0 with a continuous spectrum whose eigenfunctions corresponding to the eigenvalue E are $|H_0; E\rangle$ and are chosen such that $\langle H_0; E | H_0; E' \rangle = \delta(E - E')$. Let there also be a positive definite Hermitian operator W (the weight operator) which commutes with H_0. We state the first of our theorems as follows:

Theorem 1. Assume that a solution of

$$WU^* = U_0 \tag{1.1}$$

can be found for operators U, U_0 such that

$$U_0 = U^{-1} \tag{1.2}$$

(the asterisk denotes Hermitian adjoint). Then one can construct a Hermitian operator H:

$$H = UH_0U_0 \tag{1.3}$$

whose eigenfunctions $|H; E\rangle$ are given by

$$|H; E\rangle = U|H_0; E\rangle\,. \tag{1.4}$$

Furthermore, the eigenfunctions $|H, E\rangle$ of H have the normalization

$$\langle H; E' | H; E\rangle = [\omega(E)]^{-1}\delta(E - E') \tag{1.5}$$

where $\omega(E)$ is given by

$$\langle H_0; E | W | H_0; E'\rangle = \delta(E - E')\omega(E)\,. \tag{1.6}$$

The second and third of our theorems deal with the nature of the solutions of U, U_0, of (1.1).

Theorem 2. Let us write

$$U = I + K\,, \qquad U_0 = I + K_0\,. \tag{1.7}$$

Furthermore, let K, K_0 be triangular in some representation, say x-representation, for example,

$$\begin{cases} \langle x|K|x'\rangle = 0 & x' > x, \\ \langle x|K_0|x'\rangle = 0 & x' > x. \end{cases} \tag{1.8}$$

Then equation (1) can be solved for *both* U and U_0. Moreover, the solution will be unique. The equation for $\langle x|K|x'\rangle$ will be the Gelfand-Levitan equation. We can motivate the introduction of the triangularity conditions (1.8) by noting that from (1.2), (1.4), (1.7) and (1.8) we have

$$\begin{cases} \langle x|H;E\rangle = \langle x|H_0;E\rangle + \int\limits_{-\infty}^{x} \langle x|K|x'\rangle \mathrm{d}x' \langle x'|H_0;E\rangle \mathrm{d}\,, \\ \langle x|H_0;E\rangle = \langle x|H;E\rangle + \int\limits_{-\infty}^{x} \langle x|K_0|x'\rangle \mathrm{d}x' \langle x'|H_0;E\rangle\,. \end{cases} \tag{1.9}$$

Hence $\langle x|H;E\rangle$ approaches $\langle x|H_0;E\rangle$ as x approaches $-\infty$. That is, the triangularity conditions specify the boundary conditions which $\langle x|H;E\rangle$ are to satisfy.

Theorem 3 is the most surprising of the three theorems.

Theorem 3. If K, K_0 satisfy triangularity conditions (1.8), then the solution U, U_0 of equation (1.1) satisfies the condition (1.2).
The use of the three theorems is the following: one can find an operator H whose eigenfunctions $|H;E\rangle$ are prescribed to have the normalization given by (1.5) and the asymptotic form

$$\lim_{x\to-\infty} \langle x|H;E\rangle = \langle x|H_0;E\rangle\,, \tag{1.10}$$

where $|H_0;E\rangle$ are prescribed eigenfunctions of a known Hamiltonian H_0. To obtain the operator H and the eigenfunctions $|H;E\rangle$ one need merely construct the operator W and solve equation (1.1) for U, U_0 for the case where K, K_0 have the triangularity properties (1.8). Then H is given by (1.3) and $|H;E\rangle$ by (1.4).

2. – The eigenfunctions of the unperturbed Hamiltonian.

We have assumed that the unperturbed Hamiltonian H_0 has a continuous spectrum only. In most cases the spectrum will be degenerate. It will therefore be convenient to introduce a set of commuting variables, collectively denoted by A_0, each of which also commutes with H_0, such that A_0 and H_0 form a complete set of commuting variables. Using a modification of Dirac's notation we denote the state which is simultaneously an eigenstate of H_0 and A_0 corresponding to the eigenvalues E and a of H_0 and A_0 respectively by $|H_0, A_0; E, a\rangle$. An example of A_0 occurs in the case where H_0 is the kinetic energy operator. A_0 could then be chosen to be the operator whose eigenvalues give the direction of the momentum.

We have

$$H_0|H_0, A_0; E, a\rangle = E|H_0, A_0; E, a\rangle . \tag{2.1}$$

If we assume that the eigenfunctions are normalized, we have

$$\langle H_0, A_0; E, a|H_0, A_0; F, b\rangle = \delta(E-F)\delta(a, b) , \tag{2.2}$$

where $\delta(a, b)$ is a generalized δ-function defined by

$$\int f(a)\delta(a, b)\,\mathrm{d}a = \left.\begin{array}{ll} f(b) & \text{if the range of integration includes } b \\ 0 & \text{otherwise.}\end{array}\right. \tag{2.3}$$

(The integration is to be understood as summation over the values of a in the discrete spectrum.) Since the set of eigenfunctions is to be a complete set, we write

$$\iint |H_0, A_0; E, a\rangle\,\mathrm{d}E\,\mathrm{d}a\langle H_0, A_0; E, a| = I , \tag{2.4}$$

where I is identity operator.

3. – The eigenfunctions of the perturbed Hamiltonian and the transformation operator.

The perturbed Hamiltonian H is given by

$$H = H_0 + \varepsilon V \tag{3.1}$$

where V is the scattering potential and ε is a smallness parameter. We have assumed that the spectrum of H is the same as H_0. Hence the degeneracy of the spectrum of H is the same as that of H_0. Let us introduce a set of commuting operators A, analogous to A_0, which together with H form a complete set of commuting variables. We denote the eigenfunction of H and A which belongs to the eigenvalues E and a of H and A respectively by $|H, A; E, a\rangle$. Hence

$$H|H, A; E, a\rangle = E|H, A; E, a\rangle . \tag{3.2}$$

We also introduce the transformation operator U defined by

$$|H, A; E, a\rangle = U|H_0, A_0; E, a\rangle . \tag{3.3}$$

(This operator is also used, for example, by FRIEDRICHS [7]). We now proceed to develop some relations which U must satisfy. Substituting (3.3) into (3.2), we have

$$\begin{aligned} H|H, A; E, a\rangle &= EU|H_0, A_0; E, a\rangle \\ &= UE|H_0, A_0; E, a\rangle \\ &= UH_0|H_0, A_0; E, a\rangle \end{aligned} \tag{3.4}$$

or

$$HU|H_0, A_0; E, a\rangle = UH_0|H_0, A_0; E, a\rangle . \tag{3.5}$$

Hence on using (2.4),

$$HU = UH_0 . \tag{3.6}$$

For any function $f(H)$ of H we infer

$$f(H)U = Uf(H_0) . \tag{3.7}$$

It will be shown later that U has an inverse, U^{-1}. From (3.6), then,

$$H = UH_0U^{-1} \tag{3.8}$$

and we see that the transformation U is actually a canonical transformation between H and H_0. It is further seen that a relation like (3.8) can hold only when the spectrum of H and H_0 are the same.

---

[7] K. O. FRIEDRICHS: *Comm. on Pure and Applied Math.*, 1, 361 (1948).

The operator U or equivalently the eigenfunctions $|H, A; E, a\rangle = U|H_0, A_0; E, a\rangle$ are not unique. It can be shown that all possible solutions of (3.6) satisfy the integral equation

$$U = L + \varepsilon \int \frac{P}{E - H_0} V U \delta(E - H_0)\, \mathrm{d}E\,, \tag{3.9}$$

where L is an arbitrary operator which commutes with H_0. The symbol P means « principal part » in the integration, the range of integration is the range of the spectrum of E. The operator $\delta(E - H_0)$ is defined by

$$\delta(E - H_0) = \int |H_0, A_0; E, a\rangle \mathrm{d}a \langle H_0, A_0; E, a|\,, \tag{3.10}$$

and has the properties

$$H_0 \delta(E - H_0) = \delta(E - H_0) H_0 = E\, \delta(E - H_0)\,, \tag{3.11}$$

$$\int \delta(E - H_0)\, \mathrm{d}E = I\,. \tag{3.12}$$

Equation (3.12) follows from (2.4).

We note that operators which commute with H_0, are in a sense, diagonal in the H_0 representation. Let B be any operator which commutes with H_0; then

$$\langle H_0, A_0; E, a|B|H_0, A_0; F, b\rangle = \langle a|\beta(E)|b\rangle \delta(E - F)\,, \tag{3.13}$$

where $\langle a|\beta(E)|b\rangle$ is a function of a, b and E (a *proper* function, i.e., not a δ-function, in terms of E) which can be obtained from the operator B.

4. – The outgoing and incoming eigenfunctions; the scattering operator.

There are two transformation operators which play a particularly important role. We denote them by $U_\pm$: The integral equations for these operators are obtained by choosing the operator L properly. The operator U_- is required to satisfy the condition

$$\lim_{t \to -\infty} \exp[iH_0 t] \exp[-iHt] U_- |\varphi\rangle = |\varphi\rangle \tag{4.1}$$

where $|\varphi\rangle$ is a state in Hilbert space. The meaning of (4.1) is that if we choose the solution $|\varphi(t)\rangle$ of the perturbed Schrödinger equation

$$H|\varphi(t)\rangle = i \frac{\partial}{\partial t} |\varphi(t)\rangle \tag{4.2}$$

to behave like

$$|\varphi(t)\rangle = \exp[-iH_0t]|\varphi\rangle \quad \text{when } t \to -\infty, \tag{4.3}$$

then at finite times

$$|\varphi(t)\rangle = \exp[-iHt]U_-|\varphi\rangle\,. \tag{4.4}$$

Since the Schrödinger equation has only a first derivative in time, only one initial condition is required to obtain a unique solution. Equation (4.4) may be regarded as the solution of (4.2) with the initial condition (4.3).

In Appendix I it is shown that condition (4.1) fixes the operator L as

$$L = I - i\pi\varepsilon\int\delta(E-H_0)VU_-\delta(E-H_0)\mathrm{d}E\,. \tag{4.5}$$

It is to be noted that L as given by (4.5) commutes with H_0, as required. Substituting (4.5) into (3.9) we find

$$U_- = I + \varepsilon\int\gamma_-(E-H_0)VU_-\delta(E-H_0)\,\mathrm{d}E\,, \tag{4.6}$$

where

$$\gamma_-(x) = -i\pi\delta(x) + \frac{P}{x} = \lim_{\zeta\to+0}\frac{1}{x+i\zeta}\,. \tag{4.7}$$

The eigenfunctions $|H, A; E, a\rangle_-$ of H given by

$$|H, A; E, a\rangle_- = U_-|H_0, A_0; E, a\rangle \tag{4.8}$$

are generally called the « outgoing waves ». (Usually, however, the plus subscript is used to denote these eigenfunctions.)

The scattering operator can be conveniently expressed in terms of U_-. It is defined in the following way: If $|\varphi(t)\rangle$ is a solution of the Schrödinger equation (4.2) such that

$$\lim_{t\to-\infty}\exp[iH_0t]|\varphi(t)\rangle = |\varphi\rangle \tag{4.9}$$

exists, then the scattering operator S is given by

$$\lim_{t\to+\infty}\exp[iH_0t]|\varphi(t)\rangle = S|\varphi\rangle\,. \tag{4.10}$$

If we identify $|\varphi(t)\rangle$ with $\exp[-iHt]U_-|\varphi\rangle$ then the scattering operator is given by

$$\lim_{t\to+\infty}\exp[iH_0t]\exp[-iHt]U_-|\varphi\rangle = S|\varphi\rangle\,. \tag{4.11}$$

In Appendix I, it is shown that

$$S = I - 2\pi i\varepsilon\int\delta(E - H_0)VU_-\delta(E - H_0)\,\mathrm{d}E\,. \tag{4.12}$$

The scattering operator takes a state which behaves like the solution $\exp[-iH_0t]|\varphi\rangle$ of the unperturbed equation at $t = -\infty$ into another $\exp[-iH_0t]S|\varphi\rangle$ of the unperturbed equation when $t = +\infty$.

The operator U_+ is used to solve the final value problem. Specifically, U_+ is defined as that transformation which satisfies

$$\lim_{t\to+\infty}\exp[iH_0t]\exp[-iHt]U_+|\varphi\rangle = |\varphi\rangle\,. \tag{4.13}$$

The state $|\varphi\rangle$ at finite times is then given by

$$|\varphi(t)\rangle = \exp[-iHt]U_+|\varphi\rangle\,, \tag{4.14}$$

where $|\varphi(t)\rangle$ behaves like $\exp[-iH_0t]|\varphi\rangle$ for $t = +\infty$. In Appendix I it is shown that in this case the operator L must be taken as

$$L = I + i\pi\varepsilon\int\delta(E - H_0)VU_+\delta(E - H_0)\,\mathrm{d}E\,. \tag{4.15}$$

Hence U_+ satisfies the equation

$$U_+ = I + \varepsilon\int\gamma_+(E - H_0)VU_+\delta(E - H_0)\,\mathrm{d}E\,, \tag{4.16}$$

where

$$\gamma_+(x) = i\pi\delta(x) + \frac{P}{x} = \lim_{\zeta\to+0}\frac{1}{x - i\zeta}\,. \tag{4.17}$$

The eigenstates $|H, A; E, a\rangle_+$ of H given by

$$|H, A; E, a\rangle_+ = U_+|H_0, A_0; E, a\rangle \tag{4.18}$$

are usually called the « incoming waves ». From the definition of the scattering operator it is clear that the inverse scattering operator S^{-1} is given by

$$\lim_{t\to-\infty}\exp[iH_0t]\exp[-iHt]U_+|\varphi\rangle = S^{-1}|\varphi\rangle\,. \tag{4.19}$$

In Appendix I it is shown that

$$S^{-1} = I + 2\pi i\varepsilon\int\delta(E - H_0)VU_+\delta(E - H_0)\,\mathrm{d}E\,. \tag{4.20}$$

Generally, the operators U are not unitary. However, the operators $U_\pm$ are. In Appendix II it is shown that

$$U_\pm^* U_\pm = I \tag{4.21}$$

where the asterisk mean Hermitian adjoint. Equation (4.21) yields orthonormality relations for $|H, A; E, a\rangle_+$, since we have

$$\begin{aligned} {}_\pm\langle H, A; E, a | H, A; F, b\rangle_\pm &= \langle H_0, A_0; E, a | U_\pm^* U_\pm | H_0, A_0; F, b\rangle \\ &= \langle H_0, A_0; E, a | H_0, A_0; F, b\rangle \\ &= \delta(E-F)\delta(a, b)\,. \end{aligned} \tag{4.22}$$

Since we have assume the spectrum of H is the same as that of H_0, each set of eigenfunctions, $|H, A; E, a\rangle_+$ and $|H, A; E, a\rangle_-$, is complete. Accordingly, any arbitrary state $|\varphi\rangle$ can be expanded in terms of either set. Using (4.22) the coefficients of the expansion can be found and it is seen that

$$|\varphi\rangle = \iint |H, A; E, a\rangle_\pm \,\mathrm{d}E\,\mathrm{d}a\, {}_\pm\langle H, A; E, a|\varphi\rangle\,, \tag{4.23}$$

or

$$\iint |H, A; E, a\rangle_\pm \,\mathrm{d}E\,\mathrm{d}a\, {}_\pm\langle H, A; E, a| = \mathrm{I}; \tag{4.24}$$

thus on using (4.8), (4.18), (2.4) we have

$$U_\pm U_\pm^* = I\,. \tag{4.25}$$

Finally conditions (4.21) and (4.25) imply, by definition of inverse, that

$$U_\pm^* = U_\pm^{-1} \tag{4.26}$$

which is the condition that $U_\pm$ be unitary.

5. – The general transformation operator and the spectral weight function.

We shall now consider properties of the general operator U which are analogous to the unitary property of $U_\pm$.

From (3.9) we see that we can write U as

$$U = M_\pm + \varepsilon\int \gamma_\pm(E-H_0) V U \delta(E-H_0)\,\mathrm{d}E\,, \tag{5.1}$$

where

$$M_{\pm} = L \mp i\pi\varepsilon \int \delta(E - H_0) V U \delta(E - H_0) \, dE \,. \tag{5.2}$$

In Appendix III we prove that as a consequence of (5.1)

$$U = U_{\pm} M_{\pm} \,. \tag{5.3}$$

(This equation also appears in reference ([7])). From equation (5.2) it is clear that the operators $M_{\pm}$ commute with H_0. From (5.3) we can show that the operators $M_{\pm}$ have inverses.

As a special case of (5.3) let, us take the positive sign for the subscripts and identify U with U_{-}. Then using (4.5) for the operator L associated with U_{-}, we have

$$M_{+} = I - 2i\pi\varepsilon \int \delta(E - H_0) V U_{-} \delta(E - H_0) \, dE = S \,. \tag{5.4}$$

Hence (5.3) gives

$$U_{-} = U_{+} S \tag{5.5}$$

or on using (4.26)

$$S = U_{+}^{*} U_{-} \,. \tag{5.6}$$

From (5.3) we have generally

$$U_{+} M_{+} = U_{-} M_{-} \tag{5.7}$$

or

$$M_{+} = U_{+}^{*} U_{-} M_{-} = S M_{-} \,.$$

Hence

$$S = M_{+} M_{-}^{-1} \,. \tag{5.8}$$

Equation (5.8) shows how one can construct the scattering operator from a general transformation U.

Let us define the weight operator W by

$$W = M_{+}^{-1} M_{+}^{*-1} = M_{-}^{-1} M_{-}^{*-1} \,. \tag{5.9}$$

It is to be noted that W is a positive definite Hermitian operator which commutes with H_0.

From (5.3), (4.21), (4.25) we have

$$W U^{*} U = I \,, \tag{5.10}$$

$$U W U^{*} = I \,. \tag{5.11}$$

Equation (5.10) gives the orthogonality relations for $|H, A; E, a\rangle$, and (5.14) gives the completeness relation for these eigenstates. From (5.10) and (5.11) it is clear that

$$U^{-1} = WU^* , \tag{5.12}$$

To justify our calling W a weight operator let us write

$$\langle H_0, A_0; E, a|W|H_0, A_0; F, b\rangle = \langle a|\omega(E)|b\rangle\,\delta(E-F) , \tag{5.13}$$

as is possible from (3.13). Then for any state $|\varphi\rangle$ we have, from (5.11) and (2.4),

$$\begin{aligned}|\varphi\rangle &= UWU^*|\varphi\rangle \\ &= \iiiint U|H_0, A_0; E, a\rangle\,\mathrm{d}E\,\mathrm{d}a\langle H_0, A_0; E, a|W|H_0, A_0; F, b\rangle\,\mathrm{d}F\,\mathrm{d}b\,\cdot \\ &\quad\cdot\langle H_0, A_0; F, b|U^*|\varphi\rangle \\ &= \iiiint |H, A; E, a\rangle\,\mathrm{d}E\,\mathrm{d}a\langle H_0, A_0; E, a|W|H_0, A_0; F, b\rangle\,\mathrm{d}F\,\mathrm{d}b\,\cdot \\ &\quad\cdot\langle H, A; F, b|\varphi\rangle ,\end{aligned} \tag{5.14}$$

or on using (5.13) we have the expansion of $|\varphi\rangle$ in terms of $|H, A; E, a\rangle$, namely

$$|\varphi\rangle = \int \mathrm{d}E \iint |H, A; E, a\rangle\,\mathrm{d}a\langle a|\omega(E)|b\rangle\,\mathrm{d}b\langle H, A; E, b|\varphi\rangle . \tag{5.15}$$

If the spectrum of H had no degeneracy, (5.15) would be written as

$$|\varphi\rangle = \int \mathrm{d}E\,|H; E\rangle\omega(E)\langle H; E|\varphi\rangle , \tag{5.15a}$$

and $\omega(E)$ would play the role of the usual weight function. In the case of degeneracy the weight function must be replaced by the operator $\langle a|\omega(E)|b\rangle$.

To summarize: In the direct problem, if H is given and a set of eigenfunctions (or equivalently a transformation operator U) is specified by choosing the appropriate boundary conditions, we can find the spectral weight operator W corresponding to those eigenfunctions.

6. – The inverse problem.

In the inverse problem we are given an operator W and seek the conditions under which we can solve the equations (cf. (5.10) and (5.11))

$$WU^*U = \mathrm{I} , \tag{6.1}$$

$$UWU^* = \mathrm{I} , \tag{6.2}$$

for U such that $|H, A; E, a\rangle = U|H_0, A_0; E, a\rangle$ is an eigenfunction of a Hamiltonian $H = UH_0U^{-1}$ corresponding to a suitable choice of boundary conditions on $|H, A; E, a\rangle$. Specifically we shall rewrite (6.1) and (6.2) as

$$WU^* = U_0 \tag{6.3}$$

$$U_0 = U^{-1} \tag{6.4}$$

and obtain theorems which will tell us under what conditions we can obtain unique solutions U, U_0 of(6.3) such that (6.4) is satisfied and such that the eigenfunctions $|H, A; E, a\rangle = U|H_0, A_0; E, a\rangle$ approach $|H_0, A_0; E, a\rangle$ asymptotically in some specified representation.

First we shall discuss the necessary conditions on W, independent of the boundary conditions on the eigenfunctions $|H, A; E, a\rangle$.

We shall now show that in order for (6.3) to have a solution for U and U_0 such that (6.4) holds, *W must be a positive definite, Hermitian operator which has an inverse.* Further, in order that H. as given by

$$H = UH_0U_0 = UH_0U^{-1}, \tag{6.5}$$

be Hermitian, *W must commute with H_0.*

Once we have proved this, we shall henceforth assume that W is a positive definite Hermitian operator which commutes with H_0 and has an inverse.

First of all let us show that W must have an inverse. From (6.3) and (6.4),

$$WU^*U = I, \tag{6.6}$$

from which by multiplication on the right by $U^{-1}U^{*-1}$ we get

$$W = U^{-1}U^{*-1} = (U^*U)^{-1}. \tag{6.7}$$

Clearly W has an inverse W^{-1} given by

$$W^{-1} = U^*U. \tag{6.8}$$

That W must be Hermitian follows immediately from (6.7). We can also show from (6.7) that W is positive definite.

We now show that W must commute with H in order that H as given by (6.5) be Hermitian. From (6.5) and (6.3) we have

$$H = UH_0U_0 = UH_0WU^*, \qquad H^* = U_0^*H_0U^* = UWH_0U^*. \tag{6.9}$$

Then if H is to equal H^* we see from (6.9) that the following equation must hold:

$$UH_0WU^* = UWH_0U^* .$$

On multiplying on the right by U^{*-1} and on the left by U^{-1} we have

$$H_0W = WH_0 . \tag{6.9a}$$

Hence, a necessary condition that H be Hermitian is that W commutes with H_0. This condition is also sufficient as can be easily seen.

The potential V which one would obtain is given by

$$\varepsilon V = H - H_0 = UH_0U^{-1} - H_0 , \tag{6.10}$$

$$\varepsilon VU = UH_0 - H_0U . \tag{6.10a}$$

To show that eigenfunctions $|H, A; E, a\rangle$ of H can be obtained from

$$|H, A; E, a\rangle = U|H_0, A_0; E, a\rangle \tag{6.11}$$

we use (6.5). Then

$$\begin{aligned} H|H, A; E, a\rangle &= HU|H_0, A_0; E, a\rangle = UH_0|H_0, A_0; E, a\rangle \\ &= EU|H_0, A_0; E, a\rangle = E|H, A; E, a\rangle . \end{aligned} \tag{6.12}$$

7. – Boundary conditions on the eigenfunctions and triangularity conditions on operators.

It is clear that for a given operator W which satisfies the conditions discussed in Section **6**, there are many operators U_0 and U which satisfy $WU^* = U_0$ and $U^{-1} = U_0$. We wish to introduce additional conditions, understandable from a physical point of view, which will enable us to get essentially unique solutions U, U_0 which satisfy $U^{-1} = U_0$. Toward this end we consider imposing boundary conditions on the eigenfunctions $|H, A; E, a\rangle$ to make them unique.

Let us introduce a Hermitian operator Q which is itself a complete set of commuting variables. Let us denote the eigenvalues of Q by q, whose range is

$$q_0 \leqslant q \leqslant q_1 .$$

In special cases q_0 or q_1 can be 0, or $\pm\infty$. The eigenvector of Q corresponding to the eigenvalue q is denoted by $|q\rangle$.

The boundary condition which can reasonably be imposed on $|H, A; E, a\rangle$ is

$$\lim_{q\to q_0} [\langle q|H, A; E, a\rangle - \langle q|H_0, A_0; E, a\rangle] = 0\,, \tag{7.1}$$

i.e., that $\langle q|H, A; E, a\rangle$ approach $\langle q|H_0, A_0; E, a\rangle$ asymptotically in the Q-representation. We shall now give conditions on U, U_0 such that (7.1) will often hold and U, U_0 will be unique. These conditions will henceforth be called the *triangularity* conditions. Let us write

$$U = I + \varepsilon K\,, \qquad U_0 = I + \varepsilon K_0\,; \tag{7.2}$$

then

$$\langle q|H, A; E, a\rangle = \langle q|H_0, A_0; E, a\rangle + \varepsilon\int_{q_0}^{q_1}\langle q|K|q'\rangle dq'\langle q'|H_0, A_0; E, a\rangle\,, \tag{7.3}$$

$$\langle q|H_0, A_0; E, a\rangle = \langle q|H, A; E, a\rangle + \varepsilon\int_{q_0}^{q_1}\langle q|K_0|q'\rangle dq'\langle q'|H, A; E, a\rangle\,. \tag{7.4}$$

Let us call any operator P triangular in the Q-representation if

$$\langle q|P|q'\rangle = 0 \qquad\qquad q' > q \tag{7.5}$$

and if $\langle q|P|q'\rangle$ is a bounded function for $q' \leqslant q$. If K and K_0 are triangular and q_0 is finite in the Q-representation, then equations (7.3) and (7.4) become

$$\langle q|H, A; E, a\rangle = \langle q|H_0, A_0; E, a\rangle + \varepsilon\int_{q_0}^{q}\langle q|K|q'\rangle dq'\langle q'|H_0, A_0; E, a\rangle\,, \tag{7.6}$$

$$\langle q|H_0, A_0; E, a\rangle = \langle q|H, A; E, a\rangle + \varepsilon\int_{q_0}^{q}\langle q|K_0|q'\rangle dq'\langle q'|H, A; E, a\rangle\,. \tag{7.7}$$

and (7.1) is clearly satisfied.

If $q_0 = -\infty$, some additional restrictions on $\langle q|K|q'\rangle$ and $\langle q|K_0|q'\rangle$ as $q \to \infty$ are needed in order that the triangularity condition guarantee the asymptotic form (7.1). An example of such a condition (a rather severe one) is that $\langle q|K|q'\rangle$ and $\langle q|K_0|q'\rangle$ vanish when $q < q_{\min}$, where $q_{\min}$ is finite. In Section **17** we shall discuss sufficient conditions which W must satisfy in order that K and K_0 be such that (7.1) holds.

In our present treatment we have not specified in which representation K, K_0 is to be triangular. Our attitude is that physical reasoning should be used to tell us in which representation the asymptotic relation (7.1) is to hold and hence in which representation K, K_0 are to be triangular. In contrast to this attitude, previous authors choose the Q-representation in such a way that the potential V is diagonal in that representation. The emphasis in their papers is on the proof that K and K_0 are triangular in this representation.

Most of the remaining sections of the present paper are devoted to theorems which tell us under what conditions U, U_0 can be obtained uniquely once the Q-representation is given in which K, K_0 is to be triangular.

A summary of the theorems is the following:

1) A necessary condition for a solution is that $\langle q|W-I|q'\rangle$ be a bounded function of q' in the neighborhood of $q'=q$.

2) A sufficient condition for a solution U, U_0 of $WU^*=U_0$ is that the operator $(W-I)$ be bounded less than 1.

3) A sufficient condition for the uniqueness of the solution U, U_0 of $WU^*=U_0$ is that $\langle q|K|q'\rangle$ and $\langle q|K_0|q'\rangle$ be bounded functions in every interval. (This condition is automatically satisfied by the definition of triangularity of K, K_0 in the Q-representation).

4) A sufficient condition that $U^{-1}=U_0$ is that either $\langle q|K|q'\rangle$ or $\langle q|K_0|q'\rangle$ be a bounded continuous function of its arguments.

5) An alternative sufficient condition that U, U_0 are unique and $U^{-1}=U_0$ is that $(W-I)$ be bounded less than 1 and $(W-I)$ have an inverse.

8. – Solution of the equation for U, U_0; the Gelfand-Levitan equation.

The triangularity conditions (7.5) as applied to K, K_0 enable us to find a relation between U and W, from which, under suitable conditions on W, we can find U explicitly in terms of W.

Let us write

$$W=I+\varepsilon\Omega\,. \tag{8.1}$$

From (6.3) we have, using the fact that W is Hermitian,

$$UW=U_0^*\,. \tag{8.2}$$

Upon substituting (7.2) and (8.1) into (8.2), we obtain

$$K=K_0^*-\Omega-\varepsilon K\Omega \tag{8.3}$$

or, in terms of the Q-representation,

$$\langle q|K|q'\rangle=\langle q|K_0^*|q'\rangle-\langle q|\Omega|q'\rangle-\varepsilon\int_{q_0}^{q}\langle q|K|q''\rangle\,dq''\langle q''|\Omega|q'\rangle\,. \tag{8.4}$$

In the integral occuring in (8.4) we have used the triangularity property of K.

Now let us take $q' < q$. Since $\langle q|K_0^*|q'\rangle = 0$, we have

$$\langle q|K|q'\rangle = -\langle q|\Omega|q'\rangle - \varepsilon\int_{q_0}^{q}\langle q|K|q''\rangle dq''\langle q''|\Omega|q'\rangle \quad \text{for } q' < q. \tag{8.5}$$

Equation (8.5) is our generalization of the Gelfand-Levitan equation. It is an integral equation of the Fredholm type (if q is considered fixed) which gives K in terms of Ω or equivalently U in terms of W.

One could iterate (8.5) to obtain a solution for $\langle q|K|q'\rangle$. However, we shall proceed in a different way by rewriting equation (8.5) in terms of an operator notation.

Let us define the operator $\delta(q-Q)$ by

$$\delta(q-Q)|q'\rangle = \delta(q-q')|q'\rangle\,, \tag{8.6}$$

from which it can be seen that on using

$$\int_{q_0}^{q_1}|q\rangle dq\langle q| = \mathrm{I}\,,$$

we have

$$\delta(q-Q) = |q\rangle\langle q|\;. \tag{8.7}$$

We define the operator $\eta(q-Q)$ by

$$\eta(q-Q)|q'\rangle = \eta(q-q')|q'\rangle\,, \tag{8.8}$$

where $\eta(q)$ is the Heaviside step function

$$\eta(q) = 1\,, \qquad q > 0, \tag{8.9}$$
$$\quad\;\; = 0\,, \qquad q < 0.$$

Any operator R, say, which is triangular in the sense

$$\langle q|R|q'\rangle = 0\,, \qquad q' > q \tag{8.10}$$

together with the boundedness condition for $q' < q$ can be represented as

$$\langle q|R|q'\rangle = \langle q|P|q'\rangle\eta(q-q')\,, \tag{8.11}$$

where

$$\langle q|P|q'\rangle = \langle q|R|q'\rangle \quad \text{for} \quad q' \leqslant q\,, \tag{8.12}$$

while

$$\langle q|P|q'\rangle \text{ is arbitrary for } q' > q .$$

Relation (8.11) can be written

$$R = \int_{q_0}^{q_1} \delta(q-Q) P \eta(q-Q)\, dq = \int_{q_0}^{q_1} \eta(Q-q) P \delta(q-Q)\, dq , \tag{8.13}$$

where as before P is an operator such that

$$\langle q|P|q'\rangle = \langle q|R|q'\rangle , \qquad \text{for } q < q' \tag{8.14}$$

and

$$\langle q|P|q'\rangle \text{ is arbitrary for } q > q' .$$

In particular P can be taken as R. Let us multiply each side of equation (8.5) by $\eta(q-q')$. Since $\langle q|K|q'\rangle\eta(q-q') = \langle q|K|q'\rangle$ we have

$$\langle q|K|q'\rangle = -\langle q|\Omega|q'\rangle\eta(q-q') - \\ -\varepsilon \int_{q_0}^{q_1} \langle q|K|q''\rangle\eta(q-q'')\, dq'' \langle q''|\Omega|q'\rangle\eta(q-q') . \tag{8.15}$$

In order that $\langle q|K|q'\rangle$ have a solution it is necessary that $\langle q|\Omega|q'\rangle\eta(q-q')$ exist as the kernel of an integral operator. We shall discuss this point more fully later. From (8.15) we get, in an operator notation,

$$K = -\int_{q_0}^{q_1} \delta(q-Q)\Omega\eta(q-Q)\, dq - \varepsilon \int_{q_0}^{q_1} \delta(q-Q)K\eta(q-Q)\Omega\eta(q-Q)\, dq . \tag{8.16}$$

We regard (8.16) as our equation for K. We maintain that if the operator $A(q)$ exists, where

$$A(q) = [I + \varepsilon\Omega\eta(q-Q)]^{-1} , \qquad A^{-1}(q) = [I + \varepsilon\Omega\eta(q-Q)] \tag{8.17}$$

then a solution of (8.16) is

$$K = -\int_{q_0}^{q_1} \delta(q-Q)A(q)\Omega\eta(q-Q)\, dq . \tag{8.18}$$

This can be shown as follows. If $A(q)$ exists we have from $A(q)A^{-1}(q) = A^{-1}(q)A(q) = I$

$$A(q) = I - \varepsilon\Omega\eta(q-Q)A(q) \\ = I - \varepsilon A(q)\Omega\eta(q-Q) . \tag{8.19}$$

On substituting K as given by (8.18) into the right side of (8.16) we obtain

$$(8.20)\quad K = -\int_{q_0}^{q_1} \delta(q-Q)\Omega\eta(q-Q)\,\mathrm{d}q + \varepsilon\int_{q_0}^{q_1}\delta(q-Q)\cdot$$
$$\left\{\int_{q_0}^{q_1}\delta(q'-Q)A(q')\Omega\eta(q'-Q)\,\mathrm{d}q'\right\}\eta(q-Q)\Omega\eta(q-Q)\,\mathrm{d}q =$$
$$= -\int_{q_1}^{q_1}\delta(q-Q)\Omega\eta(q-Q)\,\mathrm{d}q +$$
$$+\varepsilon\int_{q_0}^{q_1}\delta(q-Q)A(q)\Omega\eta(q-Q)\eta(q-Q)\Omega\eta(q-Q)\,\mathrm{d}q =$$
$$= -\int_{q_0}^{q_1}\delta(q-Q)\Omega\eta(q-Q)\,\mathrm{d}q + \varepsilon\int_{q_0}^{q_1}\delta(q-Q)A(q)\Omega\eta(q-Q)\Omega\eta(q-Q)\,\mathrm{d}q\,.$$

In deriving equation (8.20) we have used the fact that $\delta(q-Q)\delta(q'-Q) = \delta(q-Q)\,\delta(q'-q)$. On using (8.19) we have

$$(8.21)\quad K = -\int_{q_0}^{q_1}\delta(q-Q)\Omega\eta(q-Q)\,\mathrm{d}q + \int_{q_0}^{q_1}\delta(q-Q)\Omega\eta(q-Q)\,\mathrm{d}q -$$
$$-\int_{q_0}^{q_1}\delta(q-Q)A(q)\Omega\eta(q-Q)\,\mathrm{d}q\,,$$

which is just (8.18). Hence (8.18) satisfies (8.16).

Now the operator $A(q) = [I + \varepsilon\Omega\eta(q-Q)]^{-1}$ will always exist (and hence (8.18) will provide a solution for K) if Ω is bounded less than $1/|\varepsilon|$, for in this case $\Omega\eta(q-Q)$ is also bounded less than $1/|\varepsilon|$ (the bound of $\eta(q-Q)$ being less than 1). Then $A(q)$ is given by the series

$$(8.22)\qquad A(q) = \sum_{n=0}^{\infty}\varepsilon^n(-1)^n[\Omega\eta(q-Q)]^n\,.$$

Substitution of (8.22) into (8.18) gives the result for K which one would obtain from iteration.

Incidentally, the condition that Ω is bounded less than $1/|\varepsilon|$ assures us that $W = I + \varepsilon\Omega$ is positive definite.

Having obtained K we can find U, on using (8.19), to be

$$U = \mathrm{I} + \varepsilon K = \mathrm{I} - \varepsilon\int_{q_0}^{q_1} \delta(q-Q)A(q)\Omega\eta(q-Q)\,\mathrm{d}q = \int_{q_0}^{q_1}\delta(q-Q)A(q)\,\mathrm{d}q\,, \tag{8.23}$$

and U_0 to be

$$U_0 = WU^* = (\mathrm{I} + \varepsilon\Omega)\int_{q_0}^{q_1} A^*(q)\delta(q-Q)\,\mathrm{d}q = \tag{8.24}$$

$$= \int_{q_0}^{q_1}\{[\mathrm{I} + \varepsilon\eta(q-Q)\Omega] + \varepsilon\eta(Q-q)\Omega\}A^*(q)\delta(q-Q)\,\mathrm{d}q =$$

$$= \int_{q_0}^{q_1}\{A^{*-1}(q) + \varepsilon\eta(Q-q)\Omega\}A^*(q)\delta(q-Q)\,\mathrm{d}q = \mathrm{I} + \varepsilon\int_{q_0}^{q_1}\eta(Q-q)\Omega A^*(q)\delta(q-Q)\,\mathrm{d}q\,.$$

where the asterisk indicates the Hermitian adjoint, i.e.

$$A^*(q) = [I + \varepsilon\eta(q-Q)\Omega]^{-1}\,, \qquad A^{*-1}(q) = [I + \varepsilon\eta(q-Q)\Omega]\,. \tag{8.17a}$$

Hence K is given by

$$K_0 = \frac{1}{\varepsilon}[U_0 - \mathrm{I}] = \int_{q_0}^{q_1}\eta(Q-q)\Omega A^*(q)\delta(q-Q)\,\mathrm{d}q\,. \tag{8.25}$$

An alternative expression for K_0 will prove useful later. It arises from the identity

$$\Omega[I + \varepsilon\eta(q-Q)\Omega]^{-1} = [I + \varepsilon\Omega\eta(q-Q)]^{-1}\Omega\,, \tag{8.26}$$

or

$$\Omega A^*(q) = A(q)\Omega\,.$$

The identity (8.26) can be proved by multiplying on the right by $[I+\varepsilon\eta(q-Q)\Omega]$ and on the left by $[I+\varepsilon\Omega\eta(q-Q)]$. The alternative form K_0 is then

$$K_0 = \int_{q_0}^{q_1}\eta(Q-q)A(q)\Omega\delta(q-Q)\,\mathrm{d}q\,. \tag{8.27}$$

On comparing (8.27) with (8.13) it is clear that K_0 has the desired triangularity properties.

9. – Uniqueness theorems.

In this section we prove the uniqueness of the solutions (8.18) and (8.27) for K and K_0. The proof presented here depends on the positive definite character of W.

Theorem: If there is a solution U, U_0 of $WU^*=U_0$ such that

$$\langle q|U|q'\rangle = \delta(q-q') + \varepsilon\langle q|K|q'\rangle$$

and

$$\langle q|U_0|q'\rangle = \delta(q-q') + \varepsilon\langle q|K_0|q'\rangle\,,$$

where $\langle q|K|q'\rangle$ and $\langle q|K_0|q'\rangle$ are bounded functions of q and q' and are triangular in the sense that they vanish when $q'>q$, then this solution is the only solution of this character.

Proof: Assume that there is another such solution $U^{(1)} = I + K^{(1)}$ and $U_0^{(1)} = I + \varepsilon K_0^{(1)}$. Then in addition to having the relation

$$WU^* = U_0 \tag{9.1}$$

we should have

$$WU^{*(1)} = U_0^{(1)}\,. \tag{9.2}$$

On subtracting we find

$$W(K^* - K^{*(1)}) = (K_0 - K_0^{(1)})\,. \tag{9.3}$$

Let us define R and R_0 by

$$R = K - K^{(1)}\,, \qquad R_0 = K_0 - K_0^{(1)}\,. \tag{9.4}$$

From the properties of K, $K^{(1)}$, K_0, $K_0^{(1)}$ we find that $\langle q|R|q'\rangle$ and $\langle q|R_0|q'\rangle$ vanish for $q'>q$ and that they are bounded functions of q and q'. Equation (9.3) can be written as

$$\int_{q_0}^{q'}\langle q|W|q''\rangle\,dq''\langle q''|R^*|q'\rangle = \langle q|R_0|q'\rangle\,. \tag{9.5}$$

from which

$$\int_{q_0}^{r}dq\int_{q_0}^{q'}dq''\langle r|R|q\rangle\langle q|W|q''\rangle\langle q''|R^*|q'\rangle = \int_{q'}^{r}\langle r|R|q\rangle\,dq\langle q|R_0|q'\rangle\,, \tag{9.6}$$

where $r > q'$. Let us now let r approach q'. Under the boundedness conditions assumed for R and R_0, the integral on the right of (9.6) vanishes. Hence

$$\int_{q_0}^{q'} dq \int_{q_0}^{q'} dq'' \langle q' | R | q \rangle \langle q | W | q'' \rangle \langle q'' | R^* | q' \rangle = 0 . \tag{9.7}$$

If we regard $\langle q | R^* | q' \rangle$ as a vector, say $|\varphi\rangle$, insofar as it is a function of q, and if we regard q' as a fixed parameter, then equation (9.7) has the form

$$\langle \varphi | W | \varphi \rangle = 0 . \tag{9.8}$$

Since W is positive definite, it follows that $|\varphi\rangle = 0$, or, equivalently $R = 0$. Hence $U = U^{(1)}$, $U_0 = U_0^{(1)}$ and the solution is unique.

In Appendix IV we present an alternative proof of the uniqueness. However, it is not quite so useful as the present proof, because instead of using the positive definite character of W, it assumes that $\varepsilon\Omega = W - I$ has an inverse.

10. – Proof that $U^{-1} = U_0$.

Thus far we have discussed only equation

$$W U^* = U_0 \tag{10.1}$$

and have shown that under quite general conditions, this equation has a unique solution if we specify that $\varepsilon K_0 = U_0 - I$ and $\varepsilon K = U - I$ are to be triangular in a specified representation, We have not yet discussed our other requirement on U and U_0, namely

$$U^{-1} = U_0 . \tag{10.2}$$

It might be expected that the solutions U and U_0 which are obtained from (10.1) using the triangularity conditions of K and K_0 would satisfy (10.2) only under rather exceptional conditions on the weight operator W, since equations (10.1) and (10.2) appear to be quite independent of one another. Surprisingly, however, it turns out that rather general conditions on W are sufficient to insure (10.2).

We give two proofs that $U^{-1} = U_0$. One is an abstract proof (using only the equation (10.1)) and the other makes use of the explicit solutions (8.23) and (8.24) for U and U_0. We present the abstract proof here because of its simplicity. The second proof is given in Appendix V.

Theorem: If $\langle q|K|q'\rangle$ and $\langle q|K_0|q'\rangle$ vanish when $q' > q$ and are bounded continuous functions of q and q' for $q < q'$, then $U^{-1} = U_0$.

Proof: We shall prove this theorem in two parts. First we shall prove $UU_0 = I$ (i.e., U_0 is the right inverse of U); then we shall show that U^{-1} exists, from which it follows that $U_0 = U^{-1}$.

Let us first prove $UU_0 = I$ for the case where $\langle q|K|q'\rangle$ and $\langle q|K_0|q'\rangle$ are matrices (i.e., where Q has a discrete spectrum) (*). Our requirements on K and K_0 are then

$$\text{(10.3)} \qquad \begin{cases} \langle q|K|q'\rangle = 0\,, & q' \geqslant q, \\ \langle q|K_0|q'\rangle = 0\,, & q' \geqslant q. \end{cases}$$

Now from (10.1) we have

$$\text{(10.4)} \qquad UW = U_0^*$$

or

$$\text{(10.5)} \qquad UWU^* = U_0^* U^* = UU_0\,.$$

The second equation follows from (10.1). From (10.5) we have, on using $U = I + \varepsilon K$, $U_0 = I + \varepsilon K_0$,

$$\text{(10.6)} \qquad K + K_0 + \varepsilon K K_0 = K^* + K_0^* + \varepsilon K_0^* K^*\,,$$

or, in terms of the Q-representation,

$$\text{(10.7)} \qquad \langle q|K|q'\rangle + \langle q|K_0|q'\rangle + \varepsilon\langle q|KK_0|q'\rangle = \\ = \langle q|K^*|q'\rangle + \langle q|K_0^*|q'\rangle + \varepsilon\langle q|K_0^*K^*|q'\rangle\,.$$

Now from (10.3) it can be shown that $\langle q|KK_0|q'\rangle = 0$ for $q' \geqslant q$. Hence the left-hand side of (10.7) vanishes when $q' \geqslant q$. But since

$$\langle q|K^*|q'\rangle = \langle q|K_0^*|q'\rangle = \langle q|K_0^*K^*|q'\rangle = 0\,, \qquad q' \leqslant q,$$

it follows that the right-hand side vanishes when $q' \leqslant q$. Hence the right and left members of (10.7) vanish identically, i.e.,

$$\text{(10.8)} \qquad K + K_0 + \varepsilon KK_0 = 0 = K^* + K_0^* + \varepsilon K_0^* K^*\,.$$

(*) This proof was suggested to the authors by Professors J. B. KELLER and B. FRIEDMAN.

Hence

$$UU_0 = I + \varepsilon K + \varepsilon K_0 + \varepsilon K K_0 = I\,, \tag{10.9}$$

as required.

We shall now indicate the generalization of this proof to the case where Q has a continuous spectrum. The sole dffierence is that we do not want or need to require that $\langle q|K|q\rangle = \langle q|K_0|q\rangle = 0$, that is, that K and K_0 be zero on the diagonal. In this case our previous considerations lead us to the conclusion that

$$\langle q|K|q'\rangle + \langle q|K_0|q'\rangle + \varepsilon\langle q|KK_0|q'\rangle = 0 \tag{10.10}$$

for $q < q'$ or $q > q'$,

and

$$\langle q|K|q\rangle + \langle q|K_0|q\rangle + \varepsilon\langle q|KK_0|q\rangle \neq 0 \tag{10.10a}$$

generally.

This expression (10.10*a*) will under our conditions be finite, however, for all q. Let $K + K_0 + \varepsilon + KK_0$ operate on a state $|\varphi\rangle$ in Hilbert space whose Q-representative $\langle q|\varphi\rangle$ is a continuous function of q. Then the Q-representative of $(K + K_0 + \varepsilon KK_0)|\varphi\rangle$ is given by

$$\int_{q_0}^{q} [\langle q|K|q'\rangle + \langle q|K_0|q'\rangle + \varepsilon\langle q|KK_0|q'\rangle]\,\mathrm{d}q'\langle q'|\varphi\rangle\,.$$

Clearly this integral equals zero, since, for fixed q, the integrand is zero for all q' except possibly for $q' = q$, where it is finite. Hence

$$(K + K_0 + \varepsilon KK_0)|\varphi\rangle = 0 \tag{10.11}$$

for that set of states $|\varphi\rangle$ whose Q-representatives $\langle q|\varphi\rangle$ are continuous. Since this set of states is dense in Hilbert space, one can extend the definition of $K + K_0 + \varepsilon KK_0$ to the whole Hilbert space, in which case (10.11) holds for all states $|\varphi\rangle$ in the Hilbert space. Hence, by definition, (10.8) holds, and, finally so does (10.9).

We proceed now to the second part of the proof. We shall show that under the conditions of the theorem U has an inverse. The inverse $U^{-1} = (I + \varepsilon K)^{-1}$ obviously exists if the bound of K is less than $1/|\varepsilon|$. However, this condition is too restrictive and does not make use of the triangularity properties of K. A necessary and sufficient condition for U to have an inverse is that the equation $U|\varphi\rangle = 0$, when applied to any state $|\varphi\rangle$ in Hilbert space, implies that $|\varphi\rangle = 0$.

We shall now appeal to the theory of the Volterra integral equation to show that U satisfies this condition if U acts on the set of states $|\varphi\rangle$ whose Q-representatives $\langle q|\varphi\rangle$ are continuous functions of q. $U|\varphi\rangle = 0$ can be written as

$$\langle q|\varphi\rangle + \varepsilon\int_{q_0}^{q}\langle q|K|q'\rangle \mathrm{d}q'\langle q'|\varphi\rangle = 0 . \tag{10.12}$$

According to the Volterra theory [8] the only continuous function $\langle q|\varphi\rangle$ which can satisfy this equation if $\langle q|K|q'\rangle$ is a continuous, bounded function of q and q' is $\langle q|\varphi\rangle = 0$.

Hence U^{-1} exists when applied to the set of states $|\varphi\rangle$ which have continuous Q-representatives. It follows from (10.4) that $U_0U = I$ when applied to any state in this set. Since the set of continuous representatives is dense in the whole Hilbert space, the extension of U_0U into the rest of Hilbert space must also be I. Finally, therefore,

$$U_0U = I \tag{10.13}$$

over the whole Hilbert space.

In the case where q_0 equals $-\infty$, our reasoning has to be modified slightly, for the Volterra theory cannot be applied directly to this case. Instead of considering all states whose representatives $\langle q|\varphi\rangle$ are continuous, we restrict ourselves to those continuous states for which a $q_{\min}$ exists such that $\langle q|\varphi\rangle = 0$ when $q < q_{\min}$. The value of $q_{\min}$ may be arbitrarily small so long as it is finite. It may also be different for different states. Hence the equation (10.12) is replaced by

$$\langle q|\varphi\rangle + \int_{q\,\min}^{q}\langle q|K|q'\rangle \mathrm{d}q'\langle q'|\varphi\rangle = 0 . \tag{10.12a}$$

Then the Volterra theory may be applied to show that the only solution in this set is $\langle q|\varphi\rangle = 0$, and a reasoning similar to that used earlies shows that $U_0U = I$ when U_0U is applied to any state in this set. Since this set is dense in the whole Hilbert space, $U_0U = I$ in the whole Hilbert space as before.

The condition that $\langle q|K|q'\rangle$ be continuous is a little more severe than we should like; we should prefer the condition that $\langle q|K|q'\rangle$ be bounded, so that we might include the case where there are a finite umber of point dis-

---

(8) W. SCHMEIDLER: *Integralgleichungen mit Anwendungen in Physik und Tecknik*, I, *Lineare Integralgleichung* (Leipzig, 1950), p. 243.

continuities. The Volterra theory can be extended to include such cases. The proof in Appendix V is also valid for such operators, provided we assume that (8.23) and (8.24) give the solutions for U and U_0 and that the operators W^{-1} and Ω^{-1} exist.

11. – Generalizations.

Up to now we have assumed that Q by itself forms a complete set of commuting variables and looked for a triangular K in terms of this representation. Let us now assume that $Q_1, Q_2, \ldots, Q_n$ form a complete set of commuting variables. Then K and K_0, in terms of this representation, are given by

$$\langle q_1, q_2, \ldots, q_n | K | q_1', q_2', \ldots, q_n' \rangle$$

and

$$\langle q_1, q_2, \ldots, q_n | K_0 | q_1', q_2', \ldots, q_n' \rangle .$$

If we assume that K and K_0 are triangular in only one of the variables, q_1, say, i.e., if

$$\langle q_1, q_2, \ldots, q_n | K | q_1', q_2', \ldots, q_n' \rangle = 0 , \qquad q_1' > q_1$$

$$\langle q_1, q_2, \ldots, q_n | K_0 | q_1', q_2', \ldots, q_n' \rangle = 0 , \qquad q_1' > q_1$$

and if these operators obey the appropriate boundedness and continuity conditions in terms of the variables q_1 and q_1', the previous theory applies without any essential alteration.

In fact, if one writes operators, say P, and vectors $|\varphi\rangle$ in the $Q_1 \ldots Q_n$ representation as

$$\langle q_1, q_2, \ldots, q_n | P | q_1', q_2', \ldots, q_n' \rangle = \langle q_1 | P | q_1' \rangle$$

and

$$\langle q_1, q_2, \ldots, q_n | \varphi \rangle = \langle q_1 | \varphi \rangle ,$$

and regards $\langle q_1 | P | q_1' \rangle$ as an operator and $\langle q_1 | \varphi \rangle$ as a vector in terms of the remaining variables, one can go through the previous discussion and see that the formulas remain unaltered. One need merely interpret continuity and boundedness in a somewhat more general fashion to preserve the various theorems. We refrain from doing so explicitly, since the procedure is almost obvious.

The following much greater generalization appears possible. If we require that K and K_0 have the same invariant subspace in Hilbert space, and that K^* and K_0^* have invariant subspaces orthogonal to those of K, K_0, the theory appears to be valid. If one works in a proper representation one can generalize the concept of triangularity in terms of a suitable Q-representation and carry out the previous procedure. Perhaps the writers will sketch this method in later papers.

12. – Relation to the Wiener-Hopf integral equation.

The Wiener-Hopf integral equation for *two* unknown functions $f(z)$ and $g(z)$ is written

$$f(z) = g(z) + \gamma(z) + \int_{-\infty}^{z} f(z-\tau)\beta(\tau)\,d\tau\,. \tag{12.1}$$

Here z ranges from $-\infty$ to $+\infty$. The functions $\gamma(z)$ and $\beta(z)$ are given, and $f(z)$ and $g(z)$, the unknown functions, satisfy the boundary conditions

$$\begin{cases} f(z) = 0 & z < 0\,, \\ g(z) = 0 & z > 0\,. \end{cases} \tag{12.2}$$

It has been shown (e.g., ref. ([9])) that with certain restrictions of $\gamma(z)$ and $\beta(z)$ one may solve for both unknown functions explicitly. The usual procedure makes use of the analytic properties of these functions. Now we shall show that our equations for K and K_0 reduce in special cases to the standard Wiener-Hopf equation (12.1). In terms of the Q-representation, we have (cf. (8.4))

$$\langle q|K|q'\rangle = \langle q|K_0^*|q'\rangle - \langle q|\Omega|q'\rangle - \varepsilon\int_{q_0}^{q} \langle q|K|q''\rangle\,dq''\langle q''|\Omega|q'\rangle\,. \tag{13.2}$$

This is an equation for both K and K_0^*. Let us now take q_0 and q_1 as being $-\infty$ and $+\infty$ respectively. Furthermore, let us assume $\langle q|\Omega|q'\rangle$, $\langle q|K|q'\rangle$ and $\langle q|K_0^*|q'\rangle$ are functions of the difference $q-q'$ only. We shall write

$$\langle q|K|q'\rangle = K(q-q')$$

$$\langle q|K_0^*|q'\rangle = K_0^*(q-q')$$

$$\langle q|\Omega|q'\rangle = \Omega(q-q')$$

([9]) R. E. A. C. Paley and N. Wiener: *Fourier Transforms in the complex domain* (New York, 1934), Chapter IV.

where, in accordance with the triangularity conditions,

$$\begin{cases} K(q-q') = 0 & q' > q, \\ K_0^*(q-q') = 0 & q' < q. \end{cases} \tag{12.4}$$

Equation (12.3) is then

$$K(q-q') = K_0^*(q-q') - \Omega(q-q') - \varepsilon\int_{-\infty}^{q} K(q-q'')\Omega(q''-q')\,dq''. \tag{12.5}$$

If we now write

$$q - q' = z, \qquad q'' - q' = \tau,$$

equation (12.5) becomes

$$K(z) = K_0^*(z) - \Omega(z) - \varepsilon\int_{-\infty}^{z} K(z-\tau)\Omega(\tau)\,d\tau. \tag{12.5a}$$

where in accordance with (12.4)

$$\begin{aligned} K(z) &= 0 & z < 0, \\ K_0^*(z) &= 0 & z > 0. \end{aligned}$$

Identifying $K(z)$ with $f(z)$, $K_0^*(z)$ with $g(z)$, $-\Omega(z)$ with $\gamma(z)$, and $-\varepsilon\Omega(z)$ with $\beta(z)$, we see that (12.5) is of the form (12.1). In a sense, then, equation (12.3) is a generalization of the Wiener-Hopf equation. In another sense, however, it is more special, since for our applications we insist that Ω be Hermitian and have other appropriate properties.

Our procedure bears another resemblance to the Wiener-Hopf techniques, that is, to the factorization process. In the Wiener-Hopf approach, one factors an operator analogous to W into a product of two exponentials. With our formalism we are able to carry out this factorization explicitly.

From

$$WU^* = U_0 \tag{12.6}$$

and

$$U_0 = U^{-1} \tag{12.7}$$

it is easy to show

$$W = U_0 U_0^*. \tag{12.8}$$

We can write

$$W = e^{B} e^{B^*}. \tag{12.9}$$

If K_0 is triangular in the Q-representation, we have

$$B = \log U_0 = \log\left\{I + \varepsilon \int_{\omega_0}^{\omega_1} \eta(Q-q)A(q)\Omega\delta(q-Q)\,\mathrm{d}q\right\}, \tag{12.10}$$

and hence B also is triangular in this representation. From (12.7) we may write

$$U = U_0^{-1} = e^{-B}. \tag{12.11}$$

Equations (12.9) and (12.11) suggest the following alternative procedure for obtaining triangular solutions of (12.6) which satisfy (12.7). Since W is positive definite, $\log W$ can be defined. Equation (12.9) can then be written

$$e^{\log W} = e^{B}e^{B^*}. \tag{12.12}$$

If we can find an operator B which is triangular in the Q-representation such that (12.12) holds, then it can be shown that a solution of (12.6) which satisfies both the triangularity requirement and equation (12.7) is

$$U_0 = e^{B}, \qquad U = e^{-B}. \tag{12.13}$$

13. – The case where the spectra of H and H_0 do not coincide.

From our theory the operators U and U_0 can provide a Hamiltonian H, where

$$H = UH_0U_0 = UH_0U^{-1} \tag{13.1}$$

such that the spectrum of H coincides with that of H_0. This fact follows from the result that H and H_0 are related by a canonical transformation.

If we wish to obtain Hamiltonians H only part of whose spectrum coincides with that of H_0, and if we wish to preserve at least an analogue of (13.1) and the equation

$$WU^* = U_0, \qquad U_0 = U^{-1}, \tag{13.2}$$

then we must extend the domain of definition of H_0 and H such that the spectra of the extended operators H_0 and H coincide and consist of the sum of that part of the spectrum of H which coincides with the spectrum of H_0, that part of the spectrum of H which lies outside of the spectrum of H_0, and that part of the spectrum of H_0 which lies outside of the spectrum of H. Then for the extended operators H and H_0 it is possible that operators U, $U_0 = U^{-1}$ exist such that

$$\tilde{H} = U\tilde{H}_0 U^{-1}\,. \tag{13.3}$$

Generally, one is required to work in a vector space larger than Hilbert space. The way of extending the space is discussed in Part II where H is taken to have a point spectrum in addition to the continuous spectrum.

14. – A necessary condition for the triangularity of K, K_0.

For many weight operators W there are generally many representations such that we can find unique operators $U=I+\varepsilon K$ and $U_0=I+\varepsilon K_0$ and such that K and K_0 are triangular in these representations. To get unique solutions for U and U_0, one must specify which of these representations one wishes to use; one should choose from physical considerations the one in which the perturbed eigenfunctions approach the unperturbed eigenfunctions asymptotically.

It should not be assumed, however, that once a positive definite operator $W = I + \varepsilon\Omega$ has been given it is possible to choose *any* representation for triangularity of K, K_0. Generally, once W has been given, there is a restricted class of representations in which K, K_0 can be triangular.

A useful necessary condition, which we have discussed earlier, for the triangularity of K and K_0 is the following: In order that K, K_0 be triangular in the Q-representation, the operator

$$\int_{q_0}^{q_1} \delta(q-Q)\Omega\eta(q-Q)\,\mathrm{d}q\,,$$

must exist.

This condition, which follows from equation (8.16), enables us to discard representations in which K, K_0 cannot be triangular. For example, if Ω is a constant, then we have

$$\int_{q_0}^{q_1} \delta(q-Q)\Omega\eta(q-Q)\,\mathrm{d}q = \Omega\int_{q_0}^{q} \delta(q-Q)\eta(q-Q)\,\mathrm{d}q\,.$$

But the operator $\int_{q_0}^{q} \delta(q-Q)\eta(q-Q)\,\mathrm{d}q$ is not defined for *any* Q-representation. Hence if Ω, or equivalently W, is a constant, K, K_0 cannot be triangular in *any* representation. At most we can require that K, K_0 be diagonal in the

Q-representation, i.e., singular along the diagonal. In this case it can be shown that the most general solution for U and U_0 which satisfies $WU^*=U_0$, $U_0=U^{-1}$ is

$$U=(W)^{-\frac{1}{2}}\exp[i\theta(Q)]\,,\qquad U_0=(W)^{\frac{1}{2}}\exp[-i\theta(Q)]\,, \tag{14.1}$$

where $\theta(Q)$ is an arbitrary real function of Q. In the case, for example, where H_0 in the Q-representation is given by $-i(\partial/\partial q)(-\infty<q<\infty)$, we have

$$H=-i\frac{\partial}{\partial q}+\theta'(q)\,. \tag{14.2}$$

where the prime indicates the derivative.

An example of a representation in which K, K_0 cannot be triangular for *any* Ω is the H_0-representation, for in this case, since Ω always commutes with H_0, we have

$$\int_{q_0}^{q_1}\delta(q-H_0)\Omega\eta(q-H_0)\,\mathrm{d}q=\Omega\int_{q_0}^{q_1}\delta(q-H_0)\eta(q-H_0)\,\mathrm{d}q\,, \tag{14.3}$$

which again is undefined. In (14.3) we have labelled the eigenvalues of H_0 by q instead of E, as in earlier sections of the paper. As before, the requirement nearest to triangularity which we can impose is that K, K_0 be diagonal in the H_0 representation. In this case the general solution for U, U_0 is similar to that obtained for the previous case, namely

$$U=(W)^{-\frac{1}{2}}\exp[i\theta(H_0)]\,,\qquad U_0=(W)^{\frac{1}{2}}\exp[-i\theta(H_0)]\,, \tag{14.4}$$

where $\theta(H_0)$ is an arbitrary real function of H_0. Now we have

$$H=H_0 \tag{14.5}$$

and the effect of the transformation U is merely to change the normalization of the eigenfunctions of H_0.

15. – A simple example.

An example which provides us with exact, unique solutions U, U_0 of $WU^*=U_0$ such that $U=U_0^{-1}$ and K, K_0 are triangular in a specific representation will now be given. Let us take the Q-representation to be the usual x-representation, i.e., let the eigenvalues q lie in a continuum between $-\infty$

and $+\infty$. Further let H_0 be

$$H_0^Q = -i\frac{d}{dq}, \tag{15.1}$$

where the superscript Q denotes that H_0 is expressed in the Q-representation. The eigenfunctions in the Q-representation are given by

$$\langle q|H_0; E\rangle = \frac{1}{\sqrt{2\pi}}\exp[iEq]. \qquad -\infty < E < +\infty \tag{15.2}$$

Let us choose an Ω such that $W = I + \varepsilon\Omega$ is positive definite and $\int\delta(q-Q)\cdot\Omega\eta(q-Q)\,dq$ exists. Such an Ω is given by

$$\Omega = \frac{2\alpha}{H_0^2+\alpha^2}, \qquad \alpha > 0, \tag{15.3}$$

and

$$W = I + 2\varepsilon\frac{\alpha}{H_0^2+\alpha^2}. \tag{15.4}$$

It is clear that W is positive definite for $\varepsilon \geqslant -\alpha/2$. We shall restrict ourselves to such values of ε. It is easy to show that

$$\langle q|\Omega|q'\rangle = 2\alpha\int_{-\infty}^{+\infty}\langle q|H_0; E\rangle\frac{1}{E^2+\alpha^2}\langle H_0; E|q'\rangle\,dE = \exp[-\alpha|q-q'|]. \tag{15.5}$$

To show that $\int\delta(q-Q)\,\Omega\eta(q-Q)\,dq$ exists we write it in the Q-representation:

$$\langle q|\int\delta(q''-Q)\Omega\eta(q''-Q)|q'\rangle = \langle q|\Omega|q'\rangle\eta(q-q') = \exp[-\alpha(q-q')]\eta(q-q'). \tag{15.6}$$

The integral equation for $\langle q|K|q'\rangle$ is

$$\langle q|K|q'\rangle = -\exp[-\alpha(q-q')] - \varepsilon\int_{-\infty}^{q}\langle q|K|q''\rangle\exp[-\alpha|q''-q'|]\,dq''. \tag{15.7}$$

A solution of this equation is

$$\langle q|K|q'\rangle = \frac{1}{\varepsilon}(\alpha-\gamma)\exp[-\gamma(q-q')]\eta(q-q'), \tag{15.8}$$

where

$$\gamma = \sqrt{\alpha^2 + 2\alpha\varepsilon}\,, \tag{15.8a}$$

or

$$\langle q|U|q'\rangle = \delta(q-q') + (\alpha-\gamma)\exp[-\alpha(q-q')]\,\eta(q-q')\,. \tag{15.9}$$

Hence

$$\langle q|U_0|q'\rangle = \delta(q-q') + (\gamma-\alpha)\exp[-\alpha(q-q')]\,\eta(q-q')\,. \tag{15.10}$$

K, K_0 satisfy the conditions for the uniqueness theorem of Section 9. Hence (15.9) and (15.10) are the only solutions of $WU^* = U_0$ in which K and K_0 have the desired triangularity properties.

Actually, (15.7) has another solution for $\varepsilon < 0$. It is

$$\langle q|K|q'\rangle = \frac{1}{\varepsilon}(\alpha+\gamma)\exp[\gamma(q-q')]\eta(q-q')\,. \tag{15.11}$$

However, it can be shown that this solution has the property

$$\int_{-\infty}^{q'}\int_{-\infty}^{q'}\langle q'|K|q\rangle\,dq\langle q|W|q''\rangle\,dq''\langle q''|K^*|q'\rangle = \frac{i(\alpha+\gamma)}{\varepsilon(\alpha-\gamma)} < 0\,, \tag{15.12}$$

since $\varepsilon < 0$, and $\alpha > \gamma$. Hence $\langle q|K^*|q'\rangle$ considered as a function of q, with q' fixed, is outside the domain where W is positive definite. But since we prescribed W to be positive definite, we reject this solution. On the other hand, the first-mentioned solutions for K and K_0 are in the domain where W is positive definite. This example illustrates how the positive definite character of W is used to select unique solutions for K.

It is easy to show that $U^{-1} = U_0$ by using the explicit forms (15.9) and (15.10).

The example that we have discussed here has the virtue of illustrating the theorems discussed earlier. However, it is somewhat trivial in the sense that

$$H = UH_0U_0 = H_0\,, \tag{15.13}$$

as follows from the fact that U commutes with H_0. In fact, with our choice of H_0 and q, U will always commute with H_0 for any W which is such that U, U_0 can be obtained uniquely. This result is a consequence of the fact that $\langle q|U|q'\rangle$ will always be a function of the difference $q - q'$ and from this it follows that $\langle H_0; E|U|H_0; E'\rangle$ is diagonal and thus commutes with H_0. We refrain from giving details. With these choices of W, however, the perturbation $V = 0$ is unique.

16. – Triangularity of operators and asymptotic conditions on eigenfunctions when $q_0 = -\infty$.

In Section 7 it was indicated that triangularity of the operators K and K_0 is not sufficient to guarantee that $\langle q|H, A; E, a\rangle$ approaches $\langle q|H_0, A_0; E, a\rangle$ as $q \to q_0$ in the case where $q_0 = -\infty$ (or $+\infty$). The triangularity conditions are usually sufficient, however, to permit one to obtain unique solutions U, U_0, and hence a potential V, from a given weight operator W, as seen from the previous work.

In particular, the problem of Section 15 provides us with an example where the triangularity conditions yield unique operators U, U_0 and V, but where the asymptotic conditions fail. To show how the conditions fail we note that on using K as obtained from Section 15 we have

$$(16.1)\qquad \int_{-\infty}^{q} \langle q|K|q'\rangle\, dq' \langle q'|H_0; E\rangle = \frac{(\alpha-\gamma)}{\varepsilon}(\gamma + iE)^{-1}\exp[iEq]\,,$$

Hence the expression $\int_{-\infty}^{q}\langle q|K|q'\rangle dq'\langle q'|H_0; E\rangle$ clearly does not approach zero as $q \to -\infty$. We have in fact for finite q:

$$(16.2)\qquad \langle q|H; E\rangle = \langle q|H_0; E\rangle[1 + (\alpha-\gamma)(\gamma + iE)^{-1}]\,.$$

Hence the eigenfunctions $\langle q|H; E\rangle$ are also eigenfunctions of H_0, which was to be expected, since $H = H_0$.

We shall now give two sets of sufficient conditions on W (or equivalently Ω) to guarantee that the triangularity condition implies the asymptotic condition (7.1).

The first sufficient condition is based on the observation that if there is a $q_{\min}$ such that $\langle q|K|q'\rangle$ vanishes when $q < q_{\min}$, then

$$(16.3)\qquad \lim_{q\to-\infty}\int_{-\infty}^{q}\langle q|K|q'\rangle\, dq'\langle q'|H_0, A_0; E, a\rangle = 0\,.$$

Theorem: If there is a $q_{\min}$ such that $\langle q|\Omega|q'\rangle$ vanishes when $q < q_{\min}$, then

$$\lim_{q\to-\infty}[\langle q|H, A; E, a\rangle - \langle q|H_0, A_0; E, a\rangle] = 0\,.$$

Proof: Since Ω is Hermitian, the condition

$$(16.4)\qquad \langle q|\Omega|q'\rangle = 0 \qquad\qquad q < q_{\min}\,,$$

implies

(16.5) $$\langle q|\Omega|q'\rangle = 0 \qquad q' < q_{\min}\,.$$

The equation for $\langle q|K|q'\rangle$ is

(16.6) $$\langle q|K|q'\rangle = -\langle q|\Omega|q'\rangle - \varepsilon\int_{q\,\min}^{q}\langle q|K|q''\rangle\,dq''\langle q''|\Omega|q'\rangle\,, \qquad (q'<q).$$

Now when $q < q_{\min}$ and $q' < q$,

(16.7) $$\langle q|K|q'\rangle = 0\,,$$

as follows from (16.4), (16.5) and (16.6). Hence (16.3) is valid and the asymptotic condition holds.

The sufficient condition which we have stated is somewhat severe in that we may expect (16.3) to hold even if $q_{\min}$ does not exist, provided that $\langle q|\Omega|q'\rangle$ approaches zero sufficiently rapidly as $q \to -\infty$. This leads us to our second sufficient condition which we shall state without proof.

Theorem: Let $\langle q|H_0, A_0; E, a\rangle$ be a bounded function of q for sufficiently small q, and let $\langle q|K|q'\rangle$ be an absolutely integrable function of q'; further, let $|\langle q|\Omega|q'\rangle| \leqslant r(q')$ for all q, where $r(q')$ is a function such that

$$\int_{-\infty}^{q} r(q')\,dq' < \infty\,.$$

Then

$$\lim_{q\to-\infty}\int_{-\infty}^{q}\langle q|K|q'\rangle\,dq'\langle q'|H_0, A_0; E, a\rangle = 0\,,$$

and hence

$$\lim_{q\to-\infty}\left[\langle q|H, A; E, a\rangle - \langle q|H_0, A_0; E, a\rangle\right] = 0\,.$$

APPENDIX I

Expressions for $U_\mp$, S, S^{-1}.

Using equation (3.7) we have

(I.1) $$\exp[iH_0t]\exp[-iHt]U_- = \exp[iH_0t]U_-\exp[-iH_0t]\,.$$

Using (3.9) with $U = U_-$, and using the fact that L commutes with H_0, we obtain

(I.2) $\exp[iH_0 t]U_- \exp[-iH_0 t] =$

$$= L + \varepsilon \int \frac{P}{E - H_0} \exp[-i(E - H_0)t] V U_- \delta(E - H_0)\, dE .$$

Now it can be shown that

$$\text{(I.3)} \qquad \lim_{t\to\pm\infty} \frac{P}{E - H_0} \exp[-i(E - H_0)t] = \mp i\pi\delta(E - H_0) ,$$

when both sides of the equation are applied to appropriate states in Hilbert space. Hence

(I.4) $\lim_{t\to-\infty} \exp[iH_0 t] \exp[-iHt] U_- |\varphi\rangle = L|\varphi\rangle +$

$$+ i\pi \int \delta(E - H_0) V U_- \delta(E - H_0)\, dE\, |\varphi\rangle = |\varphi\rangle ,$$

from which (4.5) follows. Equation (4.15) is obtained similarly. To get the expression for S one shows, using (I.3) and (4.5), that

(I.5) $\lim_{t\to+\infty} \exp[iH_0 t] \exp[-iHt] U_- |\varphi\rangle =$

$$= L|\varphi\rangle - i\pi \int \delta(E - H_0) V U_- \delta(E - H_0)\, dE\, |\varphi\rangle =$$

$$= \left[I - 2i\pi \int \delta(E - H_0) V U_- \delta(E - H_0)\, dE\right] |\varphi\rangle = S|\varphi\rangle .$$

The expression for S^{-1} is obtained similarly.

Appendix II

Proof of $U_\pm^* U_\pm = I$.

Let us consider the equation with the negative subscript. From (4.6) we have

(II.1) $U_-^* U_- = I + \varepsilon \int \gamma_-(E - H_0) V U_- \delta(E - H_0)\, dE +$

$$+ \varepsilon \int \delta(E - H_0) U_-^* V \gamma_-^*(E - H_0)\, dE +$$

$$+ \varepsilon^2 \iint \delta(E - H_0) U_-^* V \gamma_-^*(E - H_0) \gamma_-(E' - H_0) V U_- \delta(E' - H_0)\, dE\, dE' ,$$

where

$$\gamma_-^*(E-H_0) = +i\pi\delta(E-H_0) + \frac{P}{E-H_0} = -\gamma_-(H_0-E)\,. \tag{II.2}$$

Now, it can be shown that the functions $\gamma_\pm(x)$ satisfy the identity

$$\gamma_\pm(x)\gamma_\pm(y) = \gamma_\pm(x+y)[\gamma_\pm(x)+\gamma_\pm(y)]\,. \tag{II.3}$$

Hence

$$\begin{aligned}\gamma_-^*(E-H_0)\gamma_-(E'-H_0) &= -\gamma_-(H_0-E)\gamma_-(E'-H_0)\\ &= -\gamma_-(E'-E)[\gamma_-(H_0-E)+\gamma_-(E'-H_0)]\\ &= \gamma_-(E'-E)[\gamma_-^*(E-H_0)-\gamma_-(E'-H_0)]\,.\end{aligned} \tag{II.4}$$

Then the fourth term on the right of (II.1) is given by

$$\begin{aligned}&\varepsilon^2\iint\delta(E-H_0)U_-^*V\gamma_-^*(E-H_0)\gamma_-(E'-H_0)VU_-\delta(E'-H_0)\,\mathrm{d}E\,\mathrm{d}E' =\\ &= \varepsilon^2\iint\gamma_-(E'-E)\delta(E-H_0)U_-^*V\gamma_-^*(E-H_0)VU_-\delta(E'-H_0)\,\mathrm{d}E\,\mathrm{d}E' -\\ &-\varepsilon^2\iint\gamma_-(E'-E)\delta(E-H_0)U_-^*V\gamma_-(E'-H_0)VU_-\delta(E'-H_0)\,\mathrm{d}E\,\mathrm{d}E' =\\ &= \varepsilon^2\int\gamma_-(E'-H_0)\left[\int\delta(E-H_0)U_-^*V\gamma_-^*(E-H_0)\,\mathrm{d}E\right]VU_-\delta(E'-H_0)\,\mathrm{d}E' -\\ &-\varepsilon^2\int\delta(E-H_0)U_-^*V\left[\int\gamma_-(E'-H_0)VU_-\delta(E'-H_0)\,\mathrm{d}E'\right]\gamma_-(H_0-E)\,\mathrm{d}E\,.\end{aligned} \tag{II.5}$$

In equation (II.5) we have used the fact that

$$f(E)\delta(E-H_0) = f(H_0)\delta(E-H_0)\,, \tag{II.6}$$

where $f(E)$ is any function of E. In the last equation of (II.5), we write

$$\left[\varepsilon\int\delta(E-H_0)U_-^*V\gamma_-^*(E-H_0)\,\mathrm{d}E\right] = U_-^* - I\,,$$

$$\left[\varepsilon\int\gamma_-(E-H_0)VU_-\delta(E-H_0)\,\mathrm{d}E\right] = U_- - I\,,$$

so that after rearrangement of terms the fourth term on the right of (II.1) becomes

$$\begin{aligned}&\varepsilon^2\iint\delta(E-H_0)U_-^*V\gamma_-^*(E-H_0)\gamma_-(E'-H_0)VU_-\delta(E'-H_0)\,\mathrm{d}E\,\mathrm{d}E' =\\ &= -\varepsilon\int\gamma_-(E-H_0)VU_-\delta(E-H_0)\,\mathrm{d}E - \varepsilon\int\delta(E-H_0)U_-^*V\gamma_-^*(E-H_0)\,\mathrm{d}E +\\ &+\varepsilon\int\gamma_-(E-H_0)U_-^*VU_-\delta(E-H_0)\,\mathrm{d}E - \varepsilon\int\delta(E-H_0)U_-^*VU_-\gamma_-(H_0-E)\,\mathrm{d}E\,.\end{aligned} \tag{II.7}$$

The first two terms on the right of (II.6) cancel the second and third terms on the right of (II.1).

We can write the last two terms on the right on (II.7) as

$$\text{(II.8)}\quad \varepsilon\int\gamma_-(E'-H_0)U_-^*VU_-\delta(E'-H_0)\,\mathrm{d}E'-\varepsilon\int\delta(E-H_0)U_-^*VU_-\gamma_-(H_0-E)\,\mathrm{d}E=$$

$$=\varepsilon\iint\gamma_-(E'-E)\delta(E-H_0)U_-^*VU_-\delta(E'-H_0)\,\mathrm{d}E\,\mathrm{d}E'-$$

$$-\varepsilon\iint\delta(E-H_0)U_-^*VU_-\gamma_-(E'-E)\delta(E'-H_0)\,\mathrm{d}E\,\mathrm{d}E'=0\,.$$

Hence the fourth term on the right of (II.1) merely cancels the second and third terms and we are left with our desired result

$$U_-^*U_-=I\,.$$

The relation $U_+^*U_+=I$ is proved similarly.

APPENDIX III

Proof of $U=U_\pm M_\pm$.

Let us prove $U=U_-M_-$. From equations (4.6) and (5.1) we have

$$\text{(III.1)}\quad U_-^*U=M_-+\varepsilon\int\gamma_-(E-H_0)VU\delta(E-H_0)\,\mathrm{d}E+$$

$$+\varepsilon\int\delta(E-H_0)U_-^*V\gamma_-^*(E-H_0)\,\mathrm{d}E\,M_-+$$

$$+\varepsilon^2\iint\delta(E-H_0)U_-^*V\gamma_-^*(E-H_0)\gamma_-(E'-H_0)VU\delta(E'-H_0)\,\mathrm{d}E\,\mathrm{d}E'\,.$$

By a procedure similar to that used in Appendix II, one can show that the fourth term on the right of (III.1) cancels the second and third term. Hence

$$\text{(III.2)}\qquad U_-^*U=M_-\,.$$

Finally, on using the completeness relation (4.25) we have the desired result. The relation $U=U_+M_+$ is proved similarly.

Appendix IV

Alternative uniqueness theorem.

Theorem: If the inverse of $[I + \varepsilon\Omega\eta(q-Q)]$, namely $A(q)$, exists and is unique and if the inverse of Ω exists, then the only possible solution of (8.16) for K is given by (8.18).

Proof: From the triangularity properties of K we may write

$$K = \int_{q_0}^{q_1} \delta(q-Q) K \eta(q-Q)\, dq \,. \tag{IV.1}$$

Equation (8.16) may then be written

$$\int_{q_0}^{q_1} \delta(q-Q) K \eta(q-Q)[I + \varepsilon\Omega\eta(q-Q)]\, dq = -\int_{q_0}^{q_1} \delta(q-Q)\Omega\eta(q-Q)\, dq \,. \tag{IV.2}$$

Now K is to be a triangular operator such that $\langle q|K|q'\rangle = 0$ for $q' > q$. Accordingly it must always be possible to represent K in the form (see (8.13))

$$K = \int_{q_0}^{q_1} \delta(q-Q) P \eta(q-Q)\, dq \,, \tag{IV.3}$$

where $\langle q|P|q'\rangle = \langle q|K|q'\rangle$ for $q' < q$. If Ω has an inverse we can always find an operator T such that

$$T = P\Omega \,. \tag{IV.4}$$

Hence the general expression for a triangular operator K can be written

$$K = \int_{q_0}^{q_1} \delta(q-Q) T\Omega\eta(q-Q)\, dq \,. \tag{IV.5}$$

It is our objective to obtain the most general operator T which will allow K to satisfy (IV.2). Having obtained this operator we shall obtain from (IV.5) the most general K which can satisfy this equation. Let us substitute (IV.5) into (IV.2) to obtain an equation for T. The left-hand side becomes

$$\int_{q_0}^{q_1} \delta(q-Q)\left[\int_{q_0}^{q_1} \delta(q'-Q) T\Omega\eta(q'-Q)\, dq'\right] \eta(q-Q)[I + \varepsilon\Omega\eta(q-Q)]\, dq = \tag{IV.6}$$

$$= \int_{q_0}^{q_1} \delta(q-Q) T\Omega\eta(q-Q)\eta(q-Q)[I + \varepsilon\Omega\eta(q-Q)]\, dq =$$

$$= \int_{q_0}^{q_1} \delta(q-Q) T \Omega \eta(q-Q)[I + \varepsilon \Omega \eta(q-Q)]\, dq =$$

$$= \int_{q_0}^{q_1} \delta(q-Q) T [I + \varepsilon \Omega \eta(q-Q)] \Omega \eta(q-Q)\, dq .$$

In (IV.6) we have used the relations

$$\delta(q-Q)\delta(q'-Q) = \delta(q-Q)\delta(q-q') , \tag{IV.7}$$

$$\eta(q-Q)\eta(q-Q) = \eta(q-Q) . \tag{IV.8}$$

Hence we have from (IV.2) an equation for T, namely

$$\int_{q_0}^{q_1} \delta(q-Q) T [I + \varepsilon \Omega \eta(q-Q)] \Omega \eta(q-Q)\, dq = - \int_{q_0}^{q_1} \delta(q-Q) \Omega \eta(q-Q)\, dq . \tag{IV.9}$$

Equation (IV.9) is of the form

$$\int_{q_0}^{q_1} \delta(q-Q) A \eta(q-Q)\, dq = \int_{q_0}^{q_1} \delta(q-Q) B \eta(q-Q)\, dq , \tag{IV.10}$$

where A and B are two operators. In terms of the Q-representation (IV.10) reads

$$\langle q | A | q' \rangle \eta(q-q') = \langle q | B | q' \rangle \eta(q-q') , \tag{IV.11}$$

which means

$$\langle q | A | q' \rangle = \langle q | B | q' \rangle , \qquad \text{for } q' < q; \tag{IV.12}$$

$$\langle q | A | q' \rangle = \langle q | C | q' \rangle , \qquad \text{for } q' > q, \tag{IV.13}$$

where C is an arbitrary integrable operator. Let us take $C = B + D$ where D is arbitrary; then

$$\langle q | A | q' \rangle = \langle q | B | q' \rangle + \langle q | D | q' \rangle \eta(q'-q) , \tag{IV.14}$$

or

$$A = B + \int_{q_0}^{q_1} \delta(q-Q) D \eta(Q-q)\, dq . \tag{IV.15}$$

Equation (IV.15) is the most general solution of (IV.10) for A in terms of B Let us identify A with $T[I + \varepsilon \Omega \eta(q-Q)]\Omega$ of equation (IV.9) and B with $-\Omega$. Then we have

$$T[I + \varepsilon \Omega \eta(q-Q)]\Omega = -\Omega + \int_{q_0}^{q_1} \delta(q-Q) D \eta(Q-q)\, dq , \tag{IV.16}$$

from which

$$(\text{IV}.17)\qquad T = -A(q) + \int_{q_0}^{q_1} \delta(q' - Q) D\eta(Q - q')\,dq'\,\Omega^{-1}A(q)\,.$$

This is the most general solution one can obtain for T. With this T one obtains the most general solution for K, on using (IV.5), namely

$$(\text{IV}.18)\quad K = \int_{q_0}^{q_1} \delta(q - Q) T\Omega\eta(q - Q)\,dq = -\int_{q_0}^{q_1} \delta(q - Q) A(q)\Omega\eta(q - Q)\,dq +$$
$$+ \int_{q_0}^{q_1} \delta(q - Q)\left[\int_{q_0}^{q_1} \delta(q' - Q) D\eta(Q - q')\,dq'\right]\Omega^{-1}A(q)\Omega\eta(q - Q)\,dq\,.$$

Now the first term on the right is the solution which we have obtained previously. The second term will be shown to be zero: it becomes

$$(\text{IV}.19)\quad \int_{q_0}^{q_1} \delta(q - Q) D\eta(Q - q)\Omega^{-1}\Omega\eta(q - Q) A(q)\,dq =$$
$$= \int_{q_0}^{q_1} \delta(q - Q) D\eta(Q - q)\eta(q - Q) A(q)\,dq = 0\,,$$

since

$$(\text{IV}.20)\qquad \eta(Q - q)\eta(q - Q) = 0\,.$$

In (IV.19) we have used the fact that $\Omega\eta(q - Q)$ commutes with $A(q)$, a result obtained from equation (8.19).

We have thus shown that under our assumptions $A(q)$ and Ω^{-1} exist and are unique, and that hence the only solution for K is given by (8.18).

APPENDIX V

Alternative proof that $U^{-1} = U_0$.

Theorem: If the conditions on Ω are such that the equation $WU^* = U_0$ has the solutions U and U_0 given by (8.23) and (8.24), and such that Ω^{-1} exists, then $U^{-1} = U_0$.

We shall first prove that

$$(\text{V}.1)\qquad UU_0 = I\,,$$

or, what is the same, that

$$(\text{V}.2)\qquad K + K_0 = -\varepsilon K K_0\,.$$

We write (8.18) and (8.27)

$$K = -\int_{q_0}^{q_1} \delta(q - Q) A(q) \Omega \eta(q - Q)\, dq \,, \tag{V.3}$$

$$K_0 = \int_{q_0}^{q_1} \eta(Q - q) A(q) \Omega \delta(q - Q)\, dq \,. \tag{V.4}$$

Hence

$$KK_0 = -\int_{q_0}^{q_1}\int_{q_0}^{q_1} dq\, dq' \delta(q - Q) A(q) \Omega \eta(q - Q) \eta(Q - q') A(q') \Omega \delta(q' - Q) \,. \tag{V.5}$$

Now using the facts that

$$\int_{q_0}^{q_1} \delta(q - Q)\, dq = I \,, \tag{V.6}$$

and

$$\begin{cases} \delta(q - Q)\eta(q' - Q) = \eta(q' - q)\delta(q - Q) \,, \\ \delta(q - Q)\eta(Q - q') = \eta(q - q')\delta(q - Q) \,, \end{cases} \tag{V.7}$$

we see that from (V.3) and (V.4) we can write

$$K = -\int_{q_0}^{q_1}\int_{q_0}^{q_1} dq\, dq' \delta(q - Q) \eta(q - q') A(q) \Omega \delta(q' - Q) \,, \tag{V.8}$$

$$K_0 = \int_{q_0}^{q_1}\int_{q_0}^{q_1} dq\, dq' \delta(q - Q) \eta(q - q') A(q') \Omega \delta(q' - Q) \,. \tag{V.9}$$

Hence

$$K + K_0 = \int_{q_0}^{q_1}\int_{q_0}^{q_1} dq\, dq' \delta(q - Q) \eta(q - q') [A(q') - A(q)] \Omega \delta(q' - Q) \,. \tag{V.10}$$

Now

$$\begin{aligned} A(q') - A(q) &= A(q)[A^{-1}(q)A(q') - I] \\ &= A(q)[A(q') - I + \varepsilon\Omega\eta(q - Q)A(q')] \,. \end{aligned} \tag{V.11}$$

From (8.19)

$$A(q') - I = -\varepsilon\Omega\eta(q' - Q)A(q') \,. \tag{V.12}$$

Hence

$$A(q') - A(q) = \varepsilon A(q)\Omega[\eta(q - Q) - \eta(q' - Q)]A(q') \,. \tag{V.11a}$$

Furthermore,

(V.13) $$\eta(q-q')[\eta(q-Q)-\eta(q'-Q)] = \eta(q-Q)\eta(Q-q') .$$

Hence, after some obvious rearrangements we have

(V.14) $$\eta(q-q')[A(q')-A(q)] = \varepsilon A(q)\Omega\eta(q-Q)\eta(Q-q')A(q') .$$

Substituting this expression in (V.10) we have, on comparing with (V.5),

(V.15) $$K+K_0 =$$
$$= \varepsilon\int_{q_0}^{q_1}\int_{q_0}^{q_1} \mathrm{d}q\,\mathrm{d}q'\,\delta(q-Q)A(q)\Omega\eta(q-Q)\eta(Q-q')A(q')\Omega\delta(q'-Q) = -\varepsilon K K_0 ,$$

as required.

We shall now prove

(V.16) $$U_0 U = I .$$

Equation (V.16) is equivalent to

(V.17) $$W U^* U = I .$$

Since we are assuming that W has an inverse W^{-1} equation (V.17) is equivalent to

(V.18) $$W^{-1} = U^* U .$$

From equation (8.23) we have

(V.19) $$U = \int_{q_0}^{q_1} \delta(q-Q)A(q)\,\mathrm{d}q , \qquad U^* = \int_{q_0}^{q_1} A^*(q')\delta(q'-Q)\,\mathrm{d}q' .$$

Thus, on using $\delta(q-Q)\delta(q'-Q) = \delta(q-Q)\delta(q-q')$ we obtain

(V.20) $$U^* U = \int_{q_0}^{q_1} A^*(q)\delta(q-Q)A(q)\,\mathrm{d}q .$$

At this point it will be convenient to introduce derivatives of operators. We shall do so formally, although undoubtedly it is possible to justify the steps rigorously. On taking the derivative with respect to q of

(V.21) $$A(q)A^{-1}(q) = I ,$$

one obtains

(V.22) $$\left[\frac{\mathrm{d}}{\mathrm{d}q}A(q)\right]A^{-1}(q) + A(q)\left[\frac{\mathrm{d}}{\mathrm{d}q}A^{-1}(q)\right] = 0 ,$$

or

$$\frac{\mathrm{d}}{\mathrm{d}q} A(q) = - A(q) \left[\frac{\mathrm{d}}{\mathrm{d}q} A^{-1}(q)\right] A(q) . \tag{V.23}$$

But from (8.17)

$$\frac{\mathrm{d}}{\mathrm{d}q} A^{-1}(q) = \varepsilon\Omega \frac{\mathrm{d}}{\mathrm{d}q}\eta(q - Q) = \varepsilon\Omega\delta(q - Q) . \tag{V.24}$$

Thus on substituting (V.24) into (V.23) and integrating we get

$$A(q) = C - \varepsilon\int_{q_0}^{q} A(q')\Omega\delta(q'-Q)A(q')\,\mathrm{d}q' , \tag{V.25}$$

where C is an operator to be determined by boundary conditions on $A(q)$. We note that

$$\eta(q_1 - Q) = 1, \qquad \eta(q_0 - Q) = 0 , \tag{V.26}$$

which follows from the fact that

$$\eta(q_1 - Q)|\varphi\rangle = \int_{q_0}^{q_1} |q\rangle\eta(q_1 - q)\,\mathrm{d}q\langle q|\varphi\rangle ,$$

for any state $|\varphi\rangle$. Since $q_0 \leqslant q \leqslant q_1$, $\eta(q_1 - q) = 1$. Hence $\eta(q_1 - Q)|\varphi\rangle = \int_{q_0}^{q_1}|q\rangle\,\mathrm{d}q\langle q|\varphi\rangle = |\varphi\rangle$, from which the first of equations (V.26) follows. The second is obtained similarly. Therefore,

$$A^{-1}(q_1) = W, \qquad A^{-1}(q_0) = I , \tag{V.27}$$

from which it follows that

$$A(q_1) = W^{-1}, \qquad A(q_0) = I . \tag{V.28}$$

Using the second boundary condition of (V.28) in (V.25), we have

$$A(q) = I - \varepsilon\int_{q_0}^{q} A(q')\Omega\delta(q'-Q)A(q')\,\mathrm{d}q' , \tag{V.29}$$

and on using the first of equations (V.28) we get

$$W^{-1} = I - \varepsilon\int_{q_1}^{q_1} A(q)\Omega\delta(q - Q)A(q)\,\mathrm{d}q . \tag{V.30}$$

Bur from (8.26)

$$A(q)\Omega = \Omega A^*(q)\,. \tag{V.31}$$

Hence on using (V.30) and (V.20) we have

$$W^{-1} = I - \varepsilon\Omega\int_{q_0}^{q_1} A^*(q)\delta(q-Q)A(q)\,\mathrm{d}q = I - \varepsilon\Omega U^* U\,. \tag{V.32}$$

Now

$$WW^{-1} = (I + \varepsilon\Omega)W^{-1} = I\,, \tag{V.33}$$

or

$$W^{-1} = I - \varepsilon\Omega W^{-1}\,. \tag{V.34}$$

Comparing (V.34) and (V.32), we see that

$$\Omega W^{-1} = \Omega U^* U\,, \tag{V.35}$$

or, if Ω has an inverse,

$$W^{-1} = U^* U, \tag{V.36}$$

or finally

$$WU^*U = U_0 U = I\,, \tag{V.37}$$

which was to be shown.

RIASSUNTO (*)

Si mostra come dalla teoria degli operatori in uno spazio vettoriale si possa derivare una forma generalizzata del procedimento di Gelfand-Levitan per ottenere i potenziali di scattering dalle funzioni di misura spettrale. La Parte I tratta il caso in cui le Hamiltoniane perturbate e imperturbate hanno lo stesso spettro che si considera puramente continuo. Lo spazio vettoriale che si considera è lo spazio di Hilbert.

(*) *Traduzione a cura della Redazione.*

IL NUOVO CIMENTO Vol. III, N. 1 1° Gennaio 1956

The Determination of the Scattering Potential from the Spectral Measure Function.

II. Point Eigenvalues and Proper Eigenfunctions (*).

I. Kay and H. E. Moses

Institute of Mathematical Sciences, New York University - New York

(ricevuto il 27 Ottobre 1955)

Summary. — In Part I of this paper [1] we restricted ourselves to the consideration of weight operators which led to Hamiltonians H whose spectra were identical to the spectrum of a given unperturbed Hamiltonian H_0. In this part of the paper we propose to show how one may choose weight operators which lead to Hamiltonians H having spectra different from that of H_0. To make the discussion more concrete we shall take the case where H_0 has a purely continuous spectrum extending from 0 to ∞ and where H has a spectrum which has a continuous part which coincides with the spectrum of H_0 and, in addition, has negative point eigenvalues.

1. – Introduction.

In Part I of this paper [1] we restricted ourselves to the consideration of weight operators which led to Hamiltonians H whose spectra were identical to the spectrum of a given unperturbed Hamiltonian H_0. In this part of the paper we propose to show how one may choose weight operators which lead to Hamiltonians H having spectra different from that of H_0. To make the

(*) The research reported in this article was done at the Institute of Mathematical Sciences, New York University, and has been made possible through support and sponsorship extended by Geophysics Research Directorate, Air Force Cambridge Research Center under Contract No. AF 19(122)-463.

(1) I. Kay and H. E. Moses: *Nuovo Cimento*, 2, 917 (1955).

discussion more concrete we shall take the case where H_0 has a purely continuous spectrum extending from 0 to ∞ and where H has a spectrum which has a continuous part which coincides with the spectrum of H_0, and, in addition, has negative point eigenvalues. However, the procedure which is discussed is capable of being generalized considerably, at least in a formal fashion.

It is our objective to find operators U and U_0 such that

$$WU^* = U_0 \tag{1.1}$$

$$U_0 = U^{-1}, \tag{1.2}$$

where W is the weight operator. We wish H to be obtained from

$$H = UH_0U_0 = UH_0U^{-1}, \tag{1.3}$$

Clearly (1.3) cannot hold if H_0 and H have a different spectrum, which is the case being considered here. We shall have to define a vector space larger than Hilbert space such that the spectrum of H_0 in this larger space is the same as that of H in this space.

It will be possible to carry out the extension of the Hilbert space to the larger space in terms of the Q-representation which is the representation in which K is triangular in the sense described in Part I, i.e.,

$$\langle q|U|q'\rangle = \delta(q-q') + \varepsilon\langle q|K|q'\rangle, \tag{1.4}$$

where

$$\langle q|K|q'\rangle = 0 \qquad \text{for } q' > q. \tag{1.5}$$

Like Part I this paper has two main divisions. In the first (Sections 2 and 3) we shall characterize the weight operator assuming the operator H and its spectrum given. In the second division (Section 5) we shall show how H can be obtained from the weight operator. We shall see that in addition to prescribing the weight operator which gives the weight function of the eigenfunction of the continuous spectrum and prescribing the boundary conditions which these eigenfunctions must satisfy, we shall have to give the normalization constants and eigenvalues of the proper eigenfunctions. Furthermore, we shall have to specify a set of « eigenfunctions » of H_0 which span the vector space which must be added to the Hilbert space to assure us that the spectrum of H_0 coincides with that of H.

2. – The eigenfunctions of the unperturbed Hamiltonian and the extended vector space.

Since we shall work with a vector space larger than Hilbert space, it will be useful to introduce a projection operator which shall be denoted by $\eta(H_0)$. This operator is to have the property that

$$\eta(H_0)|\varphi\rangle = |\varphi\rangle \tag{2.1}$$

if the state $|\varphi\rangle$ is in the Hilbert space, and that

$$\eta(H_0)|\varphi\rangle = 0 \tag{2.2}$$

if $|\varphi\rangle$ is in that part of the vector space which is orthogonal to the Hilbert space.

As in Part I we shall denote the eigenstates of H_0, A_0 belonging to the eigenvalues E, a, respectively, by $|H_0, A_0; E, a\rangle$. Since we are assuming that the spectrum of H_0 is continuous and ranges from zero to ∞, we shall express the fact that the $|H_0, A_0; E, a\rangle$ for $E > 0$ are a complete set in the Hilbert space by

$$\iint |H_0, A_0; E, a\rangle \eta(E)\,\mathrm{d}E\,\mathrm{d}a\langle H_0, A_0; E, a| = \eta(H_0)\,, \tag{2.3}$$

$\eta(E)$ being the usual Heaviside step function.

We shall extend the definition of the operator H_0 so that it has eigenvalues which correspond to the negative point eigenvalues E_i of the operator H. Toward this end we introduce « eigenfunctions » $|H_0, A_0; E, a\rangle$ which are defined for $E < 0$ in the vicinity ΔE_i of each eigenvalue E_i of H. These eigenfunctions are to span a vector space orthogonal to the Hilbert space. We shall characterize this extended space by working in the Q-representation. Every operator A defined in Hilbert space can be represented as an integral operator with the kernel $\langle q|A|q'\rangle$. If the operator A as defined in the extended space has the same form as an integral operator in the Q-representation when operating on vectors in the extended space, we shall call the operator A « a Q-extended operator ». We shall discuss Q-extended operators in more detail in Section 4.

In particular we shall take H_0 as a Q-extended operator. Hence for $E < 0$, the eigenfunctions $|H_0, A_0; E, a\rangle$ are defined by

$$\int_{q_0}^{q_1} \langle q|H_0|q'\rangle \mathrm{d}q'\langle q'|H_0, A_0; E, a\rangle = E\langle q|H_0, A_0; E, a\rangle\,, \tag{2.4}$$

for E in the vicinity ΔE_i of each point eigenvalue E_i of H. If, for example, $H_0^Q = -\mathrm{d}^2/\mathrm{d}q^2$ (where the superscript Q means that H_0 is expressed in the Q-representation) we should solve for $\langle q|H_0, A_0; E, a\rangle$ from

$$-\frac{\mathrm{d}^2}{\mathrm{d}q^2}\langle q|H_0, A_0; E, a\rangle = E\langle q|H_0, A_0; E, a\rangle .$$

Not every solution of (2.4) will be used to define an eigenfunction of H_0 in the extended space, for we shall require that $\langle q|H_0, A_0; E, a\rangle$ be quadratically integrable functions of q in the vicinity of q_0. If $q_0 = -\infty$ then a necessary condition is

$$\lim_{q\to-\infty}\langle q|H_0, A_0; E, a\rangle = 0 . \tag{2.5}$$

As we shall see in Section 6, this condition is a necessary condition for the bound states of H to be quadratically integrable functions of q.

We shall designate the function $\langle q|H_0, A_0; E, a\rangle$ which is a solution of (2.4) subject to the above condition by the ket $|H_0, A_0; E, a\rangle$. The corresponding bra $\langle H_0, A_0; E, a|$ will be used to denote the function $\langle H_0, A_0; E, a|q\rangle$ where

$$\langle H_0, A_0; E, a|q\rangle = \langle q|H_0, A_0; E, a\rangle^* \tag{2.6}$$

where the asterisk denotes complex conjugate.

Let us denote by ${}_B\langle H_0, A_0; E, a,|q\rangle^*$ another set of functions which satisfy both (2.4) and

$$\int_{q_0}^{q_l} {}_B\langle H_0, A_0; E, a|q\rangle \mathrm{d}q\langle q|H_0, A_0; E', a'\rangle = \delta(E-E')\delta(a, a') . \tag{2.7}$$

We also require that the eigenfunctions of H_0 belonging to the positive spectrum be orthogonal to the eigenfunctions of the extended space, i.e.,

$$\eta(-E)\eta(E')\int_{q_0}^{q_l} {}_B\langle H_0, A_0; E, a|q\rangle \mathrm{d}q\langle q|H_0, A_0; E', a'\rangle = 0 . \tag{2.7a}$$

$$\eta(-E)\eta(E')\int_{q_0}^{q_l} \langle H_0, A_0; E, a|q\rangle \mathrm{d}q\langle q|H_0, A_0; E', a'\rangle = 0 . \tag{2.7b}$$

We can write

$$\int_{q_0}^{q_l} \langle q|H_0|q'\rangle\langle q'|H_0, A_0; E, a\rangle_B = E\langle q|H_0, A_0; E, a\rangle_B , \tag{2.8}$$

where $\langle q | H_0, A_0; E, a\rangle_B$ is defined by

$$\langle q | H_0, A_0; E, a\rangle_B = {}_B\langle H_0, A_0; E, a | q\rangle^* . \tag{2.9}$$

The bra ${}_B\langle H_0, A_0; E, a|$ is used to denote ${}_B\langle H_0, A_0; E, a | q\rangle$ and the ket $|H_0, A_0; E, a\rangle_B$ is used to denote $\langle q | H_0, A_0; E, a\rangle_B$. The subscript B is used to denote the fact that ${}_B\langle H_0, A_0; E, a|$ is the « bi-orthogonal » bra to $|H_0, A_0, E, a\rangle$.

If I is the identity operator of the entire vector space, we write

$$\int_{q_0}^{q_1} |q\rangle \mathrm{d}q \langle q| = I , \tag{2.10}$$

$$\langle q | q'\rangle = \delta(q - q') . \tag{2.10a}$$

Then (2.7) can be written abstractly as

$$\begin{cases} {}_B\langle H_0, A_0; E, a | H_0, A_0; E', a'\rangle = \delta(E - E')\delta(a, a') , \\ \langle H_0, A_0; E, a | H_0, A_0; E', a'\rangle_B = \delta(E - E')\delta(a, a') , \quad \text{for } (E, E' < 0); \end{cases} \tag{2.11}$$

$$\eta(-E)\eta(E')\, {}_B\langle H_0, A_0; E, a | H_0, A_0; E', a'\rangle = 0 \tag{2.11a}$$

$$\eta(-E)\eta(E')\langle H_0, A_0; E, a | H_0, A_0; E', a'\rangle = 0 . \tag{2.11b}$$

We can define the projection operator $\eta(-H_0)$ by

$$\eta(-H_0) = \iint |H_0, A_0; E, a\rangle \eta(-E) \mathrm{d}E \,\mathrm{d}a \, {}_B\langle H_0, A_0, E, a| , \tag{2.12}$$

where the integration is carried out over the intervals ΔE_i. The extended space is defined by

$$\eta(H_0) + \eta(-H_0) = I . \tag{2.13}$$

The operator $\eta(-H_0)$ can be shown to be a projection operator which maps the Hilbert space into zero and the space added to the Hilbert space to form the full space into itself. A general vector $|\Phi\rangle$ in the full space can be written as

$$\langle q | \Phi\rangle = \iint \langle q | H_0, A_0; E, a\rangle \eta(E) \mathrm{d}E \,\mathrm{d}a \langle H_0, A_0; E, a | \Phi\rangle + \\ + \iint \langle q | H_0, A_0; E, a\rangle \, \eta(-E) \mathrm{d}E \,\mathrm{d}a \, {}_B\langle H_0, A_0; E, a | \Phi\rangle , \tag{2.14}$$

where the first integral goes over the positive E axis and the second integral goes over the intervals ΔE_i.

In (2.14) the H_0 representative of $|\Phi\rangle$ which is $\langle H_0, A_0; E, a|\Phi\rangle$ is a quadratically integrable function of E, a for $E > 0$. The function ${}_s\langle H_0, A_0; E, a|\Phi\rangle$ is an integrable function of E, a in the intervals ΔE_i for $E < 0$. The definition of ${}_s\langle H_0, A_0; E, a|\Phi\rangle$ for $E < 0$ outside the intervals ΔE_i is immeterial.

Let us now consider some properties of the projection operators $\eta(H_0)$, $\eta(-H_0)$. First of all we note

$$(2.15)\qquad \begin{cases} \eta(H_0)|H_0, A_0; E, a\rangle = \eta(E)|H_0, A_0; E, a\rangle \\ \eta(-H_0)|H_0, A_0; E, a\rangle = \eta(-E)|H_0, A_0; E, a\rangle . \end{cases}$$

The projection operator $\eta(-H_0)$ maps the Hilbert space into zero and maps that part of the vector space orthogonal to the Hilbert space into itself. By definition (2.12) and properties (2.7), (2.7*a*) and (2.7*b*)

$$(2.16)\qquad \eta(-H_0)\eta(H_0)=0,\quad \eta^2(H_0)=\eta(H_0),\quad \eta^2(-H_0)=\eta(-H_0),\quad \eta(H_0)\eta(-H_0)=0.$$

If T is any operator defined in the whole extended space then

$$T = T\eta(H_0) + T\eta(-H_0)\,,$$

where $T\eta(H_0)$ is an operator which is zero when it operates on vectors orthogonal to the Hilbert space and where $T\eta(-H_0)$ is zero when it operates on the Hilbert space. All the operators defined in Part I may be considered to be of the form $T\eta(H_0)$ and hence most of the properties of the various operators discussed in Part I will have analogues to operators of this character appearing here. The principal purpose of the present paper is to define operators over the full space, particularly the operator U, $U_\pm$, and W, i.e., to define $U\eta(-H_0)$, etc. It will not be necessary, however, to extend $L\eta(H_0)$, $M_\pm(H_0)$, as will be clear from the subsequent work.

The formal Hermitian adjoint of $T\eta(H_0)$ is $\eta(H_0)T^*$. If T commutes with H_0, it can be shown formally that

$$(2.17)\qquad \eta(H_0)T^* = T^*\eta(H_0)\,.$$

Hence if such an operator operates on Hilbert space, its adjoint will still operate on the Hilbert space and maps vectors orthogonal to that space into zero.

If R is another operator which commutes with H_0, one can show

$$R\eta(H_0)T\eta(H_0) = RT\eta(H_0) . \tag{2.18}$$

Hence products of such operators R and T acting on Hilbert space can be obtained by projecting on Hilbert space the product of the operators defined on the complete space. We shall use the properties (2.17) and (2.18) often without referring to them explicitly.

3. – The eigenfunctions of the perturbed Hamiltonian and the transformation operator.

3.1. *The eigenfunctions of the continuous spectrum; the operators* $U\eta(H_0)$, $U_{\pm}\eta(H_0)$, $W\eta(H_0) = W_c$. – Let us consider the eigenfunctions $|H, A; E, a\rangle$ of $H = H_0 + \varepsilon V$. As in Part I we introduce the operator U such that

$$|H, A; E, a\rangle = U|H_0, A_0; E, a\rangle . \tag{3.1}$$

When $E > 0$ the eigenfunctions $|H_0, A_0; E, a\rangle$ are those which span the Hilbert space. But when $E = E_i$, where E_i is one of the point eigenvalues of H, we have

$$|H, A; E_i, a\rangle = U|H_0, A_0; E_i, a\rangle , \tag{3.2}$$

where $|H_0, A_0; E_i, a\rangle$ is one of the vectors added to form the extended space. In fact these elements were added in order that (3.1) should hold for all values of E in the spectrum of H.

Let us consider the continuous spectrum of H. We shall obtain expressions for $U\eta(H_0)$, $U_{\pm}\eta(H_0)$, S, etc., analogous to those obtained in Part I where the continuous part of the spectrum which we are considering was the entire spectrum. The various expressions that are given below are obtained in an almost identical manner as those for the analogous operators of Part I. Therefore, instead of carrying out the derivations in detail, we shall indicate only the more important relationships.

For the case of the continuous spectrum, (3.1) may be written

$$\begin{aligned} |H, A; E, a\rangle &= U\eta(E)|H_0, A_0; E, a\rangle = \\ &= U\eta(H_0)|H_0, A; E, a\rangle \qquad (E > 0). \end{aligned} \tag{3.2}$$

In a manner similar to that used in Section 3 of Part I it can be shown that

$U\eta(H_0)$ satisfies the equation

$$U\eta(H_0) = L\eta(H_0) + \int \frac{P}{E - H_0} V U\eta(H_0)\delta(E - H_0)\mathrm{d}E, \tag{3.3}$$

where L is an arbitrary operator which commutes with H_0, and where as in Part I,

$$\delta(E - H_0) = \int |H_0, A_0; E, a\rangle \mathrm{d}a \langle H_0, A_0; E, a| . \tag{3.3a}$$

The integral in (3.3) is formally taken over the whole spectrum of the extended operator H_0. The factor $\eta(H_0)$ in the integral, however, in effect cuts out the negative part.

Now, as in Part I, there are two operators, $U_{\pm}\eta(H_0)$, which are particularly interesting and whose integral equations can be obtained by selecting $L\eta(H_0)$ properly. These operators are defined by the conditions

$$\lim_{t\to-\infty} \exp[iH_0t]\exp[-iHt]U_-\eta(H_0)|\varphi\rangle = \eta(H_0)|\varphi\rangle , \tag{3.4}$$

$$\lim_{t\to+\infty} \exp[iH_0t]\exp[-iHt]U_+\eta(H_0)|\varphi\rangle = \eta(H_0)|\varphi\rangle , \tag{3.4a}$$

where $\eta(H_0)|\varphi\rangle$ is an arbitrary state in Hilbert space. As in Part I, $U_{\pm}\eta(H_0)$ satisfy

$$U_{\pm}\eta(H_0) = \eta(H_0) + \varepsilon\int \gamma_{\pm}(E - H_0)VU_{\pm}\eta(H_0)\delta(E - H_0)\mathrm{d}E, \tag{3.5}$$

The scattering operator S is defined by

$$\lim_{t\to+\infty} \exp[iH_0t]\exp[-iHt]U_-\eta(H_0)|\varphi\rangle = S\eta(H_0)|\varphi\rangle , \tag{3.6}$$

where, as before, $\eta(H_0)|\varphi\rangle$ is an arbitrary state in Hilbert space. It can be shown that

$$S = \eta(H_0) - 2\pi i\varepsilon \int \delta(E - H_0)VU_-\eta(H_0)\delta(E - H_0)\mathrm{d}E , \tag{3.7}$$

and that its inverse is given by

$$S^{-1} = \eta(H_0) + 2\pi i\varepsilon \int \delta(E - H_0)VU_+\eta(H_0)\delta(E - H_0)\mathrm{d}E . \tag{3.8}$$

From (3.7) and (3.8) it is clear that S and S^{-1} can be written

$$S = S\eta(H_0)\,, \tag{3.9}$$

$$S^{-1} = S^{-1}\eta(H_0) \tag{3.10}$$

and therefore that S and S^{-1} maps the space orthogonal to the Hilbert space into zero.

The general operator $U\eta(H_0)$ may be expressed in terms of $U_\pm\eta(H_0)$ as follows:

$$U\eta(H_0) = U_\pm\eta(H_0)M_\pm\eta(H_0), \tag{3.11}$$

where

$$M_\pm\eta(H_0) = L\eta(H_0) \pm i\varepsilon\int\delta(E-H_0)V\,U\eta(H_0)\delta(E-H_0)\,\mathrm{d}E\,. \tag{3.12}$$

From (3.12) it is clear that $M_\pm\eta(H_0)$ commutes with H_0. It can be shown that $M_\pm\eta(H_0)$ has an inverse in Hilbert space which commutes with H_0 and which we shall denote by $M_\pm^{-1}\eta(H_0)$:

$$M_\pm^{-1}\eta(H_0)M_\pm\eta(H_0) = M_\pm\eta(H_0)M_\pm^{-1}\eta(H_0) = \eta(H_0); \tag{3.13}$$

alternatively, noting that $M_\pm^{-1}$ commutes with H_0 and using (2.13), we can write

$$M_\pm^{-1}M_\pm\eta(H_0) = M_\pm M_\pm^{-1}\eta(H_0) = \eta(H_0). \tag{3.13a}$$

It can also be shown that

$$\eta(H_0)U_\pm^*U_\pm\eta(H_0) = \eta(H_0)\,, \tag{3.14}$$

which is a generalization of (4.21) of Part I. Equation (3.14) is essentially the normalization condition on the eigenfunctions

$$|H, A; E, a\rangle_\pm = U_\pm\eta(H_0)|H_0, A_0; E, a\rangle \tag{3.14a}$$

(see Part I), namely

$$\eta(E)\eta(E')_\pm\langle H, A; E, a|H, A; E', a'\rangle_\pm = \eta(E)\delta(E-E')\delta(a, a'). \tag{3.14b}$$

From (3.11) and (3.13) the correspondùng relation for U is

$$W_c\,\eta(H_0)U^*U\eta(H_0) = \eta(H_0) \tag{3.15}$$

where

$$W_c\,\eta(H_0) = M_\pm^{-1}\eta(H_0)M_\pm^{-1*}\eta(H_0) = M_\pm^{-1}M_\pm^{*-1}\eta(H_0) = \eta(H_0)W_c\,. \tag{3.16}$$

It is to be noted that $W_c\,\eta(H_0)$ is a positive definite operator which has an inverse in Hilbert space. Equation (3.15) expresses the orthogonality relations between the eigenfunctions of the continuous spectrum of H associated with the weight operator $W_c\eta(H_0)$.

Two expressions for S whose analogues appear in Part I are

$$S = \eta(H_0)U_+^*U_-\,\eta(H_0)\,, \tag{3.17}$$

$$S = M_+\eta(H_0)M_-^{-1}\eta(H_0) = M_+\,M_-^{-1}\eta(H_0). \tag{3.18}$$

3'2. ***The eigenfunctions of the discrete spectrum; the operator*** $U\eta(-H_0)$. – Let us now consider the eigenfunctions of the discrete spectrum of H, namely $|H, A;\, E_i, a\rangle$. We should first note that the nature of the degeneracy operator A and its eigenvalues may be completely different for the discrete spectrum of H and for the continuous spectrum. In the case of the continuous spectrum the operator A may be chosen so that its eigenvalues are not countable (i.e., A may have a continuous spectrum). However, in the case of point eigenvalues E_i of H, the degeneracy is always countable, and in most applications is finite for a given value of E_i. Hence the operator A, when operating on that subspace of Hilbert space spanned by the eigenstates of the discrete spectrum of H, will have to be defined so as to have countable eigenvalues.

In order that the spectrum of the extended operator H_0 have the same spectrum as H, it is necessary that the degeneracy operator A_0 and its eigenvalues of additional eigenvectors $|H_0, A_0;\, E, a\rangle$ which are introduced for values of E in the neighborhood ΔE_i of each eigenvalue E_i of H, have the same point spectrum as the eigenvalues of A in the eigenstate $|H, A;\, E_i\, a\rangle$.

We may then write, since $E_i < 0$,

$$\begin{aligned} |H, A;\, E_i, a\rangle &= U|H_0, A_0;\, E_i, a\rangle = \\ &= U\eta(-E)|H_0, A_0;\, E_i, a\rangle = \\ &= U\eta(-H_0)|H_0, A_0;\, E_i, a\rangle\,. \end{aligned} \tag{3.19}$$

For each value of a and E_i the operator maps an « eigenvector » of the extended operator H_0 into an eigenvector H with the same value of a as a degeneracy label.

Generally the eigenfunctions of H corresponding to the discrete spectrum satisfy the following orthogonality relation:

$$\langle H, A; E_i, a | H, A; E_j, b\rangle = \delta(i, j)\langle a|\omega_d^{-1}(E_i)|b\rangle, \tag{3.20}$$

where $\delta(i, j)$ is the Kronecker δ. The matrix $\langle a|\omega^{-1}(E_i)|b\rangle$ is a positive definite Hermitian matrix in the space of eigenfunctions of the operator A corresponding to a fixed eigenvalue E_i of H. In the case where the eigenfunctions belonging to the same eigenvalue E_i but having different degeneracy labels have been made orthogonal to each other, $\langle a|\omega^{-1}(E_i)|b\rangle$ would have the

$$\langle a|\omega_d^{-1}(E_i)|b\rangle = C_{ia}\delta(a, b).$$

The constant C_{ia} (which is always positive) is the normalization constant for the eigenfunction $|H, A; E_i, a\rangle$, that is,

$$\langle H, A; E_i, a | H, A; E_i, a\rangle = C_{ia}.$$

The operator $\langle a|\omega_d^{-1}(E_i)|b\rangle$ has a positive-definite inverse which we denote by $\langle a|\omega_d(E_i)|b\rangle$; the latter, by definition, satisfies the relation

$$\sum_b \langle a|\omega_d(E_i)|b\rangle\langle b|\omega_d^{-1}(E_i)|c\rangle = \sum_b \langle a|\omega_d^{-1}(E_i)|b\rangle\langle b|\omega_d(E_i)c\rangle = \delta(a, c). \tag{3.21}$$

There is one other orthogonality relation which we should note, namely the relation which expresses the orthogonality between the eigenfunctions of the discrete and continuous spectrum of H. It is

$$\eta(E)\eta(-E_i)\langle H, A; E, a | H, A; E_i, b\rangle = 0. \tag{3.22}$$

3·3. *The completeness theorem for H; definition of H in the extended space.* – The principal difference between the results of Parts I and II arises from the fact that in the present case the Hilbert space is spanned by the eigenfunctions belonging to the discrete as well as the continuous eigenfunctions of H.

Let us consider any state $\eta(H_0)|\varphi\rangle$ in Hilbert space. Further, let us take for the eigenfunctions of the continuous spectrum either the outgoing or incoming eigenstates $|H, A; E, a\rangle_\pm = U_\pm \eta(H_0)|H_0, A_0; E, a\rangle$. The arbitrary state $\eta(H_0)|\varphi\rangle$ can then be expanded in the following way:

$$\eta(H_0)|\varphi\rangle = \iint |H, A; E, a\rangle_\pm \eta(E)\,dE\,da\; {}_\pm\langle H, A; E, a|\eta(H_0)|\varphi\rangle + \sum_i\sum_{a,b}|H, A; E_i, a\rangle\langle a|\omega_d(E_i)|b\rangle\langle H, A; E_i, b|\eta(H_0)|\varphi\rangle. \tag{3.23}$$

The coefficients of $|H, A; E, a\rangle_{\pm}$ and $|H, A; E_i, a\rangle$ in the expansion (3.23) follow from the normalization conditions (3.20) and (3.14*b*). Since $\eta(H_0)|\varphi\rangle$ is an arbitrary state in Hilbert space, (3.23) is equivalent to

$$\iint |H, A; E, a\rangle_{\pm}\eta(E)\,\mathrm{d}E\,\mathrm{d}a\,{}_{\pm}\langle H, A; E, a|\eta(H_0) + \\ + \sum_i \sum_{a,b} |H, A; E_i, a\rangle\langle a|\omega_d(E_i)|b\rangle\langle H, A; E_i, b|\eta(H_0) = \eta(H_0)\,. \tag{3.24}$$

It is useful to define the positive definite matrices $\langle a|\omega_d(E)|b\rangle$ in the previously described intervals ΔE_i so that for $E = E_i$ the operator $\langle a|\omega_d(E)|b\rangle$ are the matrices $\langle a|\omega_d(E_i)|b\rangle$ introduced earlier. On using (3.19), (3.14*a*) and (2.3) we have

$$U_{\pm}\eta(H_0)U_{\pm}^*\eta(H_0) + \sum_i \sum_{a,b} U\eta(-H_0)\delta(E_i - H_0)\cdot \\ \cdot\int |H_0, A_0; E, a\rangle\langle a|\omega_d(E)|b\rangle\,\mathrm{d}E\langle H_0, A_0; E, b|\eta(-H_0)U^*\eta(H_0) = \eta(H_0)\,, \tag{3.25}$$

where the integral is taken over each of the intervals $\Delta_i E$. Let us introduce the operator

$$W_d\eta(-H_0) = \sum_{a,b}\int |H_0, A_0; E, a\rangle\,\langle a|\omega_d(E)|b\rangle\,\mathrm{d}E\langle H_0, A_0; E, b|\eta(-H_0)\,. \tag{3.26}$$

It is clear that $W_d\eta(-H_0)$ is a positive definite operator in the space orthogonal to the Hilbert space and that it commutes with H_0.

On using (3.11) and (3.16), Eq. (3.25) becomes

$$UWU^*\eta(H_0) = \eta(H_0)\,, \tag{3.27}$$

where

$$W = W_c\eta(H_0) + \sum_i \delta(E_i - H_0)W_d\,\eta(-H_0)\,. \tag{3.28}$$

Eq. (3.28) is the weight operator for the case where point eigenvalues exist. As can be seen, the weight operator is positive definite and has an inverse in the vector space. Eq. (3.28) provides a decomposition of W into two parts: W_c, which characterizes the orthogonality relations of the continuous spectrum of H (hence the subscript c for « continuous »); and W, which characterizes the orthogonality relations of the discrete spectrum (hence the subscript d for « discrete »).

We have not yet discussed how H is to be defined in the extended space. The extension of H can be carried out by giving the vectors which span the extended space. Since we want the spectrum of the extended operator H to have the same spectrum as the spectrum of the extended operator H_0 and hence of the original operator H, it is seen that the eigenstates of H which span the Hilbert space must also span the extended space. It is clear that the spectrum of H is thus preserved.

The effect of the extension of the definition of H is that equation (3.27) is replaced by

$$UWU^* = I . \tag{3.29}$$

4. – More detailed discussion of the Q-representation and Q-extended operators.

As in Part I we shall introduce the Q-representation and require that the operators K and K_0 be triangular in terms of this representation, i.e.,

$$\begin{cases} \langle q|K|q'\rangle = 0 , & q' > q , \\ \langle q|K_0|q'\rangle = 0 , & q' > q , \end{cases} \tag{4.1}$$

where K and K_0 are given by

$$U = I + \varepsilon K , \qquad U_0 = WU^* = I + \varepsilon K_0 . \tag{4.2}$$

Generally, the Q is defined only in the Hilbert space (i.e., only $Q\eta(H_0)$ is defined). Its eigenstates $|q\rangle$ satisfy the completeness relation

$$\int_{q_0}^{q_1} |q\rangle \mathrm{d}q \langle q|\eta(H_0) = \eta(H_0) , \tag{4.3}$$

where q_1 and q_0 are the upper and lower limits of the eigenvalues of Q. To extend the definition of Q into the whole vector space we write the completeness relation for the eigenfunctions $|q\rangle$ as

$$\int_{q_0}^{q_1} |q\rangle \mathrm{d}q \langle q| = I . \tag{4.4}$$

From (4.4) we have

$$\langle q|q'\rangle = \delta(q - q') = \langle q|I|q'\rangle . \tag{4.5}$$

Eq. (4.4) means that any vector $|\varphi\rangle$ in the extended space has q-representative namely $\langle q|\varphi\rangle$. If the state $|\varphi\rangle$ is in Hilbert space it is quadratically integrable in the function $\langle q|\varphi\rangle$. If $\eta(-H_0)|\varphi\rangle \neq 0$, i.e., if $|\varphi\rangle$ has the component outside the Hilbert space, the functions $\langle q|\varphi\rangle$ are not quadratically integrable.

As usual, every operator A defined in the extended space can be given as an integral operator with the kernel $\langle q|A|q'\rangle$, i.e., if $|\varphi\rangle$ is a state in the extended space $\langle q|A|\varphi\rangle$ can be written as (Using (4.4))

$$\langle q|A|\varphi\rangle = \int_{q_0}^{q_1} \langle q|A|q'\rangle \, dq' \langle q'|\varphi\rangle \, . \tag{4.6}$$

As in ordinary Hilbert space theory, the operator A is defined if the kernel $\langle q|A|q'\rangle$ is given.

Let us consider now an operator of the form $A\eta(H_0)$. This operator operates in a non-trivial way only on the Hilbert space. It maps vectors orthogonal to the Hilbert space into zero. The kernel $\langle q|A\eta(H_0)|q'\rangle$ is, of course, a known function of q and q'.

One might wish to extend the definition of the operator $A\eta(H_0)$ to an operator A which can be applied in a non-trivial way to the whole vector space and which can be applied in a non-trivial way to the whole vector space and which equals $A\eta(H_0)$ when applied to the Hilbert space. One possible way of defining the extended operator A is to define the kernel $\langle q|A|q'\rangle$ as a function of q and q' as being equal to the kernel $\langle q|A\eta(H_0)|q'\rangle$,

$$\langle q|A|q'\rangle = \langle q|A\eta(H_0)|q'\rangle \, . \tag{4.7}$$

It should be noted that Eq. (4.7) does *not* imply that $A = A\eta(H_0)$. An operator A obtained from $A\eta(H_0)$ by this means vill be called a Q-extended operator. In Section 2 we took H_0 to be a Q-extended operator.

The simplest example of a Q-extended operator is the identity I, which is obtained from the operator $\eta(H_0)$ because we have

$$\langle q|\eta(H_0)q'\rangle = \delta(q-q') \tag{4.8}$$

when $\eta(H_0)$ acts on the Hilbert space. Furthermore,

$$\langle q|I|q'\rangle = \delta(q-q') = \langle q|\eta(H_0)|q'\rangle \, . \tag{4.9}$$

We shall require that the operator U be a Q-extended operator obtained from $U\eta(H_0)$. A consequence of this requirement is that

$$\langle q|K|q'\rangle = \langle q|K\eta(H_0)|q'\rangle . \tag{4.10}$$

Since every operator can be written as an integral operator with a kernel, any operator which is given in terms of the Q-representation when acting on the Hilbert space has the same form as a Q-extended operator.

It should not be thought, however, that every operator acting on the full space is a Q-extended operator. For example, the weight operator W is not a Q-extended operator, since generally

$$\langle q|W\eta(H_0)|q'\rangle = \langle q|W_c\eta(H_0)|q'\rangle$$

does *not* equal

$$\langle q|W|q'\rangle = \langle q|W_c\eta(H_0)|q'\rangle + \langle q|W_d\eta(-H_0)|q'\rangle .$$

5. – Theorems on the inverse problem. The equation for K, K_0.

We shall now present theorems analogous to those of Part I. We shall not go through the proof because the proofs of Part I are essentially the same if the Hilbert space is extended properly.

Theorem I. Let there be given a vector space spanned by the eigenfunctions of the extended operators H_0, A_0 as described above, the negative point eigenvalues of H, E_i and the degeneracy of these eigenvalues being given. Further, let us prescribe the positive definite operator

$$W = W_c\eta(H_0) + \sum_i \delta(E_i - H_0)W_d\eta(-H_0)$$

in the vector space; W commutes with H_0 and has an inverse. Then if the equations

$$U_0 = WU^* \tag{5.1}$$

$$U_0 = U^{-1} \tag{5.2}$$

can be solved for U, U_0, the operator H given by

$$H = UH_0U_0 \tag{5.3}$$

is a Hermitian operator whose eigenstates corresponding to the continuous spectrum are

$$|H, A; E, a\rangle = U\eta(E)|H_0, A_0; E, a\rangle \qquad \text{for } 0 \leqslant E < \infty \tag{5.4}$$

and whose eigenstates corresponding to the point eigenvalues E_i with the prescribed degeneracies are

$$|H, A; E_i, a\rangle = U|H_0, A_0; E_i, a\rangle . \tag{5.5}$$

The eigenstates of H will satisfy the completeness relation

$$\iiint |H, A; E, a\rangle da\langle a|\omega_c(E)|b\rangle\eta(E)dE db\langle H, A; E, b| + \\ + \sum_i \sum_{a,b} |H, A; E_i, a\rangle\langle a|\omega_d(E_i)|b\rangle\langle H, A; E_i, b| = I , \tag{5.6}$$

where

$$\langle a|\omega_c(E)|b\rangle\eta(E)$$

and

$$\langle a|\omega_d(E)|b\rangle\eta(-E)$$

are defined by

$$\langle H_0, A_0; E, a|W_c\eta(H_0)|H_0, A_0; E', b\rangle = \delta(E-E')\langle a|\omega_c(E)|b\rangle\eta(E) , \tag{5.7}$$

$$_s\langle H_0, A_0; E, a|W_s\eta(-H_0)|H_0, A_0; E', b\rangle_s = \delta(E-E')\langle a|\omega_d(E)|b\rangle\eta(-E). \tag{5.8}$$

Furthermore, the eigenfunctions will satisfy the normalization conditions

$$\int\langle a|\omega_c(E)|b\rangle db\langle H, A; E, b|H, A; F, c\rangle = \delta(E-F)\delta(a, c) , \tag{5.9}$$

$$\langle H, A; E, b|H, A; E_i, c\rangle = 0 , \qquad (E > 0), \tag{5.10}$$

$$\sum_i \langle a|\omega_d(E_i)|b\rangle\langle H, A; E_i, b|H, A; E_j, c\rangle = \delta(i, j)\delta(a, c) . \tag{5.11}$$

The condition that W be a positive definite operator which commutes with H_0 and which has an inverse is equivalent to requiring the same properties of $W_c\eta(H_0)$ and $W_s\eta(-H_0)$ in the Hilbert space and the vector space orthogonal to the Hilbert space respectively.

Theorem II. If we write $U = I + \varepsilon K$, $U_0 = I + \varepsilon K_0$ and require K, K_0 to be triangular in the Q-representation then $U_0 = WU^*$ has a unique solution for U, U_0 if W satisfies conditions analogous to those required in Part I.

Theorem III. It can be shown that if the extension of the Hilbert space is carried out properly then $U_0 = U^{-1}$.

We shall now write the generalized Gelfand-Levitan equation for K, From (5.1) we have

(5.12) $$UW = U_0^* .$$

Let us write

(5.13) $$\begin{cases} U = I + \varepsilon K \\ U_0 = I + \varepsilon K_0 \\ W = \eta(H_0) + \varepsilon\Omega . \end{cases}$$

Then (5.12) becomes

(5.14) $$\eta(H_0) + \varepsilon\Omega + \varepsilon K\eta(H_0) + \varepsilon^2 K\Omega = I + \varepsilon K_0^*$$

or in terms of the Q-representation

(5.15) $$\langle q|\eta(H_0)|q'\rangle + \varepsilon\langle q|\Omega|q'\rangle + \varepsilon\langle q|K\eta(H_0)|q'\rangle + \\ + \varepsilon^2\langle q|K\Omega|q'\rangle = \delta(q-q') + \varepsilon\langle q|K_0^*|q'\rangle .$$

Now

(5.16) $$\langle q|\eta(H_0)|q'\rangle = \delta(q-q')$$

when $\eta(H_0)$ acts on Hilbert space; otherwise

$$\langle q|\eta(H_0)|q'\rangle = 0 .$$

Hence

(5.17) $$\langle q|\eta(H_0)|q'\rangle \equiv 0 \qquad (q > q').$$

From the triangularity condition (4.1) on K_0

(5.18) $$\langle q|K_0^*|q'\rangle = 0 \qquad (q > q').$$

We now use the fact that K is a Q-extended operator and write

(5.19) $$\langle q|K\eta(H_0)|q'\rangle = \langle q|K|q'\rangle.$$

Hence, finally we obtain our equation for K, namely

$$\langle q|K|q'\rangle = -\langle q|\Omega|q'\rangle - \varepsilon\int_{q_0}^{q}\langle q|K|q''\rangle \mathrm{d}q''\langle q''|\Omega|q'\rangle\,, \qquad (q > q'), \tag{5.20}$$

which is the same equation for $\langle q|K|q'\rangle$ as obtained in Part I.

6. – Conditions on the eigenfunctions $|H_0, A_0; E_i, a\rangle$.

The condition that $|H, A; E_i, a\rangle$ be in Hilbert space and hence that the q-representative $\langle q|H, A; E_i, a\rangle$ be a quadratically integrable function of q leads to a condition on $\langle q|H_0, A_0; E_i, a\rangle$ if we assume K as triangular in the Q-representation. We can write

$$\langle q|H, A; E_i, a\rangle = \langle q|H_0, A_0; E_i, a\rangle + \varepsilon\int_{q_0}^{q}\langle q|K|q'\rangle \mathrm{d}q'\langle q'|H_0, A_0; E_i, a\rangle\,. \tag{6.1}$$

Now since the second term approaches zero as q approaches q_0, we see that the quadratic integrability of $\langle q|H, A; E, a\rangle$ implies that $\langle q|H_0, A_0; E_i, a\rangle$ be a quadratically integrable function of q in the neighborhood of $q = q_0$:

It appears, in fact, that a sufficient condition on the extension of the Hilbert space to guarantee the validity of the three theorems of Section 5 is the condition that $\langle q|H_0, A_0; E_i, a\rangle$ satisfy

$$\int_{q_0}^{q_a}\langle H_0, A_0; E_i, a|q'\rangle \mathrm{d}q'\langle q'|H_0, A_0; E_i, a\rangle < \infty \qquad \text{for } q < q_1\,. \tag{6.2}$$

That is, the added eigenfunctions $\langle q|H_0, A_0; E_i, a\rangle$ are « essentially » in Hilbert space if one replaces these functions by zero in the vicinity of $q = q_1$. The functions $\langle q|H_0, A_0; E_i, a\rangle$ will have to be singular at $q = q_1$ in order that these eigenfunctions span a space bigger than Hilbert space. Then the q-representative $\langle q|\varphi\rangle$ of any state $|\varphi\rangle$ as a function of q is quadratically integrable

$$\int_{q_0}^{q_a}\langle\varphi|q\rangle \mathrm{d}q\langle q|\varphi\rangle < \infty \qquad \text{for } q < q_1\,. \tag{6.3}$$

Those states $|\varphi\rangle$ which are in Hilbert space are, of course, quadratically integrable in $q_0 \leqslant q \leqslant q_1$ be definition:

$$\int_{q_0}^{q_1} \langle\varphi|q\rangle \,\mathrm{d}q \langle q|\varphi\rangle < \infty . \tag{6.4}$$

States which are not in Hilbert space will have representatives $\langle q|\varphi\rangle$ which have singularities for $q = q_1$ which are so severe that the square $\langle q|\varphi\rangle$ does not exist:

$$\int_{q_0}^{q_1} \langle\varphi|q\rangle \,\mathrm{d}q \langle q|\varphi\rangle = \infty . \tag{6.5}$$

Under the conditions of these extension of the Hilbert space, one can say that the Hilbert space is « dense » in the vector space; that is, every state in the vector space $\langle q|\varphi\rangle$ can be considered as being approximated by an element of Hilbert space for $q < q_1$.

The vector space of Ref. ([2]) has essentially this property.

([2]) I. KAY: *On the determination of a linear system from the reflection coefficient*; EM-74, *Institute of Mathematical Sciences, Division of Electromagnetic Research, New York University*.

RIASSUNTO (*)

Nella Parte I di questo lavoro ci siamo limitati alla considerazione di operatori di ponderazione che conducono ad Hamiltoniane H i cui spettri sono identici allo spettro di una data Hamiltoniana imperturbata H_0. In questa parte del lavoro ci proponiamo di mostrare come si possano scegliere degli operatori di ponderazione che conducano ad Hamiltoniane H dotate di spettri differenti di quelli di H_0. Per rendere la discussione più concreta tratteremo il caso in cui H_0 ha spettro puramente continuo estendentesi da 0 a ∞ e in cui H ha uno spettro dotato di una parte continua coincidente con lo spettro di H_0 e, inoltre, ha autovalori discreti negativi.

(*) *Traduzione a cura della Redazione.*

I. KAY and H. E. MOSES
1956, Febbraio
Il Nuovo Cimento
Serie X, Vol. 3, pag. 276-304

The Determination of the Scattering Potential from the Spectral Measure Function.

III. Calculation of the Scattering Potential from the Scattering Operator for the One-Dimensional Schrödinger Equation (*).

I. KAY and H. E. MOSES

Institute of Mathematical Sciences, New York University - New York

(ricevuto il 27 Ottobre 1955)

Summary. — A procedure discussed in previous papers for the calculation of the scattering potentials from the spectral measure functions associated with certain eigenfunctions of the continuous spectrum of the perturbed Hamiltonian, has been adapted to the problem of obtaining the scattering potential from the scattering operator for the one-dimensional Schrödinger equation ($-\infty < x < \infty$). Examples are given to show how the procedure may be used.

1. - Introduction and Summary of Results.

Several writers (1,2,3) have recently discussed the problem of obtaining the scattering potential from the scattering phases for a variety of radial wave equations. Their work is characterized by the fact they require the perturbed eigenfunctions to approach the unperturbed eigenfunctions at the origin. From this boundary condition and from specific properties of the unperturbed Hamil-

(*) The research reported in this article was done at the Institute of Mathematical Sciences, New York University, and has been made possible through support and sponsorship extended by Geophysics Research Directorate, Air Force Cambridge Research Center under Contract No. AF19(122)-463.

(1) R. JOST and W. KOHN: *Det. Kgl. Danske Videnskabernes Selskab, Mathematisk-fysiske Meddeleleser*, **27**, nr. 9 (1953).

(2) E. CORINALDESI: *Nuovo Cimento*, **11**, 468 (1954).

(3) R. G. NEWTON and R. JOST: *Nuovo Cimento*, **1**, 590 (1955).

tonians and of the perturbation, they then obtain the Gelfand-Levitan equation ([4,5]). In the special cases discussed in ([1-3]) this equation was used to obtain the scattering potential in terms of the spectral measure function. By showing how the spectral measure function can be obtained from the scattering phases, it is then possible to derive the potential in terms of the phases.

In a previous paper ([6]) the authors provided a more abstract formulation of the problem of obtaining the scattering potential from the spectral measure function. It is the objective of the present paper to illustrate the use of the theorems presented there by considering the one-dimensional wave equation $(-\infty < x < \infty)$. By showing how the spectral measure function can be obtained from the scattering operator S one can then calculate the potential from suitable elements of S. Actually this one-dimensional problem has already been treated by KAY ([7]) using properties of the one-dimensional wave equation discussed by FRIEDRICHS ([8]). The advantage of the more general approach of the present paper is that it seems to indicate a way to tackle the more interesting problem of three-dimensional scattering from a general (i.e., non-symmetric) potential.

For conciseness we shall summarize the results below.

First, let us define the reflection coefficient $b(k)$. Let us consider solutions $\Psi_k(x)$ of the equation

$$\left[-\frac{d^2}{dx^2} + V(x)\right]\Psi_k(x) = k^2\Psi_k(x)\,, \tag{1.1}$$

which satisfy the condition that they represent the sum of an incident plane wave and an «outgoing wave», i.e., which satisfy the integral equation

$$\Psi_k^{\text{out}}(x) = \exp[ikx] - \frac{i}{2|k|}\int_{-\infty}^{+\infty}\exp[i|k|\,|x-x'|\,]V(x')\Psi_k^{\text{out}}(x')\,dx'\,. \tag{1.2}$$

The reflection coefficient $b(k)$ is defined by considering the following asymptotic form for $\Psi_k(x)$ when $k > 0$

$$\lim_{x\to-\infty}\Psi_k^{\text{out}}(x) = \exp[ikx] + b(k)\exp[-ikx] \qquad (k>0). \tag{1.3}$$

([4]) I. M. GELFAND and B. M. LEVITAN: *Iszv. Akad. Nauk. SSSR, Math. Series*, **15**, 309 (1951).

([5]) N. LEVINSON: *Phys. Rev.*, **89**, 755 (1953).

([6]) I. KAY and H. E. MOSES: *Nuovo Cimento*, part I, **2**, 917 (1955); part II, **3**, 66 (1956).

([7]) I. KAY: *On the determination of a linear system from the reflection coefficient*: New York University, Institute of Mathematical Sciences, Division of Electromagnetic Research Report No. EM-74.

([8]) K. O. FRIEDRICHS: *Spectral representation of linear operators*; Lecture Notes, New York University, Institute of Mathematical Sciences, (Summer 1948).

The reflection coefficient gives the amplitude of the plane wave which is reflected toward the left when an incident wave of unit amplitude moves toward the right.

Though we have defined $b(k)$ for $k > 0$ only, one can define this function for $k < 0$. One can show by analytic continuation

$$b(-k) = b^*(k) \qquad \text{for } k \text{ real} \tag{1.4}$$

where the asterisk means complex conjugate.

We shall show that the potential $V(x)$ can be obtained from the knowledge of the reflection coefficient $b(k)$ alone if it satisfies certain sufficient conditions. (Some of these conditions are also necessary.)

Let $b(k)$ have the form

$$b(k) = g(k) \exp[-2ik\alpha] \qquad (\alpha > 0) \tag{1.5}$$

where

$$g(k) = 0(k^{-1}) \qquad \text{for } |k| \to \infty \tag{1.6}$$

Further let the analytic continuation of $g(k)$ have positive imaginary residues r_i at a finite number of poles $k_j = i\tau_j$, $(j=1, 2, ..., n)$ where $\tau_j > 0$. Then if the equation

$$\langle x|K|x'\rangle = -\langle x|\Omega|x'\rangle - \eta(x+x'+2\alpha)\int_{-(2\alpha+x')}^{x}\langle x|K|x''\rangle \mathrm{d}x''\langle x''|\Omega|x'\rangle , \tag{1.7}$$

can be solved for $\langle x|K|x'\rangle$ $(x' < x)$, where

$$\langle x|\Omega|x'\rangle = \frac{1}{2\pi}\int_{-\infty}^{+\infty} b(k)\exp[-ik(x+x')]\,\mathrm{d}k - i\sum_j r_j \exp[\tau_j(x+x'+2\alpha)] , \tag{1.8}$$

and $\eta(x)$ is the usual Heaviside unit function

$$\eta(x) = \begin{cases} 1 & \text{for } x > 0, \\ 0 & \text{for } x < 0, \end{cases} \tag{1.8a}$$

then the scattering potential $V(x)$ is given by

$$V(x) = 2\frac{\mathrm{d}}{\mathrm{d}x}\langle x|K|x\rangle . \tag{1.9}$$

Furthermore

$$\langle x|K|x'\rangle \equiv V(x) \equiv 0 \qquad \text{when } x < -\alpha \tag{1.10}$$

and the point eigenvalues of $-(d^2/dx^2)+V(x)$ will have the values $-\tau_j^2$. The reflection coefficient associated with $V(x)$ as obtained from (1.9) will be $b(k)$. as required by a consistent theory.

One will be able to obtain a unique set of eigenfunctions of $-(d^2/dx^2)+V(x)$ corresponding to the continuous spectrum. These are given by

$$\Psi_k(x) = \exp[ikx] + \int_{-\infty}^{x} \langle x|K|x'\rangle \exp[ikx']\,dx', \tag{1.11}$$

which satisfy the boundary condition

$$\lim_{x\to-\infty} \Psi_k(x) = \exp[ikx]. \tag{1.12}$$

The eigenfunctions corresponding to the point eigenvalue $-\tau_j^2$ denoted by $\Psi_{-i\tau_j}(x)$ are given by

$$\Psi_{-i\tau_j} = \exp[\tau_j x] + \int_{-\infty}^{x} \langle x|K|x'\rangle \exp[\tau_j x']\,dx'. \tag{1.13}$$

The normalization of the eigenfunctions $\Psi_k(x)$ of the continuous spectrum given by (1.11) is

$$\frac{1}{2\pi}\int_{-\infty}^{+\infty} \Psi_k^*(x)\Psi_{k'}(x)\,dx + \frac{b(-k)}{2\pi}\int_{-\infty}^{+\infty} \Psi_{-k}^*(x)\Psi_{k'}(x)\,dx = \delta(k-k'). \tag{1.14}$$

The bound states are orthogonal to eigenfunctions of the continuous spectrum. They satisfy the following orthogonality relation with respect to each other

$$\int_{-\infty}^{+\infty} \Psi_{-i\tau_m}(x)\Psi_{-i\tau_j}(x)\,dx = i\delta_{mj}r_j^{-1}\exp[-2\tau_j\alpha]. \tag{1.15}$$

The eigenstates $\Psi_k(x)$ and $\Psi_{-i\tau_j}$ form a complete set and satisfy the completeness relationship

$$\frac{1}{2\pi}\left[\int_{-\infty}^{+\infty} \Psi_k(x)\Psi_k^*(x')\,dk + \int_{-\infty}^{+\infty} \Psi_k(x)b(-k)\Psi_{-k}^*(x')\,dk\right] - \tag{1.16}$$
$$- i\sum_j \Psi_{-i\tau_j}(x)\Psi_{-i\tau_j}(x')r_j \exp[2\tau_j\alpha] = \delta(x-x').$$

2. – The one-dimensional eigenfunctions; the scattering operator.

The unperturbed Hamiltonian H_0 is given by

$$H_0^x = -\frac{\mathrm{d}^2}{\mathrm{d}x^2}, \qquad -\infty < x < +\infty, \tag{2.1}$$

where the superscript x on H_0^x signifies that the operator is expressed in the x-representation. It is convenient to use the eigenfunction $|H_0, A_0; E, a\rangle$ given by

$$\langle x | H_0, A_0; E, a\rangle = \frac{(E)^{-\frac{1}{4}}}{2\sqrt{\pi}} \exp[ia\sqrt{E}x], \tag{2.2}$$

where a is restricted to having the values $+1$ or -1. It is easily verified from (2.2) that

$$\langle H_0, A_0; E', a' | H_0, A_0; E, a\rangle = \int_{-\infty}^{+\infty} \langle H_0, A_0; E', a' | x\rangle \mathrm{d}x \langle x | H_0, A_0; E, a\rangle = \\ = \delta(E - E')\,\delta_{a,a'}, \tag{2.3}$$

where $\delta_{a,a'}$ is the usual Kronecker δ. It can also be verified that

$$\sum_a \int_0^\infty |H_0, A_0; E, a\rangle \mathrm{d}E \langle H_0, A_0; E, a| = \eta(H_0), \tag{2.4}$$

where $\eta(H_0)$ is the identity operator in Hilbert space and maps remainder of the extended vector space into zero.

The perturbed Hamiltonian H is given by

$$H = H_0 + \varepsilon V. \tag{2.5}$$

Initially, we shall assume that $\langle x | V | x'\rangle$ is generally not diagonal. Later we shall give conditions under which $\langle x | V | x'\rangle$ is diagonal, i.e., has the form

$$\langle x | V | x'\rangle = V(x)\,\delta(x - x'). \tag{2.6}$$

We shall assume, however, that $\langle x | V | x'\rangle$ is such that a scattering operator exists.

The outgoing, and incoming eigenfunctions of H corresponding to the continuous spectrum satisfy the integral equations

$$(2.7)\quad \langle x|H, A; E, a\rangle_\pm = \langle x|H_0, A_0; E, a\rangle \pm$$

$$\frac{i}{2}\varepsilon(E)^{-\frac{1}{2}}\int_{-\infty}^{+\infty}\int_{-\infty}^{+\infty}\exp[\mp i\sqrt{E}|x-x'|]\,dx'\langle x'|V|x''\rangle\,dx''\langle x''|H, A; E, a\rangle_\pm .$$

In (2.7) we use the well-known result

$$(2.7a)\qquad \langle x|\gamma_\pm(E-H_0)|x'\rangle = \pm\frac{i}{2}(E)^{-\frac{1}{2}}\exp[\mp i\sqrt{E}|x-x'|] .$$

The scattering operator and its inverse can be represented in the H_0-representation by (see (⁶), Part I, eqs. (4.12), (4.20), and (⁶), Part. II, Eqs. (3.7), (3.8))

$$(2.8)\qquad \langle H_0, A_0; E, a|S|H_0, A_0; E', a'\rangle = \delta(E-E')\langle a|S(E)|a'\rangle$$

and

$$(2.9)\qquad \langle H_0, A_0; E, a|S^{-1}|H_0, A_0; E', a'\rangle = \delta(E-E')\langle a|S^{-1}(E)|a'\rangle ,$$

where

$$(2.10)\qquad \langle a|S(E)|a'\rangle = \delta_{a,a'} - 2\pi i\varepsilon\langle H_0, A_0; E, a|V|H, A; E, a'\rangle_- =$$

$$= \delta_{a,a'} - 2\pi i\varepsilon\int_{-\infty}^{+\infty}\int_{-\infty}^{+\infty}\langle H_0, A_0; E, a|x\rangle\,dx\langle x|V|x'\rangle\,dx'\langle x'|H, A; E, a'\rangle_- ,$$

and

$$(2.11)\qquad \langle a|S^{-1}(E)|a'\rangle = \delta_{a,a'} + 2\pi i\varepsilon\langle H_0, A_0; E, a|V|H, A; E, a'\rangle_+ =$$

$$= \delta_{a,a'} + 2\pi i\varepsilon\int_{-\infty}^{+\infty}\int_{-\infty}^{+\infty}\langle H_0, A_0; E, a|x\rangle\,dx\langle x|V|x'\rangle\,dx'\langle x'|H, A; E, a'\rangle_+ .$$

For future reference we note the reciprocity theorem

$$(2.12)\qquad \langle H_0, A_0; E, a|V|H, A; E, b\rangle_- = {}_+\langle H, A; E, a|V|H_0, A_0; E, b\rangle =$$

$$= \langle H_0, A_0; E, b|V|H, A; E, a\rangle_+^* ,$$

where the asterisk means complex conjugate.

If $\langle x|V|x'\rangle$ dies down with sufficient rapidity, the asymptotic forms of $\langle x|H, A; E, a\rangle_\pm$ are related to $\langle a|S(E)|b\rangle$ and $\langle a|S^{-1}(E)|b\rangle$ by

$$(2.13)\qquad \lim_{x\to\pm\infty}\langle x|H, A; E, a\rangle_- = \langle x|H_0, A_0; E, a\rangle + \langle x|H_0, A_0; E, \pm 1\rangle$$

$$[\langle\pm 1|S(E)|a\rangle - \delta_{\pm 1, a}]$$

$$\lim_{x\to\pm\infty} \langle x|H, A; E, a\rangle_+ = \langle x|H_0, A_0; E, a\rangle + \langle a|H_0, A_0; E, \mp 1\rangle [\langle \mp 1|S^{-1}(E)|a\rangle - \delta_{\mp 1, a}], \tag{2.14}$$

as can be shown from (2.10) and (2.11).

3. - Expression for the weight operator in terms of the scattering operator.

We shall now express in terms of S and S^{-1} the operators W_c, $M_\pm$ associated with those operators U such that the eigenfunctions of H given by

$$\langle x|H, A; E, a\rangle = \langle x|U|H_0, A_0; E, a\rangle \tag{3.1}$$

satisfy the boundary condition

$$\lim_{x\to-\infty} \langle x|H, A; E, a\rangle - \langle x|H_0, A_0; E, a\rangle = 0 \tag{3.2}$$

or, symbolically,

$$\lim_{x\to-\infty} \langle x|H, A; E, a\rangle = \langle x|H_0, A_0; E, a\rangle . \tag{3.2a}$$

Since we shall work in the continuous spectrum of H, we shall drop the factor $\eta(H_0)$ for the time being, since it will always act like the identity operator.

From $U = U_- M_-$ we can write

$$\langle x|H, A; E, a\rangle = \sum_{a'} \langle x|H, A; E, a'\rangle_- \langle a'|\mu_-(E)|a\rangle , \tag{3.3}$$

where $\langle a|\mu_-(E)|a'\rangle$ is given by (by cf. ([6]), Part I, Eq. (5.5))

$$\langle H_0, A_0; E, a|M_-|H_0, A_0; E', a'\rangle = \delta(E - E')\langle a|\mu_-(E)|a'\rangle . \tag{3.4}$$

From Eqs. (3.2a) and (2.13) we obtain an equation for $\langle a|\mu_-(E)|a'\rangle$, namely

$$\langle x|H_0, A_0; E, a\rangle = \sum_{a'} \langle x|H_0, A_0; E, a'\rangle\langle a'|\mu_-(E)|a\rangle + \langle x|H_0, A_0; E, -1\rangle \sum_{a'} [\langle -1|S(E)|a'\rangle - \delta_{-1,a}]\langle a'|\mu_-(E)|a\rangle . \tag{3.5}$$

On multiplying (3.5) through by $\langle H_0, A_0; E'', a''|x\rangle$ and integrating with respect to x, and then using the orthogonality relation (2.3), one finds that

$$\delta(E''-E)\delta_{a'',a} = \delta(E''-E)\sum_{a'}\delta_{a'',a'}\langle a'|\mu_-(E)|a\rangle + \\ + \delta(E''-E)\delta_{a'',-1}\sum_{a'}[\langle -1|S(E)|a'\rangle - \delta_{-1,a'}]\langle a'|\mu_-(E)|a\rangle , \tag{3.6}$$

which on renaming some of the variables, may be rewritten as

$$\delta_{a,a'} = \langle a|\mu_-(E)|a'\rangle + \delta_{a,-1}\sum_{a''}[\langle -1|S(E)|a''\rangle - \delta_{-1,a''}]\langle a''|\mu_-(E)|a'\rangle . \tag{3.7}$$

Eq. (3.7) may be regarded as an equation for the matrix $\langle a|\mu_-(E)|a'\rangle$. The solution of (3.7) can be shown to be

$$\langle a|\mu_-(E)|a'\rangle = \begin{pmatrix} 1 & 0 \\ \dfrac{-\langle -1|S(E)|+1\rangle}{\langle -1|S(E)|-1\rangle} & \dfrac{1}{\langle -1|S(E)|-1\rangle} \end{pmatrix}, \tag{3.8}$$

where we adopt the convention that the first row and column are labeled by $+1$ and the second by -1, i.e., $\langle +1|\mu_-(E)|-1\rangle = 0$. Eq. (3.8) gives us M_- in the H_0-representation.

From ([6]), Part I, Eq. (5.9) we have

$$W_c = M_-^{-1}M_-^{*-1} . \tag{3.9}$$

To find W_c, therefore, we shall have to obtain M_-^{-1} and M_-^{*-1}. We shall express these operators in terms of the H_0-representation. Let us define $\langle a|\mu_-^{-1}(E)|a'\rangle$ by

$$\langle H_0, A_0; E, a|M_-^{-1}|H_0, A_0; E', a'\rangle = \delta(E-E')\langle a|\mu_-^{-1}(E)|a'\rangle . \tag{3.10}$$

From $M_-^{-1}M_- = M_-M_-^{-1} = \eta(H_0)$, (where we recall $\eta(H_0)$ is the identity in Hilbert space) it can be shown that for fixed E the quantities $\langle a|\mu_-^{-1}(E)|a'\rangle$ are just the elements of the matrix which is the inverse of the matrix whose elements are $\langle a|\mu_-(E)|a'\rangle$. It is then shown that

$$\langle a|\mu_-^{-1}(E)|a'\rangle = \begin{pmatrix} 1 & 0 \\ \langle -1|S(E)|+1\rangle & \langle -1|S(E)|-1\rangle \end{pmatrix} \tag{3.11}$$

Let us now define $\langle a|M^{*-1}(E)|a'\rangle$ by

$$\langle H_0, A_0; E, a|M_-^{*-1}|H_0, A_0; E', a'\rangle = \delta(E-E')\langle a|\mu_-^{*-1}(E)|a'\rangle . \tag{3.12}$$

From the fact that $M_{-}^{*-1} = (M^*)^{-1} = (M^{-1})^*$ is the Hermitian adjoint of M^{-1} we have

$$\langle a|\mu_{-}^{*-1}(E)|a'\rangle = \langle a'|\mu_{-}^{-1}(E)a\rangle^* . \tag{3.13}$$

If, as in (6) we define $\langle a|\omega_c(E)|a'\rangle$ by

$$\langle H_0, A_0; E, a|W_c|H_0, A_0; E', a'\rangle = \delta(E-E')\langle a|\omega_c(E)|a'\rangle , \tag{3.14}$$

we have from (3.9)

$$\langle a|\omega_c(E)|a'\rangle = \sum_{a''}\langle a|\mu_{-}^{-1}(E)|a''\rangle\langle a''|\mu_{-}^{*-1}(E)|a'\rangle , \tag{3.15}$$

or, on using (3.13) and (3.11)

$$\langle a|\omega_c(E)|a'\rangle = \begin{pmatrix} 1 & \langle -1|S(E)|+1\rangle^* \\ \langle -1|S(E)|+1\rangle & 1 \end{pmatrix}. \tag{3.16}$$

In deriving (3.16) from (3.15) we have used the relation

$$|\langle -1|S(E)|-1\rangle|^2 + |\langle -1|S(E)|+1\rangle|^2 = 1 , \tag{3.17}$$

which follows from the fact that $S^* = S^{-1}$, i.e., $SS^* = \eta(H_0)$.

For the sake of completeness, we shall also give the expression for $\langle a|\mu^+(E)|a'\rangle$ defined by

$$\langle x|H, A; E, a\rangle = \sum_{a'}\langle x|H, A; E, a'\rangle_+ \langle a'|\mu_+(E)|a\rangle , \tag{3.18}$$

where $\langle x|H, A; E, a\rangle$ satisfies the boundary condition (3.2*a*). The expression can be shown to be

$$\langle a|\mu_+(E)|a'\rangle = \begin{pmatrix} \dfrac{1}{\langle +1|S^{-1}(E)|+1\rangle} & \dfrac{-\langle +1|S^{-1}(E)|-1\rangle}{\langle +1|S^{-1}(E)|+1\rangle} \\ 0 & 1 \end{pmatrix}. \tag{3.19}$$

4. – The complex energy plane.

In this section we consider the analytic continuation of various quantities introduced above, and for this purpose we introduce the complex E-plane which shall have a cut along the positive real axis. The results obtained here will be useful in later sections.

Let us first introduce the function $\langle x|H_0, A_0; E, a\rangle^+$. This function is defined as the analytic continuation of $\langle x|H_0, A_0; E, a\rangle$ when E is on the upper part of the cut. From (2.2), we see that when E is in the lower part of the cut

(4.1) $$\langle x|H_0, A_0; E, a\rangle^+ = -i\langle x|H_0, A_0; E, -a\rangle .$$

One can also introduce $\langle x|H_0, A_0; E, a\rangle^-$ which defined as being equal to $\langle x|H_0, A_0; E, a\rangle$ when E is on the lower side of the cut. Hence, from (4.1)

(4.2) $$\langle x|H_0, A_0; E, a\rangle^- = i\langle x|H_0, A_0; E, -a\rangle^+$$

and

(4.3) $$\langle x|H_0, A_0; E, a\rangle^- = i\langle x|H_0, A_0; E, -a\rangle$$

when E is on the upper side of the cut.

We can also introduce $\langle H_0, A_0; E, a|x\rangle^+$ which equals $\langle H_0, A_0; E, a|x\rangle$ when E is on the upper side of the cut. Then since

(4.4) $$\langle H_0, A_0; E, a|x\rangle = \langle x|H_0, A_0; E, -a\rangle$$

we have

(4.5) $$\langle H_0, A_0; E, a|x\rangle^+ = \langle x|H_0, A_0; E, -a\rangle^+ .$$

Similarly one can define $\langle H_0, A_0; E, a|x\rangle^-$ as being the analytic continuation of $\langle H_0, A_0; E, a|x\rangle$ when E is at the bottom of the cut, where from (4.4) it can be shown that

(4.6) $$\langle H_0, A_0; E, a|x\rangle^- = \langle x|H_0, A_0; E, -a\rangle^- .$$

Now we define $\langle x|H, A; E, a\rangle^+_\pm$ as the analytic continuation of $\langle x|H, A; E, a\rangle_\pm$ when E is on the top of the cut. From the integral equation (2.7) it is clear that when E is at the bottom of the cut we have

(4.7) $$\langle x|H, A; E, a\rangle^+_\pm = -i\langle x|H_0, A_0; E, -a\rangle \mp \frac{i}{2}\,\varepsilon(E)^{-\frac{1}{2}}\cdot$$

$$\cdot\int_{-\infty}^{+\infty}\int_{-\infty}^{+\infty} dx'\,dx''\,\exp\left[\pm i\sqrt{E}\,|x-x'|\right]\langle x'|V|x''\rangle\langle x''|H, A; E, a\rangle^+_\pm ,$$

from which it can be shown that

(4.8) $$\langle x|H, A; E, a\rangle^+_\pm = -i\langle x|H, A; E, -a\rangle_\mp$$

when E is on the lower part of the cut.

Likewise one can introduce $\langle x|H, A; E, a\rangle_{\pm}^{-}$ which is the analytic continuation of $\langle x|H, A; E, a\rangle_{\pm}$ defined on the lower part of the cut. It is clear from (4.8) that

$$\langle x|H, A; E, a\rangle_{\pm}^{-} = i\langle x|H, A; E, -a\rangle_{\mp}^{+} \tag{4.9}$$

and that when E is at the top of the cut

$$\langle x|H, A; E, a\rangle_{\pm}^{-} = i\langle x|H, A; E, -a\rangle_{\mp} . \tag{4.10}$$

The analytic continuation of $\langle a|S(E)|a'\rangle$ when E is on the top of the cut will be denoted by $\langle a|S(E)|a'\rangle^{+}$. From (2.10), (4.8), (4.5), (4.1), (2.11), (2.12) we see that when E is at the bottom of the cut

$$\begin{aligned}\langle a|S(E)|a'\rangle^{+} &= \delta_{a,a'} + 2\pi i\int_{-\infty}^{+\infty}\mathrm{d}x\int_{-\infty}^{+\infty}\mathrm{d}x' \cdot \\ &\cdot\langle H_0, A_0; E, -a|x\rangle\langle x|V|x'\rangle\langle x'|H, A; E, -a'\rangle_{+} = \\ &= \langle -a|S^{-1}(E)|-a'\rangle = \langle -a'|S(E)|-a\rangle^{*} ,\end{aligned} \tag{4.11}$$

where, as usual, the asterisk means complex conjugate. We denote by $\langle a|S(E)|a'\rangle^{-}$ the analytic continuation of $\langle a|S(E)|a'\rangle$ when E is at the bottom of the cut. It can be shown that when E is at the top of the cut

$$\langle a|S(E)|a'\rangle^{-} = \langle -a|S^{-1}(E)|-a'\rangle = \langle -a'|S(E)|-a\rangle^{*} . \tag{4.12}$$

5. – Extension of the unperturbed Hamiltonian. Diagonal form for the scattering potential.

In ([6]) we saw that it is necessary to extend the definition of the operator H_0 to a space larger than Hilbert space in order that the extended operator H_0 have the same spectrum as H. In ([6]), Part II the extension was carried out by introducing « eigenfunctions » of the extended operator H_0. Here we specialize this procedure to the present case.

In Hilbert space H_0 is defined by the way it acts in the x-representation, namely,

$$H_0^x = -\frac{\mathrm{d}^2}{\mathrm{d}x^2} . \tag{5.1}$$

We shall require that the extended space, H_0, be also given by (5.1). In the notation of (*), Part II, H_0 is then an x-extended operator. With this extension the potential V has the desirable property that it is diagonal in the x-representation, as we shall show shortly. First, however, we shall obtain the additional eigenfunctions necessary to complete the extended space, In accordance with (*), Part II we shall look for solutions of

$$H_0|H_0, A_0; E, a\rangle = E|H_0, A_0; E, a\rangle \tag{5.2}$$

for values of E in the vicinity of each E_i which is a point eigenvalue of H such that

$$\lim_{x\to-\infty} \langle x|H_0, A_0; E, a\rangle = 0 . \tag{5.3}$$

In terms of the x-representation Eq. (5.2) becomes

$$-\frac{d^2}{dx^2}\langle x|H_0, A_0; E, a\rangle = E\langle x|H_0, A_0; E, a\rangle . \tag{5.4}$$

For a fixed negative value of E, there is only one solution which satisfies (5.3). We conclude therefore that *the spectrum of the extended Hamiltonian H_0 and the point spectrum of H cannot be degenerate.*

Let us denote by $\langle x|H_0; E\rangle$ the suitable solutions of (5.4). Then we can write

$$\langle x|H_0\,; E\rangle = A(E)\frac{\exp[-i(\pi/4)]}{2\sqrt{\pi}}(-E)^{-\frac{1}{4}}\exp[\sqrt{-E}x] = \tag{5.5}$$

$$= A(E)\langle x|H_0, A_0; E, -1\rangle^+ ,$$

where $A(E)$ is an arbitrary function of E which has no zeros or singularities for $E<0$. With this choice of $|H_0; E\rangle$ we take for the bi-orthogonal vector $|H_0; E\rangle_B$ another solution of (5.2) or (5.4) which satisfies

$$_B\langle H_0, E|H_0, E'\rangle = \delta(E-E') . \tag{5.6}$$

It is clear that

$$_B\langle H_0; E|x\rangle = [A(E)]^{-1}\frac{\exp[-i(\pi/4)]}{2\sqrt{\pi}}(-E)^{-\frac{1}{4}}\exp[\sqrt{-E}x] = \tag{5.5a}$$

$$= \langle H_0, A_0; E, -1|x\rangle^+ .$$

Also

$$\langle H_0; E|x\rangle = A^*(E)\frac{\exp[i(\pi/4)]}{2\sqrt{\pi}}(-E)^{-\frac{1}{4}}\exp[\sqrt{-E}x] = \tag{5.5b}$$

$$= iA^*(E)\langle H_0, A_0; E, +1|x\rangle^+ .$$

In the general vector space all states $\langle x|\varphi\rangle$ will have the property that they are quadratically integrable from $x = -\infty$ to any finite limit.

We shall now prove V is diagonal in the x-representation. We have

$$\varepsilon V = H - H_0 = UH_0U_0 - H_0UU_0 = (UH_0 - H_0U)U_0 = \tag{5.7}$$

$$= \varepsilon(KH_0 - H_0K) + \varepsilon^2(KH_0 - H_0K)K_0 .$$

We shall first calculate $(KH_0 - H_0K)$ in terms of the x-representation. Since K is triangular in the x-representation we can always write

$$\langle x|K|x'\rangle = \eta(x-x')\langle x|P|x'\rangle , \tag{5.8}$$

where

$$\begin{cases} \langle x|P|x'\rangle = \langle x|K|x'\rangle & \text{for } x' \leqslant x, \\ \langle x|P|x'\rangle = \text{an arbitrary function,} & \text{for } x' > x, \end{cases} \tag{5.9}$$

We choose $\langle x|P|x'\rangle$ so that it is continuous at $x = x'$. Since H_0 is Hermitian, we have

$$\langle x|KH_0|x'\rangle = -\langle x|K|x'\rangle \frac{\partial^2}{\partial x'^2} = -\frac{\partial^2}{\partial x'^2}\langle x|K|x'\rangle = \tag{5.10}$$

$$= -\frac{\partial^2}{\partial x'^2}\eta(x-x')\langle x|P|x'\rangle .$$

Now

$$\frac{\partial}{\partial x'}\eta(x-x')\langle x|P|x'\rangle = -\delta(x-x')\langle x|P|x'\rangle + \eta(x-x')\frac{\partial}{\partial x'}\langle x|P|x'\rangle = \tag{5.11}$$

$$= -\delta(x-x')\langle x|P|x\rangle + \eta(x-x')\frac{\partial}{\partial x'}\langle x|P|x'\rangle ,$$

and

$$\frac{\partial^2}{\partial x'^2}\eta(x-x')\langle x|P|x'\rangle = \delta'(x-x')\langle x|P|x\rangle - \delta(x-x')\frac{\partial}{\partial x'}\langle x|P|x\rangle + \tag{5.12}$$

$$+ \eta(x-x')\frac{\partial^2}{\partial x'^2}\langle x|P|x'\rangle .$$

Likewise

$$\langle x|H_0K|x'\rangle = -\frac{\partial^2}{\partial x^2}\langle x|K|x'\rangle = -\frac{\partial^2}{\partial x^2}\eta(x-x')\langle x|P|x'\rangle . \tag{5.13}$$

But

$$(5.14)\qquad \frac{\partial}{\partial x}\eta(x-x')\langle x|P|x'\rangle = \delta(x-x')\langle x|P|x\rangle + \eta(x-x')\frac{\partial}{\partial x}\langle x|P|x'\rangle\,,$$

and

$$(5.15)\qquad \frac{\partial^2}{\partial x^2}\eta(x-x')\langle x|P|x'\rangle = \delta'(x-x')\langle x|P|x\rangle + \delta(x-x')\frac{\mathrm{d}}{\mathrm{d}x}\langle x|P|x\rangle + \\ + \delta(x-x')\frac{\partial}{\partial x}\langle x|P|x\rangle + \eta(x-x')\frac{\partial^2}{\partial x^2}\langle x|P|x'\rangle\,.$$

In (5.15) the expression $(\mathrm{d}/\mathrm{d}x)\langle x|P|x\rangle$ means that after x' has been set equal to x in $\langle x|P|x'\rangle$, one takes the derivative of $\langle x|P|x\rangle$ with respect to x. In contrast, the expression $(\partial/\partial x)\langle x|P|x\rangle$ means that one has taken the derivative of $\langle x|P|x'\rangle$ with respect to x and then set x' equal to x. In (5.12) $(\partial/\partial x')\langle x|P|x\rangle$ means that one is to take the derivative of $\langle x|P|x'\rangle$ with respect to x' and then set x' equal to x. From (5.11)-(5.15), (5.8), and the fact that

$$\frac{\partial}{\partial x}\langle x|P|x\rangle + \frac{\partial}{\partial x'}\langle x|P|x\rangle = \frac{\mathrm{d}}{\mathrm{d}x}\langle x|P|x\rangle\,,$$

one can write

$$(5.16)\qquad \langle x|KH_0-H_0K|x'\rangle = \\ = 2\delta(x-x')\frac{\mathrm{d}}{\mathrm{d}x}\langle x|K|x\rangle + \eta(x-x')\left[\frac{\partial^2}{\partial x^2}\langle x|K|x'\rangle - \frac{\partial^2}{\partial x'^2}\langle x|K|x'\rangle\right].$$

Hence from (5.7) and the fact that K_0 is triangular in the x-representation we have

$$(5.17)\qquad \langle x|V|x'\rangle = 2\delta(x-x')\frac{\mathrm{d}}{\mathrm{d}x}\langle x|K|x\rangle + \eta(x-x')\cdot \\ \cdot\left\{\frac{\partial^2}{\partial x^2}\langle x|K|x'\rangle - \frac{\partial^2}{\partial x'^2}\langle x|K|x'\rangle + 2\varepsilon\frac{\mathrm{d}}{\mathrm{d}x}\langle x|K|x\rangle\cdot\langle x|K_0|x'\rangle + \right. \\ \left. + \varepsilon\int_{x'}^{x}\left[\frac{\partial^2}{\partial x^2}\langle x|K|x''\rangle - \frac{\partial^2}{\partial x'^2}\langle x|K|x''\rangle\right]\mathrm{d}x''\langle x''|K_0|x'\rangle\right\}.$$

Now from the general theory, V is Hermitian if the generalized Gelfand-Levitan equation of (•) can be solved. Furthermore, it will later be shown that $\langle x|K|x'\rangle$ is real. Therefore the right-hand side of (5.17) is the sum of a Hermitian operator which is diagonal in the x-representation, and an operator which, because it is triangular in the x-representation, cannot be Hermitian. Since the right-hand side must be Hermitian, the expression in brackets must vanish and we are left with

$$\langle x|V|x'\rangle = \delta(x-x')V(x)\,, \tag{5.18}$$

where

$$V(x) = 2\,\frac{\mathrm{d}}{\mathrm{d}x}\langle x|K|x\rangle\,. \tag{5.19}$$

Hence V is diagonal in the x-representation and can be obtanied from K very simply, by means of (5.19).

6. – The weight operator and the Gelfand-Levitan equation.

We shall now develop expressions for the weight operator in terms of the H_0-representation. In accordance with the general theory, the equation for K is

$$\langle x|K|x'\rangle = -\langle x|\Omega|x'\rangle - \varepsilon\int_{-\infty}^{x}\langle x|K|x''\rangle\,\mathrm{d}x''\langle x''|\Omega|x'\rangle\,, \tag{6.1}$$

where

$$\varepsilon\Omega = W - \eta(H_0) = [W_c - \eta(H_0)]\eta(H_0) + W_d\sum_i\delta(E_i - H_0)\eta(-H_0)\,. \tag{6.2}$$

We have already expressed W in terms of the H_0-representation and have seen that it can be obtained in terms of the reflection coefficient of the scattering operator (Eqs. (3.14) and (3.16)). We shall now obtain the expression for W_d in terms of the H_0-representation.

From the general procedure discussed in ([6]), Part II, W_d as given in the H_0-representation has the form

$${}_B\langle H_0, A_0; E, a|W_d|H_0, A_0; E', b\rangle_B = \delta(E-E')\langle a|\omega_d(E)|b\rangle\,, \tag{6.3}$$

where $|H_0, A_0; E, a\rangle_B$ are the bi-orthogonal eigenfunctions (5.5a) of H_0 defined for negative eigenvalues E in the neighborhoods of the point eigenvalues E_i of H. Since, in the present case, these eigenfunctions have no degeneracy, we shall write

$${}_B\langle H_0, E|W_d|H_0, E'\rangle_B = \delta(E-E')\omega_d(E) = \delta(E-E')[C(E)]^{-1}\,, \tag{6.4}$$

where, in accordance with ([6]), Part II, the $C(E_i)$ are the normalization constants of the proper eigenstates of H, that is,

$$\langle H, E_i | H, E_j \rangle = C(E_i)\delta_{ij} . \tag{6.5}$$

We shall now obtain $\varepsilon\langle x|\Omega|x'\rangle$ in the form

$$\varepsilon\langle x|\Omega|x'\rangle = \langle x|(W_c - \eta(H_0))\eta(H_0)|x'\rangle + \\ + \langle x| W_d \sum_i \delta(E_i - H_0)\eta(-H_0)|x'\rangle . \tag{6.6}$$

We note that

$$\langle x|(W_c - \eta(H_0))\eta(H_0)|x'\rangle = \sum_{a,b}\int\langle x|H_0, A_0; E, a\rangle[\langle a|\omega_c(E)|b\rangle - \delta_{ab}]\cdot \\ \cdot\eta(E)\,\mathrm{d}E\langle H_0, A_0; E, b|x'\rangle . \tag{6.7}$$

On using (3.16) and (2.2) we have

$$\langle x|(W_c - \eta(H_0))\eta(H_0)|x'\rangle = \tag{6.8}$$

$$=\int_0^\infty \mathrm{d}E\langle x|H_0, A_0; E, -1\rangle\langle -1|S(E) + 1\rangle\langle H_0, A_0; E, +1|x'\rangle +$$

$$+\int_0^\infty \mathrm{d}E\,\langle x|H_0, A_0; E, +1\rangle\langle -1|S(E)|+1\rangle^*\,\langle H_0, A_0; E, -1|x'\rangle =$$

$$= 2\mathrm{Re}\int_0^\infty \mathrm{d}E\langle x|H_0, A_0; E, -1\rangle\langle -1|S(E)|+1\rangle\langle H_0, A_0; E, +1|x'\rangle ,$$

where Re means « real part ». Eq. (6.8) leads to

$$\langle x|(W_c - \eta(H_0))\eta(H_0)|x'\rangle = \tag{6.9}$$

$$= \mathrm{Re}\,\frac{1}{2\pi}\int_0^\infty \mathrm{d}E\,E^{-\frac{1}{2}}\exp\left[-i\sqrt{E}(x + x')\right]\langle -1|S(E)|+1\rangle .$$

We can simplify (6.9) by introducing the momentum variable defined by

$$E = k^2 . \tag{6.10}$$

Then

$$\langle x|(W_c - \eta(H_0))\eta(H_0)|x'\rangle = \mathrm{Re}\,\frac{1}{\pi}\int_0^\infty \mathrm{d}k\,\exp\left[-ik(x + x')\right]\langle -1|S(k^2)|+1\rangle . \tag{6.11}$$

We also have

$$(6.12)\qquad \langle x|W_d\sum_i\delta(E_i-H_0)\eta(-H_0)|x'\rangle =$$

$$=\sum_i\int dE\langle x|H_0;\,E\rangle[C(E)]^{-1}\delta(E_i-E)\langle H_0;\,E|x'\rangle =$$

$$=\sum_i\langle x|H_0;\,E_i\rangle[C(E_i)]^{-1}\langle H_0;\,E_i|x'\rangle\,,$$

where the integration is over the intervals ΔE_i. On using (5.5) we have

$$(6.13)\qquad \langle x|W_d\sum_i\delta(E_i-H_0)\eta(-H_0)|x'\rangle =$$

$$=\frac{1}{4\pi}\sum_i(-E_i)^{-\frac{1}{2}}\exp\left[\sqrt{-E_i}(x+x')\right][C(E_i)]^{-1}|A(E_i)|^2=$$

$$=i\sum_i[C(E_i)]^{-1}|A(E_i)|^2\langle x|H_0,A_0;\,E_i,-1\rangle^+\langle H_0,A_0;\,E_i,+1|x'\rangle^+.$$

Hence finally we have

$$(6.14)\qquad \varepsilon\langle x|\Omega|x'\rangle =$$

$$=2\mathrm{Re}\int_0^\infty dE\langle x|H_0,A_0;\,E,-1\rangle\langle -1|S(E)|+1\rangle\langle H_0,A_0;\,E,+1|x'\rangle +$$

$$+i\sum_i[C(E_i)]^{-1}|A(E_i)|^2\langle x|H_0,A_0;\,E_i,-1\rangle^+\langle H_0,A_0;\,E_i,+1|x'\rangle^+ =$$

$$=\mathrm{Re}\frac{1}{\pi}\int_0^\infty dk\exp\left[ik(x+x')\right]\langle -1|S(k^2)|+1\rangle +$$

$$\frac{1}{4\pi}\sum_i(-E_i)^{-\frac{1}{2}}[C(E_i)]^{-1}|A(E_i)|^2\exp\left[\sqrt{-E_i}(x+x')\right].$$

Clearly $\langle x|\Omega|x'\rangle$ is real. Then from the uniqueness theorem it can be shown that $\langle x|K|x'\rangle$ is also real, as stated in Section **5**.

7. - The conditions on the bound states and the reflection coefficient.

In accordance with the general theory of ([6]), we can generally find the operators $U=I+\varepsilon K$, $U_0=I+\varepsilon K_0$, and therefore the complete set of eigenfunctions $|H,A;\,E,a\rangle=U|H_0,A_0;\,E,a\rangle$ and $|H;E_i\rangle=U|H_0;E_i\rangle$ corresponding to the continuous and the discrete spectrum of $H=UH_0U_0$ respectively whose normalization is given by the matrix elements of $W=\eta(H_0)+\varepsilon\Omega$, where Ω is given by (6.14); these eigenfunctions satisfy the relations

$$\text{(7.1)}\quad \langle H, A; E, a | H, A; E', a' \rangle = \langle H_0, A_0; E, a | W^{-1}\eta(H_0) | H_0, A_0; E', a' \rangle = \langle a | \omega^{-1}(E) | a' \rangle \delta(E - E')\eta(E) ,$$

$$\text{(7.2)}\quad \langle H, E_i | H, E_j \rangle = \delta_{ij} C_i ,$$

$$\text{(7.3)}\quad \langle H, A; E, a | H, E_i \rangle = 0 .$$

However, in order to be interesting, our solution must be such that we are able to construct a scattering of S in which the reflection coefficient $\langle -1 | S(E) | +1 \rangle$ is the one used to form the weight operator W. Toward this end we shall obtain some necessary conditions on the reflection coefficient, point eigenvalues, and normalization constants.

The necessary conditions follow from the requirements on the rapidity with which $V(x)$ must decay as x approaches $+\infty$ or $-\infty$.

Let us first discuss consequences of the decay of $V(x)$ as $x \to -\infty$. If $V(x)$ is to decay sufficiently rapidly to ensure the existence of a scattering operator, one can show that the outgoing wave $\langle x | H, A; E, a \rangle_-$ must be unique. As a consequence of this uniqueness it follows that we must have

$$\text{(7.4)}\quad \langle x | H, A; E, a \rangle \to \langle x | H_0, A_0; E, a \rangle \qquad \text{as } x \to -\infty.$$

In ([6]), Part I, Section 16 we have already discussed two requirements on $\langle x | \Omega | x' \rangle$ which ensure the validity of (7.4). However, in the present application of the general procedure, it turns out that because $\langle x | \Omega | x' \rangle$ is a function of $x + x'$ (see (6.14)) another approach is useful. A sufficient condition for the validity of (7.4) is that a positive number α exists such that

$$\text{(7.5)}\quad \langle x | K | x' \rangle \equiv 0 , \qquad x < -\alpha.$$

It is clear that a consequence of(7.5) is that

$$\text{(7.6)}\quad V(x) = 2 \frac{\mathrm{d}}{\mathrm{d}x} \langle x | K | x \rangle \equiv 0 , \qquad \text{for } x < -\alpha,$$

i.e., that $V(x)$ is cut-off from below.

We shall now show that a necessary and sufficient condition for (7.5) to hold is

$$\text{(7.7)}\quad \langle x | \Omega | 0 \rangle \equiv 0 , \qquad x < -2\alpha,$$

and that a necessary condition for (7.7) and hence (7.5) to be valid is that the point eigenvalues E_i be the poles of $\langle -1|S(E)|+1\rangle^+$, i.e., the analytic continuation of the reflection coefficient. A sufficient condition for (7.7) and (7.5) to hold is that $\langle +1|S(E)-1\rangle^+$ be of the form $\exp[-i2\sqrt{E}\,\alpha]f(E)$ where $f(E)$ behaves like $0(E^{-\frac{1}{2}})$ for $|E|\to\infty$ in the cut complex plane.

Let us first prove that (7.7) is a necessary condition for (7.5) to hold. From (6.14) the following identity is obviuos:

$$\langle x|\Omega|x'\rangle = \langle x+x'|\Omega|0\rangle . \tag{7.8}$$

The Gelfand-Levitan equation (6.1) can then be written

$$\langle x|K|x'\rangle = -\langle x+x'|\Omega|0\rangle - \varepsilon\int_{-\infty}^{x}\langle x|K|x''\rangle dx''\langle x''+x'|\Omega|0\rangle \qquad (x>x') . \tag{7.9}$$

Now let $x<-\alpha$. From (7.5) we have

$$\langle x+x'|\Omega|0\rangle \equiv 0 \qquad \text{for } x<-\alpha,\ x'<x<-\alpha, \tag{7.10}$$

from which (7.7) follows as a necessary condition.

Let us now show the sufficiency of condition (7.7). If we use Eq. (7.7), then Eq. (7.9) becomes

$$\begin{aligned}\langle x|K|x'\rangle &= -\langle x+x'|\Omega|0\rangle - \\ &\quad -\varepsilon\eta(x+2\alpha+x')\int_{-(2\alpha+x')}^{x}\langle x|K|x''\rangle dx''\langle x''+x'|\Omega|0\rangle ; \qquad (x<x').\end{aligned} \tag{7.11}$$

When $x<-\alpha$, we have $x+x'<-2\alpha$ or, equivalently, $x<-(2\alpha+x')$; hence the right-hand side of (7.11) vanishes and we are left with (7.5).

To show the connection between the singularities of $\langle -1|S(E)|+1\rangle$ and the bound states we note that we can write the expression for $\varepsilon\langle x|\Omega|x'\rangle$ as an integral in the complex E-plane. On using (4.11), which gives the analytic continuation of the matrix elements of the scattering operator, we see that

$$\langle -1|S(E)|+1\rangle^+ = \langle -1|S(E)|+1\rangle^*$$

where E is at the bottom of the cut.

Hence, using this fact and the fact that

$$\begin{aligned}\langle x|H_0, A_0; E, -1\rangle^+ \langle H_0, A_0; E, +1|x'\rangle^+ &= \\ &= \frac{E^{-\frac{1}{2}}}{2\sqrt{\pi}}\langle x+x'|H_0, A_0; E, -1\rangle^+ ,\end{aligned} \tag{7.12}$$

we have

$$(7.13)\quad \varepsilon\langle x|\Omega|x'\rangle = = \frac{1}{2\sqrt{\pi}}\int_\gamma \langle -1|S(E)|+1\rangle^+ (E)^{-\frac{1}{2}}\langle x + x'|H_0, A_0; E, -1\rangle^+ dE + + \frac{1}{2\sqrt{\pi}} \exp[i\pi/4] \sum_i [C(E_i)]^{-1}|A(E_i)|^2(-E_i)^{-\frac{1}{2}}\langle x + x'|H_0, A_0; E_i, -1\rangle^+ ,$$

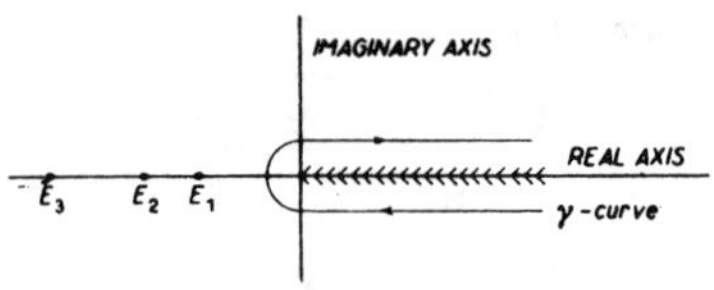

Fig. 1. – Complex E-Plane.

where the contour γ goes in a clockwise direction around the cut in the E-plane.

Let us introduce the momentum variable $k = \sqrt{E}$. The E-plane is then transformed to the upper k-plane. Let us write $b(k) = \langle -1|S(k^2)|+1\rangle$. From (4.11) it is clear that $b(-k) = b^*(k)$ for real k. We have

$$(7.14)\quad \varepsilon\langle x|\Omega|0\rangle = \frac{1}{2\pi}\int_{-\infty}^{+\infty} dk b(k) \exp[-ikx]\, dk + + \frac{1}{2\sqrt{\pi}} \exp[i\pi/4] \sum_i [C(E_i)]^{-1}|A(E_i)|^2(-E_i)^{-\frac{1}{2}} \langle x|H_0, A_0; E_i, -1\rangle^+ .$$

Let us write

$$(7.15)\quad b(k) = \exp[-i2k\alpha] g(k)$$

where $g(k) \to O(k^{-1})$ as $|k| \to \infty$.

The integral of Eq. (7.14) becomes

$$(7.16)\quad \frac{1}{2\pi}\int_{-\infty}^{+\infty} g(k) \exp[-ik(x + 2\alpha)]\, dk .$$

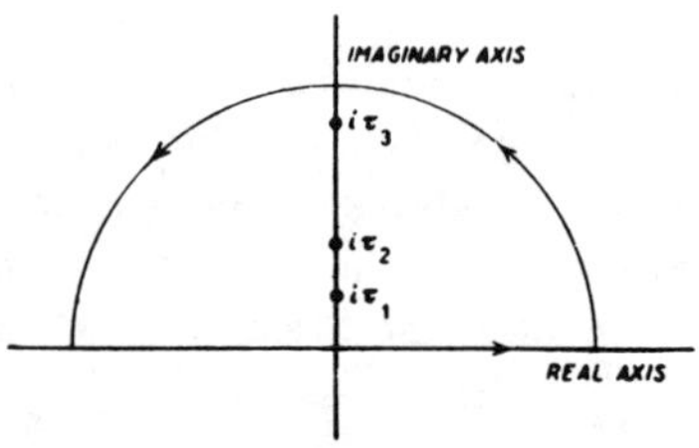

Fig. 2. – Complex k-Plane.

When $x < -2\alpha$, one can close the contour by the infinite semi-circle in the upper half of the k-plane, since the integral over this semi-circle will contribute nothing.

The integral (7.16) is equal to the sum of the residues of the integrand, namely,

$$(7.17)\quad \frac{1}{2\pi}\int_{-\infty}^{+\infty} g(k) \exp[-ik(x + 2\alpha)]\, dk = i \sum_j r_j \exp[-2ik_j\alpha] \exp[-ik_jx] ,$$

where r_j are the residues of $g(k)$ at the poles k_j. Since the integral must be real for all x, it is clear that the poles k_j must lie on the positive imaginary axis. We shall write $k_j = i\tau_j$ where τ_j are positive real numbers see (Fig. 2). Likewise, ir_j must be real.

Finally, then, we can write for $x < -2\alpha$.

$$(7.18)\qquad \varepsilon\langle x|\Omega|0\rangle = 2\sqrt{\pi}\exp[i\pi/4]i\sum_j r_j\exp[2\tau_j\alpha](-E_j)^{\frac{1}{4}}\langle x|H_0, A_0; E_j, -1\rangle^+ + \\ + \frac{1}{2\sqrt{\pi}}\exp[i\pi/4]\sum_i [C(E_i)]^{-1}|A(E_i)|^2(-E_i)^{-\frac{1}{4}}\langle x|H_0, A_0; E_i, -1\rangle^+ ,$$

where in the first sum $E_j = k_j^2 = -\tau_j^2$ (see Fig. 1).

Since $\langle x|\Omega|0\rangle$ is to vanish for $x < -2\alpha$ it is clear that it is sufficient to require that the poles k_j of $g(k)$ be equal to $i\sqrt{-E_j}$, where E_j are the point eigenvalues of H, and that

$$(7.19)\qquad [C(E_i)]^{-1}|A(E_i)|^2 = -4\pi i\tau_i r_i \exp[2\tau_i\alpha] .$$

From (7.19) it is seen that r_i must be a positive imaginary number in order that $C(E_i)$ be positive.

On using (7.19) in (7.14) we obtain

$$(7.14a)\qquad \varepsilon\langle x|\Omega|0\rangle = \frac{1}{2\pi}\int_{-\infty}^{+\infty} dk\, b(k)\exp[-ikx] - i\sum_j r_j \exp[\tau_j(x + 2\alpha)] .$$

One then obtains the results in terms of the notation of the introduction by writing

$$\langle x|H, A; E, a\rangle = \frac{1}{2\sqrt{\pi}} E^{-\frac{1}{4}}\Psi_k(x) ,$$

where $k = a\sqrt{E}$.

We have given sufficient conditions for the reproduction of the reflection coefficient $\langle -1|S(E)|+1\rangle$, in particular, the condition that $V(x) = 0$ for $x < -\alpha$. However, we are not as yet able to obtain conditions to guarantee the vanishing of $V(x)$ for x sufficiently large. This vanishing is a necessary condition for the full scattering operator to exist. It is, in fact, possible to give examples where the prescribed reflection coefficient exists, but where the full scattering operator does not exist.

Incidentally, the use of a triangular operator $\langle x|K|x'\rangle$ which vanishes when $x < -\alpha$ has the consequence that $\langle x|H, A; E, a\rangle = \langle x|U|H_0, A_0; E, a\rangle$ is an analytic function of E in the complex E-plane. Furthermore, the bound states $\langle x|H, E_i\rangle$ can be obtained from the analytic continuation of $\langle x|H, A; E, -1\rangle$. In fact

$$\langle x|H, A; E_i, -1\rangle^+ = \langle x|U|H_0, A_0; E_i, -1\rangle^+ = [A(E_i)]^{-1}\langle x|H, E_i\rangle . \tag{7.20}$$

Usually one chooses the arbitrary function $A(E_i)$ to be unity, to simplify the relation (7.20). However, to obtain the results in terms of the notation of the introduction, we write

$$A(E_i) = 2\sqrt{\pi}\exp[i\pi/4](-E_i)^{\frac{1}{4}} .$$

8. – Examples.

We shall now discuss three examples. In what follows we shall take $\varepsilon=1$ for simplicity.

8·1. *The repulsive δ-function potential.* – Let us consider the case where

$$\langle -1|S(E)|+1\rangle = \frac{-(iA)\exp[-2i\sqrt{E}\alpha]}{2E^{\frac{1}{2}} + iA}, \qquad (A > 0), \tag{8.1}$$

or equivalently

$$b(k) = \frac{-i(A/2)\exp[-2ik\alpha]}{k + (iA/2)} . \tag{8.2}$$

We see that $b(-k) = b^*(k)$ (for real k) as required. Furthermore, $b(k) = O(k^{-1})$ for $|k|\to\infty$. Since $b(k)$ has no poles on the positive imaginary axis, there will be no bound states.

From (1.8) we have

$$\langle x|\Omega|x'\rangle = \langle x + x'|\Omega|0\rangle = \tag{8.3}$$

$$= \frac{1}{2\pi}\int_{-\infty}^{+\infty} dk\, b(k)\exp[-ik(x + x')] = \frac{-iA}{4\pi}\int_{-\infty}^{+\infty}\frac{dk}{k + (iA/2)}\exp[-ik(x + x' + 2\alpha)] =$$

$$= -\frac{A}{2}\eta(x + x' + 2\alpha)\exp\left[-\frac{A}{2}(x + x' + 2\alpha)\right] .$$

The Gelfand-Levitan equation (1.7) is

$$(8.4)\quad \langle x|K|x'\rangle = \frac{A}{2}\eta(x+x'+2\alpha)\exp\left[-\frac{A}{2}(x+x'+2\alpha)\right]+\frac{A}{2}\eta(x+x'+2\alpha)\cdot$$
$$\cdot\int_{-(2\alpha+x')}^{x}\langle x|K|x''\rangle\eta(x''+x'+2\alpha)\exp\left[-\frac{A}{2}(x''+x'+2\alpha)\right]\mathrm{d}x''\,.$$

It is easily verified that the solution of (8.4) is

$$(8.5)\qquad \langle x|K|x'\rangle = \frac{A}{2}\eta(x+x'+2\alpha)\,. \qquad x'<x.$$

Hence on using the potential $V(x)$ is

$$(8.6)\qquad V(x) = 2\frac{\mathrm{d}}{\mathrm{d}x}\frac{A}{2}\eta(2x+2\alpha) = A\delta(x+\alpha)\,.$$

It is clear that $V(x)\equiv 0$ and $\langle x|K|x'\rangle = 0$ when $x<-\alpha$ as required.

The eigenfunctions of $H=-\mathrm{d}^2/\mathrm{d}x^2+A\,\delta(x+\alpha)$ are given by

$$(8.7)\quad \langle x|H,A;E,a\rangle=\langle x|H_0,A_0;E,a\rangle+\frac{A}{2}\int_{-\infty}^{+\infty}\eta(x+x'+2\alpha)\langle x'|H_0,A_0;E,a\rangle\,\mathrm{d}x'=$$
$$=\langle x|H_0,A_0;E,a\rangle+\frac{A}{2}\eta(x+\alpha)\int_{-(x+2\alpha)}^{x}\langle x'|H_0,A_0;E,a\rangle\,\mathrm{d}x'=$$
$$=\langle x|H_0,A_0;E,a\rangle-\frac{iAa}{2}\eta(x+\alpha)E^{-\frac{1}{2}}\cdot$$
$$\cdot[\langle x|H_0,A_0;E,a\rangle-\langle x+2\alpha|H_0,A_0;E,-a\rangle]\,.$$

We shall now calculate the scattering operator for the problem. Toward this end we shall use

$$(8.8)\qquad S = M_+M_-^{-1}\,.$$

(Cf. ([6]), Part I, Eq. (5.8)) or equivalently

$$(8.9)\qquad \langle a|S(E)|b\rangle = \sum_c\langle a|\mu_+(E)|c\rangle\langle c|\mu_-^{-1}(E)|b\rangle\,.$$

Hence we shall have to evaluate $\langle a|M_+|b\rangle$ and $\langle a|M_-^{-1}(E)|b\rangle$.

From (⁶), Part I, Eq. (5.1) we have

$$(8.10)\qquad M_{\pm} = U - \int \gamma_{\pm}(E - H_0) V U \delta(E - H_0) \mathrm{d}E .$$

If there were bound states, the above equation would have to be replaced by

$$(8.11)\qquad M_{\pm}\eta(H_0) = U\eta(H_0) - \int \gamma_{\pm}(E - H_0) V U \eta(H_0) \delta(E - H_0) \mathrm{d}E .$$

Equation (8.10) is equivalent to

$$(8.12)\qquad \sum_{a'} \langle x | H_0, A_0; E, a'\rangle\langle a' | \mu_{\pm}(E) | a\rangle =$$

$$= \langle x | H, A; E, a\rangle \mp \frac{i}{2}(E)^{-\frac{1}{2}} \int_{-\infty}^{+\infty} \exp[\mp i\sqrt{E}|x - x'|] V(x') \langle x' | H, A; E, a\rangle \mathrm{d}x' .$$

In obtaining (8.12) from (8.10) we used Eqs. (3.10) and (2.7a).

On using (8.6) and (8.7) we have for $x < -\alpha$

$$(8.13)\qquad \sum_{a'} \langle x | H_0, A_0; E, a'\rangle\langle a' | \mu_{\pm}(E) | a\rangle =$$

$$= \langle x | H_0, A_0; E, a\rangle \mp \frac{Ai}{2}(E)^{-\frac{1}{2}} \exp[\pm i\sqrt{E}(x + \alpha)]\langle -\alpha | H_0, A_0; E, a\rangle ,$$

or

$$(8.14)\qquad \sum_{a'} \langle x | H_0, A_0; E, a'\rangle\langle a' | \mu_{\pm}(E) | a\rangle =$$

$$= \langle x | H_0, A_0; E, a\rangle \mp \frac{Ai}{2} E^{-\frac{1}{2}} \exp[-i\alpha(a \mp 1)\sqrt{E}]\langle x | H_0, A_0; E, \pm 1\rangle .$$

Since $\langle a | \mu_{\pm}(E) | a'\rangle$ must be independent of x, we have

$$(8.15)\qquad \langle a | \mu_-(E) | a'\rangle = \begin{pmatrix} 1 & 0 \\ \frac{AiE^{-}}{2} \exp[-2i\alpha\sqrt{E}] & 1 + \frac{AiE^{-\frac{1}{2}}}{2} \end{pmatrix},$$

$$(8.16)\qquad \langle a | \mu_+(E) | a'\rangle = \begin{pmatrix} 1 - \frac{AiE^{-\frac{1}{2}}}{2} & -\frac{AiE^{-\frac{1}{2}}}{2}\exp[2i\alpha\sqrt{E}] \\ 0 & 1 \end{pmatrix}.$$

Since the matrix $\langle a | \mu_-^{-1}(E) | a'\rangle$ is just the reciprocal of $\langle a | \mu_-(E) | a'\rangle$ we see that

$$(8.17)\qquad \langle a|\mu_-^{-1}(E)|a'\rangle = \begin{pmatrix} 1 & 0 \\ \dfrac{-Ai\exp[-2i\alpha\sqrt{E}]}{2\sqrt{E}+iA} & \dfrac{2\sqrt{E}}{2\sqrt{E}+iA} \end{pmatrix}.$$

Finally on using (8.9)

$$(8.18)\qquad \langle a|S(E)|a'\rangle = \begin{pmatrix} \dfrac{2\sqrt{E}}{2\sqrt{E}+iA} & \dfrac{-Ai\exp[2i\alpha\sqrt{E}]}{2\sqrt{E}+iA} \\ \dfrac{-iA\exp[-2i\alpha\sqrt{E}]}{2\sqrt{E}+iA} & \dfrac{2\sqrt{E}}{2\sqrt{E}+iA} \end{pmatrix}.$$

The reflection coefficient

$$\langle -1|S(E)|+1\rangle = \frac{-Ai\exp[-2i\alpha\sqrt{E}]}{2\sqrt{E}+iA}$$

is just the one we started with.

8·2. *The attractive δ-function potential.* – Let us now consider the reflection coefficient

$$(8.19)\qquad \langle -1|S(E)|+1\rangle = \frac{iB\exp[-2i\alpha\sqrt{E}]}{2\sqrt{E}-iB},\qquad B>0,$$

or

$$(8.20)\qquad b(k) = \frac{i(B/2)\exp[-2i\alpha k]}{k-(iB/2)}.$$

In this case there is a bound state, since $b(k)$ has a singularity on the positive imaginary axis. In fact we have

$$(8.21)\qquad \begin{cases} \tau_1 = \dfrac{B}{2} \\ r_1 = \dfrac{iB}{2}, \end{cases}$$

and the bound state E_1 is given by

$$(8.22)\qquad E_1 = -\tau_1^2 = -\frac{B^2}{4}.$$

We have

$$(8.23)\quad \langle x|\Omega|x'\rangle = \frac{1}{2\pi}\int_{-\infty}^{+\infty} b(k)\exp[-ik(x+x')]\,dk - ir_1\exp[\tau_1(x+x'+2\alpha)] =$$

$$= \frac{iB}{4\pi}\int_{-\infty}^{+\infty}\frac{\exp[-ik(x+x'+2\alpha)]}{k-(iB/2)}\,dk + \frac{B}{2}\exp\left[\frac{B}{2}(x+x'+2\alpha)\right] =$$

$$= -\frac{B}{2}\eta(-x-x'-2\alpha)\exp\left[\frac{B}{2}(x+x'+2\alpha)\right] + \frac{B}{2}\exp\left[\frac{B}{2}(x+x'+2\alpha)\right] =$$

$$= \frac{B}{2}\eta(x+x'+2\alpha)\exp\left[\frac{B}{2}(x+x'+2\alpha)\right].$$

The solution of the Gelfand-Levitan equation is

$$(8.24)\qquad \langle x|K|x'\rangle = -\frac{B}{2}\eta(x+x'+2\alpha)\,,$$

and the potential $V(x)$ is

$$(8.25)\qquad V(x) = 2\frac{d}{dx}\langle x|K|x\rangle = -B\delta(x+\alpha)\,.$$

The eigenfunctions of the continuous spectrum are

$$(8.26)\quad \langle x|H, A; E, a\rangle = \langle x|H_0, A_0; E, a\rangle + \frac{iBa}{2}\eta(x+\alpha)(E)^{-\frac{1}{2}}\cdot$$
$$\cdot[\langle x|H_0, A_0; E, a\rangle - \langle x+2\alpha|H_0, A_0; E, -a\rangle]\,,$$

while the proper eigenfunction corresponding to the eigenvalue $E_1 = -B^2/4$ is on using (7.20) and (5.5)

$$(8.27)\quad \langle x|H, E_1\rangle = A(E_1)\langle x|H, A; E_1, -1\rangle^+ =$$
$$= \frac{A(E_1)}{2\sqrt{\pi}}\exp\left[-\frac{i\pi}{4}-\frac{B}{2}\alpha\right]\exp\left[-\left|\frac{B}{2}(x+\alpha)\right|\right],$$

where, in accordance with (7.19) and (8.21)

$$(8.28)\qquad [C(E_1)]^{-1}|A(E_1)|^2 = \frac{4\pi B^2}{4}\exp[B\alpha]\,.$$

We can choose either $A(E_1)$, in which case the normalization $C(E_1)$ is determined or $C(E_1)$ in which case $A(E_1)$ is determined to within a phase factor.

It can also be shown that

$$\langle a|\mu_-(E)|a'\rangle = \begin{pmatrix} 1 & 0 \\ \dfrac{-BiE^{-\frac{1}{2}}\exp[-2i\alpha\sqrt{E}]}{2} & 1-\dfrac{BiE^{-\frac{1}{2}}}{2} \end{pmatrix}, \tag{8.29}$$

$$\langle a|\mu_+(E)|a'\rangle = \begin{pmatrix} 1+\dfrac{BiE^{-\frac{1}{2}}}{2} & \dfrac{B}{2}iE^{-\frac{1}{2}}\exp[2i\alpha\sqrt{E}] \\ 0 & 1 \end{pmatrix}, \tag{8.30}$$

$$\langle a|\mu_-^{-1}(E)|a'\rangle = \begin{pmatrix} 1 & 0 \\ \dfrac{Bi\exp[-2i\alpha\sqrt{E}]}{2\sqrt{E}-iB} & \dfrac{2\sqrt{E}}{2\sqrt{E}-iB} \end{pmatrix}, \tag{8.31}$$

$$\langle a|S(E)|a'\rangle = \begin{pmatrix} \dfrac{2\sqrt{E}}{2\sqrt{E}-iB} & \dfrac{Bi\exp[2i\alpha\sqrt{E}]}{2\sqrt{E}-iB} \\ \dfrac{Bi\exp[-2i\alpha\sqrt{E}]}{2\sqrt{E}-iB} & \dfrac{2\sqrt{E}}{2\sqrt{E}-iB} \end{pmatrix}. \tag{8.32}$$

8·3. ***A more complicated example.*** – We shall now summarize the results of a more complicated example discussed in more detail in (7).

Let us find the potential whose reflection coefficient is given by

$$\langle -1|S(E)|+1\rangle = -\exp[-2i\alpha\sqrt{E}]\frac{1}{E+1}. \tag{8.33}$$

or

$$b(k) = -\exp[-i2\alpha k]\frac{1}{(k+i)(k-i)}. \tag{8.34}$$

There is only one pole in the upper k-plane, namely $k_1 = i$, and hence we can calculate the quantities associated with the bound states:

$$\begin{cases} \tau_1 = 1\,, \\ r_1 = \dfrac{i}{2}\,, \\ E_1 = -1\,, \\ C^{-1}(E_1)|A(E_1)|^2 = 2\pi\exp[2\alpha]\,. \end{cases} \tag{8.35}$$

Also

$$\langle x|\Omega|x'\rangle = \eta(x+x'+2\alpha)\sinh(x+x'+2\alpha)\,. \tag{8.36}$$

Then it can be shown that the solution of the Gelfand-Levitan equation is

$$(8.37)\qquad \langle x|K|x'\rangle = -\frac{1}{\sqrt{2}}\,\eta(x+x'+2\alpha)\,\frac{\sinh\sqrt{2}(x+\alpha)+\sinh\sqrt{2}(x'+\alpha)}{\cosh\sqrt{2}(x+\alpha)}\,.$$

The potential $V(x)$ is

$$(8.38)\qquad V(x) = 2\,\frac{\mathrm{d}}{\mathrm{d}x}\langle x|K|x\rangle = -4\eta(x+\alpha)\,\mathrm{sech}^2\sqrt{2}\,(x+\alpha)\,.$$

The eigenfunctions of the continuous spectrum are given by

$$(8.39)\qquad \langle x|H, A;\, E, a\rangle = \frac{(E)^{-\frac{1}{4}}}{2\sqrt{\pi}}\exp[ia\sqrt{E}x] - \frac{\eta(x+\alpha)}{\sqrt{2\pi}}\,\frac{(E)^{-\frac{1}{4}}}{(E+2)}\exp[-ia\sqrt{E}\alpha]\cdot$$
$$\cdot\Big\{\mathrm{tgh}\,\sqrt{2}(x+\alpha)[a(E+2)\sin a\sqrt{E}(x+\alpha) - iaE\cos a\sqrt{E}(x+\alpha)] +$$
$$+\sqrt{2}i\sqrt{E}\sin a\sqrt{E}(x+\alpha)\Big\}\,.$$

The single eigenstate corresponding to a point eigenvalue is given by

$$(8.40)\qquad \langle x|H, E_1\rangle = A(E_1)\exp[-i\pi/4]\cdot$$
$$\cdot\left\{\frac{\exp[x]}{2\sqrt{\pi}} + \frac{\eta(x+\alpha)}{\sqrt{2\pi}}\exp[-\alpha]\,[\mathrm{tgh}\,\sqrt{2}(x+\alpha)\exp[-(x+\alpha)] - \sqrt{2}\sinh(x+\alpha)]\right\}.$$

It is readily verified that $\lim_{x\to\pm\infty}\langle x|H, E_1\rangle = 0$.

We also derive using the methods of Part A

$$(8.41)\qquad \langle a|\mu_-(E)|a'\rangle = \begin{pmatrix} 1 & 0 \\ \dfrac{\exp[-2i\sqrt{E}\alpha]}{\sqrt{E}(\sqrt{E}+i\sqrt{2})} & \dfrac{E+1}{\sqrt{E}(\sqrt{E}+i\sqrt{2})} \end{pmatrix},$$

$$(8.42)\qquad \langle a|\mu_-^{-1}(E)|a'\rangle = \begin{pmatrix} 1 & 0 \\ \dfrac{-\exp[-2i\sqrt{E}\alpha]}{(E+1)} & \dfrac{\sqrt{E}(\sqrt{E}+i\sqrt{2})}{(E+1)} \end{pmatrix},$$

$$(8.43)\qquad \langle a|\mu_+(E)|a'\rangle = \begin{pmatrix} \dfrac{(E+1)(\sqrt{E}+i\sqrt{2})}{\sqrt{E}(E+2)} & \dfrac{\exp[2i\alpha\sqrt{E}](\sqrt{E}+i\sqrt{2})}{\sqrt{E}(E+2)} \\ 0 & 1 \end{pmatrix},$$

$$\langle a|S(E)|a'\rangle = \begin{pmatrix} \dfrac{\sqrt{E}(\sqrt{E}+i\sqrt{2})}{(E+1)} & \dfrac{\exp[2i\alpha\sqrt{E}](\sqrt{E}+i\sqrt{2})^2}{(E+2)(E+1)} \\ \dfrac{\exp[-2i\alpha\sqrt{E}]}{(E+1)} & \dfrac{\sqrt{E}(\sqrt{E}+i\sqrt{2})}{(E+1)} \end{pmatrix}. \tag{8.44}$$

In (7) it is shown how one may solve the Gelfand-Levitan equation when one is given any reflection coefficient which is a rational function of $\sqrt{E}$. In such a case, the Gelfand-Levitan equation reduces to a system of algebraic equations for which a procedure for solving them can be found. The method is analogous to that used by V. BARGMANN for the radial equation.

RIASSUNTO (*)

Il procedimento per il calcolo dei potenziali di scattering dalle funzioni spettrali di misura dell'Hamiltoniana perturbata discusso in precedenti lavori, è stato adattato al problema di ottenere il potenziale di scattering dall'operatore di scattering per l'equazione di Schrödinger unidimensionale ($-\infty < x < \infty$). Si dànno esempi dell'uso del procedimento.

(*) *Traduzione a cura della Redazione.*

SUPPLEMENTO AL VOLUME V, SERIE X
DEL NUOVO CIMENTO

N. 1, 1957
1° Semestre

The Determination of the Scattering Potential from the Spectral Measure Function.

PART IV. – « Pathological » Scattering Problems in One Dimension (*).

I. KAY and H. E. MOSES

New York University, Institute of Mathematical Sciences - New York, N.Y.

(ricevuto il 30 Settembre 1956)

CONTENTS. — 1. Introduction. – 2. The general Gelfand-Levitan equation for one-dimensional scattering. – 3. Asymptotic form for pathological wave functions. – 4. Case where there is no bound state corresponding to a pole of the reflection coefficient. – 5. Case where bound states exist but the reflection coefficient does not have a corresponding pole: reflectionless potentials. – 6. Pathological character of the reflection coefficient $\langle +1|S(E)|-1\rangle$. – 7. Generalizations.

1. – Introduction.

In a previous paper (1) we have given the Gelfand-Levitan equation which enables one to obtain the potential from the reflection coefficient and the eigenvalues and normalization of the bound states. The equation we obtained was quite general. However, we emphasized the conditions on the reflection coefficient and the point eigenvalues which would lead to potentials $V(x)$ which vanished identically for $x < -\alpha$, where α is some positive number. This requirement was used because such potentials might be regarded as

(*) The research reported in this article has been made possible through support and sponsorship extended by the Geophysics Research Directorate, Air Force Cambridge Research Center, under Contract No. AF-19(604)1705.

(1) I. KAY and H. E. MOSES: *Nuovo Cimento*, **3**, 276 (1956). Also *Institute of Mathematical Sciences, Division of Electromagnetic Research, Research Report* N. CX-20 (1955).

having the characteristics of more general potentials which vanished very strongly as $x \to -\infty$. Furthermore, such potentials would guarantee that we would obtain from our theory the same reflection coefficient that we started with. It was found that in order to obtain potentials $V(x)$ which vanished identically for $x < -\alpha$ it was sufficient that 1) the reflection coefficient have the form $b(k) = g(k)\exp[-2ik\alpha]$, where $g(k)$ is an analytic function of k such that $g(k) \to 0(k^{-1})$ as $|k| \to \infty$, and $g(-k) = g^*(k)$ for real k; and that 2) the poles of $b(k)$ in the upper k-plane lie on the positive imaginary axis in such a way that the point eigenvalues are the squares of the poles and the normalization is obtained from the residues of $b(k)$ at the poles. Reflection coefficients which satisfied both conditions 1) and 2) were shown to determine $V(x)$ uniquely.

We shall call scattering problems in which the reflection coefficients satisfy 2) « normal scattering problems ». Those scattering problems for which 2) is not satisfied will be called « pathological scattering problems ».

The motivation for distinguishing between normal and pathological scattering arises from the consideration of the integral equations which the « outgoing eigenfunctions » of the continuous spectrum of H and the proper eigenfunctions corresponding to point eigenvalues of H must satisfy, namely:

$$\Psi_k^{\text{out}}(x) = \exp[ikx] - \frac{i}{2|k|}\int_{-\infty}^{+\infty}\exp[i|k||x-x'|]V(x')\Psi_k^{\text{out}}(x')\,dx', \tag{1.1}$$

$$\Psi_{E_i}(x) = -\frac{1}{2(-E_i)^{\frac{1}{2}}}\int_{-\infty}^{+\infty}\exp[-\sqrt{-E_i}|x-x'|]V(x')\Psi_{E_i}(x')\,dx', \tag{1.2}$$

where E_i is a point eigenvalue of H ($E_i < 0$). Now let us consider equation (1.1) for $k > 0$, so that $|k| = k$. If we continued the solution as a function of k in the complex k-plane to $k = i\sqrt{-E_i}$, then we should have

$$\Psi_k^{\text{out}}(x) = \exp[-\sqrt{E_i}x] - \frac{1}{2(-E_i)^{\frac{1}{2}}}\cdot \int_{-\infty}^{+\infty}\exp[-\sqrt{-E_i}|x-x'|]V(x')\Psi_k^{\text{out}}(x')\,dx'. \tag{1.3}$$

But (1.2) is a homogeneous equation for the function Ψ_{E_i} of the form $A\Psi_{E_i} = 0$ whereas (1.3) is an inhomogeneous equation of the form $A\Psi_k = \exp[-\sqrt{E_i}x]$, where the operator A is the sum of the identity and the integral operator with the kernel $(-1/2(-E_i)^{\frac{1}{2}})\exp[-\sqrt{-E_i}|x-x'|]V(x')$.

Now in usual algebraic cases, if the homogeneous equation $Ax = 0$ has a non-trivial solution, then in general the inhomogeneous equation $Ax = y$ cannot have one. Thus since the homogeneous equation (1.2) for Ψ_{E_i} is assumed to have a solution, we may expect that there is no solution of the inhomogeneous equation (1.3). Thus the solution of (1.1) for $k > 0$ should « blow up » or have a singularity in the k-plane when continued to values of $k = i\sqrt{-E_i}$, where E_i is any point eigenvalue of H. In particular, let us consider the asymptotic form of $\Psi_k^{\text{out}}(x)$ for $k > 0$. Then

$$\lim_{x \to -\infty} \Psi_k^{\text{out}}(x) = \exp[ikx] + b(k)\exp[-ikx]\,, \qquad k > 0, \tag{1.4}$$

where $b(k)$ is, by definition, the reflection coefficient. If the analytic continuation of $\Psi_k^{\text{out}}(x)$ has a singularity at $k = i\sqrt{-E_i}$, we might expect the asymptotic form (1.4) to have such a singularity also. This singularity clearly would have to appear in the reflection coefficient $b(k)$.

Now, in what we have called the normal scattering problem, this condition is satisfied. However, in the case of the pathological scattering problem this condition is not satisfied and the usual intuitive arguments given above are not valid.

We shall consider in the present paper the special class of reflection coefficients which lead to pathological scattering potentials, namely those reflection coefficients which satisfy condition 1), p. 231, but not condition 2). Instead of imposing condition 2) we shall assume that the reflection coefficient $b(k)$ has poles on the positive imaginary axis which do not correspond to point eigenvalues, and conversely that there are point eigenvalues which do not correspond to such poles (*).

It will be shown that the potentials $V(x)$ obtained from such bound states and reflection coefficients will behave like the sum of decaying exponentials as $x \to -\infty$. Hence even in this case the original reflection coefficient will be obtained when one considers the asymptotic forms of the outgoing waves.

We shall consider two special examples.

1) The reflection coefficient has a pole on the positive imaginary axis to which there corresponds no bound state:

In this example, we shall take the reflection coefficient arising from the use of an attractive δ-function potential. Such a reflection coefficient will have a pole corresponding to the bound state which the δ-function potential sup-

(*) These situations are analogous but not identical to the case of « redundant zeros » which occur in the discussion of the radial equation (e.g. (2)).

(2) S. T. Ma: *Phys. Rev.*, **69**, 668 (1946).

ports. However, we shall obtain another potential which has the same reflection coefficient, but not the bound state.

2) There exist bound states, but the reflection coefficient has no poles corresponding to such bound states:

In this case we shall take $b(k) \equiv 0$. The potentials which we shall obtain in this manner will be perfectly reflectionless, i.e., every particle will be transmitted whatever its initial energy. These curious potentials are discussed elsewhere ([3]) in more detail than will be given here. In ([3]) all such « transparent » potentials are given explicitly and their properties discussed.

2. – The General Gelfand-Levitan Equation for One-Dimensional Scattering.

The most general Gelfand-Levitan equation for the one-dimensional scattering problem is

$$\langle x|K|x'\rangle = -\langle x|\Omega|x'\rangle - \int_{-\infty}^{x}\langle x|K|x''\rangle\,\mathrm{d}x''\langle x''|\Omega|x'\rangle\,, \qquad (x > x'), \tag{2.1}$$

where

$$\langle x|\Omega|x'\rangle = \frac{1}{2\pi}\int_{-\infty}^{+\infty} b(k)\exp[-ik(x+x')]\,\mathrm{d}k + \\ + \frac{1}{4\pi}\sum_i(-E_i)^{-\frac{1}{2}}\exp[\sqrt{-E_i}(x+x')][C(E_i)]^{-1}|A(E_i)|^2\,, \tag{2.2}$$

(see ([1]), equations (6.1) and (6.14). For simplicity we have taken and shall take $\varepsilon = 1$. This procedure is equivalent to solving the Gelfand-Levitan equation for $\varepsilon\langle x|K|x'\rangle$ rather than for $\langle x|K|x'\rangle$).

In equation (2.1), $b(k) = \langle -1|S(k^2)|+1\rangle$ is the reflection coefficient. Although the reflection coefficient is defined only for $k > 0$, one can show by analytic continuation that

$$b(-k) = b^*(k)\,, \qquad \text{for } k \text{ real.} \tag{2.3}$$

Hence $b(k)$ is defined along the whole real axis.

The quantities E_i are the point eigenvalues of the total Hamiltonian. The constants $C(E_i)$ are positive and are the normalization constants of the eigen-

([3]) I. KAY and H. E. MOSES: *Institute of Mathematical Sciences, Division of Electromagnetic Research, Research Report* No. EM-91 (1956); also, *Jour., of Applied Phys.* (in press).

states of H corresponding to the eigenvalues E_i in a sense to be described shortly. The constants $A(E_i)$ are essentially arbitrary and can be chosen to have any convenient value.

The sense in which $C(E_i)$ are the normalization constants is that the bound states $\langle x|H; E_i\rangle$ of H, obtained from

$$\langle x|H; E_i\rangle = \langle x|U|H_0; E_i\rangle = \langle x|H_0; E_i\rangle + \int_{-\infty}^{x} \langle x|K|x'\rangle \, dx' \langle x'|H_0; E_i\rangle , \tag{2.4}$$

where

$$\langle x|H_0; E_i\rangle = A(E_i) \frac{\exp[-i(\pi/4)]}{2\sqrt{\pi}} (-E_i)^{-\frac{1}{4}} \exp[\sqrt{-E_i}\, x] , \tag{2.4a}$$

(cfr. ([1]), equation (5.5)) will have the normalization $C(E_i)$, i.e.

$$\langle H; E_i|H; E_i\rangle = C(E_i) . \tag{2.5}$$

3. – Asymptotic Form for Pathological wave Functions.

As in ([1]) where normal scattering was considered we shall assume that the reflection coefficient $b(k)$ has the form

$$b(k) = g(k) \exp[-2ik\alpha] , \tag{3.1}$$

where

$$\lim_{|k|\to\infty} g(k) \to 0(k^{-1}) . \tag{3.1a}$$

Hence, from equation (7.17) of ([1]) we have

$$\frac{1}{2\pi}\int_{-\infty}^{+\infty} b(k) \exp[-ik(x+x')] \, dk = \sum_j M_j \exp[\tau_j(x+x')] , \quad \text{for } x+x' < -2\alpha. \tag{3.2}$$

In (3.2) the constants τ_j are real and positive and

$$\tau_j = -ik_j , \tag{3.2a}$$

the quantities k_j being the poles of $g(k)$, or equivalently $b(k)$, on the positive imaginary axis. (As was shown in ([1]), the only possible positions of the poles of $g(k)$ in the upper half-plane are on the positive imaginary axis). Furthermore, M_j are real constants given by

$$M_j = ir_j \exp[2\tau_j\alpha]\,, \tag{3.2b}$$

where r_j are the residues of $g(k)$ at the poles k_j. Hence, from (2.2), we see that for $x + x' < -2\alpha$ we have

$$\langle x|\Omega|x'\rangle = \sum_j M_j \exp[\tau_j(x + x')] + \sum_j N_j \exp[\sqrt{-E_j}(x + x')]\,, \qquad (x + x' < -2\alpha), \tag{3.3}$$

where

$$N_j = \frac{1}{4\pi}(-E_j)^{-\frac{1}{2}}[C(E_j)]^{-1}|A(E_j)|^2\,. \tag{3.3a}$$

In ([1]) the eigenvalues E_j and coefficients N_j were chosen so that the first sum of exponentials cancelled the second sum, term by term, and hence $\langle x|\Omega|x'\rangle \equiv 0$ for $x + x' < -2\alpha$. In the pathological case, however, we shall assume that such cancellations, if they occur at all, are not complete. Hence, generally, we shall write

$$\langle x|\Omega|x'\rangle = \sum_{j=1}^{N} P_j \exp[\nu_j(x + x')]\,, \qquad (x + x' < -2\alpha), \tag{3.4}$$

where the constants ν_j are real and positive, and P_j are real and of any sign.

The equation for $\langle x|K|x'\rangle$ becomes the following, when $x' < x < -\alpha$:

$$\langle x|K|x'\rangle = -\sum_{j=1}^{N} P_j \exp[\nu_j(x + x')] - \sum_{j=1}^{N} P_j \cdot \int_{-\infty}^{x} \langle x|K|x''\rangle \exp[\nu_j(x'' + x')]\,\mathrm{d}x''\,. \tag{3.5}$$

We shall indicate how (3.5) is to be solved. On obtaining the solution one will then be able to use

$$V(x) = 2\frac{\mathrm{d}}{\mathrm{d}x}\langle x|K|x\rangle\,, \tag{3.6}$$

to find $V(x)$ explicitly when $x < -\alpha$.

Let us make the Ansatz

$$\langle x|K|x'\rangle = \sum_{j=1}^{N} P_j^{\frac{1}{2}} g_j(x) \exp[\nu_j x']\,. \tag{3.7}$$

One obtains the following equations for $g_j(x)$ by substituting (3.7) into (3.5) and identifying the coefficients of the exponents $\exp[\nu_j x']$:

$$\sum_j B_{ij} g_j = -P_i^{\frac{1}{2}} \exp[\nu_i x], \tag{3.8}$$

where

$$B_{ij} = (P_i P_j)^{\frac{1}{2}} \frac{\exp[(\nu_i + \nu_j)x]}{\nu_i + \nu_j} + \delta_{ij}. \tag{3.9}$$

One can then show that the potential $V(x)$ is given by

$$V(x) = -2 \frac{d^2}{dx^2} \log \det B, \qquad x < -\alpha, \tag{3.10}$$

where B is the matrix whose elements are B_{ij}. By examining (3.9) and (3.10) carefully (see ([3])), we see that $V(x)$ decays exponentially as $x \to -\infty$. Therefore the reflection coefficient $b(k)$ which we used originally will be reproduced by the potential.

4. – Case where there is no Bound State Corresponding to a Pole of the Reflection Coefficient.

As an especially simple case, let us take as a reflection coefficient

$$b(k) = \frac{iB}{2} \frac{1}{k - (iB/2)}, \qquad B > 0. \tag{4.1}$$

If we require that there be a bound state whose normalization and eigenvalue correspond to the pole and residue of $b(k)$ in the upper half plane as in ([1]), we should obtain as the scattering potential $V(x) = -B\delta(x)$. This potential was, in fact, derived in ([1]).

In the present case, however, we shall require that the potential have no bound states. Then, on using

$$\langle x|\Omega|x'\rangle = -\frac{B}{2}\eta(-x-x')\exp[(B/2)(x+x')], \tag{4.2}$$

we have as the Gelfand-Levitan equation

$$\langle x|K|x'\rangle = \frac{B}{2}\eta(-x-x')\exp[(B/2)(x+x')] + \frac{B}{2}\int_{-\infty}^{\infty}\langle x|K|x''\rangle\eta(-x''-x')\exp[(B/2)(x''+x')]\,dx''. \tag{4.3}$$

The solution of this equation is

$$\langle x|K|x'\rangle = \frac{B\eta(-x)\exp[(B/2)(x+x')]}{(2-\exp[Bx])} + \tag{4.4}$$
$$+\frac{B}{2}\eta(x)\eta(x+x') + B\eta(x)\eta(-x-x')\exp[(B/2)(x+x')],$$

where, as usual, $\eta(x)$ is defined by

$$\begin{cases} \eta(x) = 1, & x > 0, \\ = 0. & x < 0. \end{cases} \tag{4.4a}$$

Then, since

$$\langle x|K|x\rangle = \frac{B\eta(-x)\exp[Bx]}{(2-\exp[Bx])} + \frac{B}{2}\eta(x), \tag{4.5}$$

we have for the scattering potential

$$V(x) = 2\frac{\mathrm{d}}{\mathrm{d}x}\langle x|K|x\rangle = -B\delta(x) + \frac{4B^2\eta(-x)}{(2-\exp[Bx])^2}\exp[Bx]. \tag{4.6}$$

This scattering potential differs from the « normal » scattering potential $V(x) = -B\delta(x)$, which has a bound state, by the exponential tail when $x < 0$. One might think that such a change in potential is so slight that it could not greatly change the ability of the potential to support a bound state. However, we note that the potential $V(x)$ given by (4.6) has as its integral

$$\int_{-\infty}^{+\infty} V(x)\,\mathrm{d}x = B > 0, \tag{4.7}$$

whereas the integral $\int_{-\infty}^{+\infty} V(x)\,\mathrm{d}x$ for the normal potential has the value $-B$. The fact that one integral is positive whereas the other is negative is essential and indicates why one potential can have a bound state and the other cannot.

For the sake of completeness we shall give the elements of the matrices $\langle a|\mu_-(E)|a'\rangle$, $\langle a|\mu_-^{-1}(E)|a'\rangle$, $\langle a|\mu_+(E)|a'\rangle$ and the elements of the scattering operator $\langle a|S(E)|a'\rangle$. These quantities are obtained in a manner similar to that used in the examples of (1). We have

$$\langle a|\mu_-(E)|a'\rangle = \begin{pmatrix} 1 & 0 \\ \dfrac{-iB}{2\sqrt{E}}\,\dfrac{\sqrt{E}+iB/2}{\sqrt{E}-iB/2} & \dfrac{\sqrt{E}+iB/2}{\sqrt{E}} \end{pmatrix}, \tag{4.8}$$

$$\text{(4.9)}\qquad \langle a|\mu_-^{-1}(E)|a'\rangle = \begin{pmatrix} 1 & 0 \\ \dfrac{iB}{2}\dfrac{1}{\sqrt{E}-iB/2} & \dfrac{\sqrt{E}}{\sqrt{E}+iB/2} \end{pmatrix},$$

$$\text{(4.10)}\qquad \langle a|\mu_+(E)|a'\rangle = \begin{pmatrix} \dfrac{\sqrt{E}-iB/2}{\sqrt{E}} & \dfrac{iB}{2\sqrt{E}}\dfrac{\sqrt{E}-iB/2}{\sqrt{E}+iB/2} \\ 0 & 1 \end{pmatrix},$$

$$\text{(4.11)}\qquad \langle a|S(E)|a'\rangle = \begin{pmatrix} \dfrac{\sqrt{E}}{\sqrt{E}+iB/2} & \dfrac{iB}{2}\dfrac{\sqrt{E}-iB/2}{(\sqrt{E}+iB/2)^2} \\ \dfrac{iB}{2}\dfrac{1}{\sqrt{E}-iB/2} & \dfrac{\sqrt{E}}{\sqrt{E}+iB/2} \end{pmatrix}.$$

It is, perhaps, interesting to note that none of the elements of the scattering operator, except the reflection coefficient, $\langle -1|S(E)|+1\rangle$ (which was given initially), shows the presence of a pole in the upper k-plane.

5. – Case where Bound States Exist but the Reflection Coefficient Does not Have a Corresponding Pole: Reflectionless Potentials.

The case which we shall consider is one in which the reflection coefficient is *identically* zero for all energies, i.e.

$$\text{(5.1)}\qquad b(k) \equiv \langle -1|S(E)|+1\rangle \equiv 0\,.$$

The potentials which give rise to such reflection coefficients will be called reflectionless potentials. In the presence of such potentials, particles which start out at $x = -\infty$ will always go to $+\infty$, whatever the initial energy.

If we did not want the potential to have bound states, it is clear from (2.1) and (2.2) that $\langle x|K|x'\rangle \equiv 0$, and hence $V(x) = 2(\mathrm{d}/\mathrm{d}x)\langle x|K|x\rangle = 0$. Therefore, in order to have a non-trivial answer we shall assume that the potential supports a finite number of bound states. Under the requirement that there are N bound states, the Gelfand-Levitan equation becomes (see equations (2.1) and (2.2))

$$\text{(5.2)}\qquad \langle x|K|x'\rangle = -\sum_{j=1}^{N} N_j \exp[\sqrt{-E_j}(x+x')] - \\ -\sum_{j=1}^{N} N_j \int_{-\infty}^{x} \langle x|K|x''\rangle \exp[\sqrt{-E_j}(x''+x')]\,\mathrm{d}x''\,,$$

where

$$N_j = \frac{1}{4\pi}(-E_j)^{\frac{1}{2}}[C(E_j)]^{-1}[A(E_j)]^{-1}, \tag{5.2a}$$

and hence the N_j are positive. The quantities E_j are the (negative) point eigenvalues which we require the potential to support. Since the reflection coefficient is identically zero, the reflection coefficient, of course, has no poles corresponding to these point eigenvalues.

Equation (5.2) can be solved by the method used to solve equation (3.7). In fact, the equations (5.2) and (3.5) are essentially the same, the latter being valid only for $x < -\alpha$, whereas the former is valid for all x.

The potentials which are obtained are discussed in detail in ([3]). They are shown to be given by

$$V(x) = -2\frac{d^2}{dx^2}\log\det B\,, \tag{5.3}$$

where B is the matrix whose elements are

$$B_{ij} = (N_i N_j)^{\frac{1}{2}}\frac{\exp[(\sqrt{-E_i}+\sqrt{-E_j})x]}{\sqrt{-E_i}+\sqrt{-E_j}} + \delta_{ij}\,. \tag{5.4}$$

In ([3]), it is shown that because the quantities N_j are positive, the potentials $V(x)$ are everywhere finite. Furthermore, it is shown that they decay exponentially as $|x| \to \infty$ and that *all* reflectionless potentials are given by (5.3) and (5.4). Hence we can choose reflectionless potentials by picking any finite set of *positive* numbers N_i and corresponding *negative* numbers E_i.

To be more explicit, let us assume the potential supports only one bound state. Hence only N_1 and E_1 are given. Then the solution of (5.2) is

$$\langle x|K|x'\rangle = -\frac{N_1\exp[\sqrt{-E_1}(x+x')]}{\left[1+\dfrac{N_1}{2\sqrt{-E_1}}\exp[2\sqrt{-E_1}x]\right]}\,. \tag{5.5}$$

Furthermore, $V(x)$ is given by

$$V(x) = \frac{-4\sqrt{-E_1}\,N_1\exp[2\sqrt{-E_1}x]}{\left[1+\dfrac{N_1}{2\sqrt{-E_1}}\exp[2\sqrt{-E_1}x]\right]^2} = 2E_1\,\mathrm{sech}^2\sqrt{-E_1}(x-x_0)\,, \tag{5.6}$$

where

$$x_0 = \frac{\log(2\sqrt{-E_1}/N_1)}{2\sqrt{-E_1}}\,. \tag{5.6a}$$

In order to find the elements of the scattering operator we shall proceed in a simpler way than heretofore. The simplicity of the present procedure arises from the fact that $b(k) \equiv 0$, or equivalently

$$\langle -1|S(E)|+1\rangle = 0\,. \tag{5.7}$$

Hence on using (5.7) above and equations (8.12) and (3.8) of ([1]), we have

$$\langle x|H, A; E, +1\rangle_- = \langle x|H, A; E, +1\rangle\,, \tag{5.8}$$

where

$$\begin{aligned}\langle x|H, A; E, +1\rangle &= \\ &= \langle x|H_0, A_0; E, +1\rangle + \int_{-\infty}^{x}\langle x|K|x'\rangle\langle x'|H_0, A_0; E, +1\rangle = \\ &= \frac{(E)^{-\frac14}}{2\sqrt{\pi}}\left[1 - \frac{N_1 \exp[2\sqrt{-E_1}x]}{\left(1+\dfrac{N_1\exp[2\sqrt{-E_1}x]}{2\sqrt{-E_1}}\right)(\sqrt{-E_1}+i\sqrt{E})}\right]\exp[i\sqrt{E}x]\,.\end{aligned} \tag{5.9}$$

On taking the limit

$$\lim_{x\to+\infty}\langle x|H, A; E, +1\rangle_- = \lim_{x\to+\infty}\langle x|H, A; E, +1\rangle\,, \tag{5.10}$$

we have for the transmission coefficient, on using (5.9) above and eq. (2.13) of ([1]),

$$\langle +1|S(E)|+1\rangle = \frac{\sqrt{E}+i\sqrt{-E_1}}{\sqrt{E}-i\sqrt{-E_1}}\,. \tag{5.11}$$

(Incidentally, if we write $k=\sqrt{E}$, for $k>0$ and $E>0$, then when $\langle +1|S(E)|+1\rangle$ is continued analytically as a function of k, it has poles in the upper half-plane corresponding to the point eigenvalue).

On using the fact that generally

$$\langle a|S(E)|a'\rangle = \langle -a'|S(E)|-a\rangle\,, \tag{5.12}$$

we obtain

$$\langle -1|S(E)|-1\rangle = \langle +1|S(E)|+1\rangle = \frac{\sqrt{E}+i\sqrt{-E_1}}{\sqrt{E}-i\sqrt{-E_1}}\,. \tag{5.13}$$

If we then use the fact that the scattering operator is unitary, we obtain the remaining element, which is the reflection coefficient at $+\infty$:

$$\langle +1|S(E)|-1\rangle = 0\,. \tag{5.14}$$

Hence,

$$\langle a|S|a'\rangle = \begin{pmatrix} \dfrac{\sqrt{E}+i\sqrt{-E_1}}{\sqrt{E}-i\sqrt{-E_1}} & 0 \\ 0 & \dfrac{\sqrt{E}+i\sqrt{-E_1}}{\sqrt{E}-i\sqrt{-E_1}} \end{pmatrix}. \tag{5.15}$$

It should be mentioned that BARGMANN (4) has constructed analogous potentials for the radial equation, corresponding to zero angular momentum, which give zero phase shift. He uses what appears to be a special method to construct such potentials and it is not clear whether all potentials which give zero phase shift can be obtained by his procedure. It would appear to be possible that the use of the Gelfand-Levitan equation for the radial equation (5) might lead to all such potentials.

6. – Pathological Character of the Reflection Coefficient $\langle +1|S(E)|-1\rangle$.

One can ask whether the reflection coefficient from the other side, namely $c(k) = \langle +1|S(E)|-1\rangle$, is «pathological» when $b(k)$ is a «normal» reflection coefficient. The third example of (1), where $c(k)$ has a pole in the upper half-plane which does not correspond to a bound state, is just such a case.

In fact, it appears that $c(k)$ is generally pathological when $b(k)$ is normal and hence the potential generally will have an exponential tail for $x \to +\infty$ even when the potential is identically zero for sufficiently small x. In particular, it has been possible to prove that if $b(k)$ has the form

$$b(k) = \frac{1}{A(k)}\,,$$

where $A(k)$ is a polynomial in k of degree higher than one, then $c(k)$ will have poles in the upper half-plane which do not correspond to bound states. When $A(k)$ is linear in k, the only possible potential which does not lead to pathological reflection coefficients $c(k)$ is the δ-function potential. These points will be discussed in subsequent papers.

(4) V. BARGMANN: *Rev. Mod. Phys.*, **21**, 488 (1949).

(5) R. JOST and W. KOHN: *Det. Kgl. Danske Vidensk. Selskab, Mat.-fys. Medd.*, **27**, 9 (1953).

Incidentally, the example of Sect. **4** of the present paper offers a case where $c(k)$ is normal but $b(k)$ is pathological. The example of Sect. **5** gives reflection coefficients $b(k)$ and $c(k)$ which are both pathological.

7. – Generalizations.

In [1] and in the present paper we have required that the reflection coefficient satisfy conditions (3.1) and (3.1a). These conditions are merely sufficient conditions to assure us that the potential which is obtained will give back the reflection coefficient with which we started. The Gelfand-Levitan equation, however, is a necessary condition which must be satisfied for all potentials for which a reflection coefficient can be defined. Hence there are far more reflection coefficients which yield meaningful results than those satisfying (3.1) and (3.1a) and it would appear that potentials leading to even more « pathological » or bizarre types of scattering are possible.

IL NUOVO CIMENTO VOL. XXII, N. 4 16 Novembre 1961

The Determination of the Scattering Potential from the Spectral Measure Function.

V. – The Gelfand-Levitan Equation for the Three-Dimensional Scattering Problem.

I. KAY (*)

Institute for Mathematical Sciences, New York University - New York

H. E. MOSES (**)

Department of Physics, Polytechnic Institute of Brooklyn - Brooklyn

(ricevuto il 29 Luglio 1961)

Summary. — A Gelfand-Levitan equation is introduced which enables one to calculate scattering potentials from part of the scattering amplitude in the three-dimensional scattering problem. The Gelfand-Levitan equation is a generalization of the one-dimensional Gelfand-Levitan equation of Kay and Moses. The part of the scattering amplitude which is required is an analogue of the reflection coefficient in one dimension. The potential which is obtained is more general than that usually assumed in the three-dimensional scattering problem in that it is diagonal in the radius variable but is a real integral operator in terms of the angular variables. The purely local potential (*i.e.* the potential which is diagonal in both the radial and angular variables) appears as a special case. In the present paper we assume that the perturbed Hamiltonian has a purely continuous spectrum which coincides with that of the unperturbed Hamiltonian. The case where there are also point eigenvalues is omitted in order to present the main ideas without unnecessary complications. We shall derive the Gelfand-Levitan equation from the very general Gelfand-Levitan method of Kay and Moses. This formalism already guarantees that the potential which is

(*) The work of I. KAY was supported by the National Science Foundation, NSF G-14000.

(**) The work of H. E. MOSES was supported by the Office of Naval Research, contract Nonr-839(30), Project no. NR-013-106. Present address: Geophysics Corporation of America, Bedford, Mass.

4339

obtained will reproduce the reflection coefficient and that the equation is correct. In a later paper, however, we shall prove that the Gelfand-Levitan equation which we derive is correct by an elementary formalism which makes no use of the general Gelfand-Levitan technique at all; the advantage of this procedure being that one is immediately convinced that the equation is the right one.

---

1 – Introduction and summary of results: the Gelfand-Levitan equation in three dimensions.

Various forms of the Gelfand-Levitan equation have been introduced for the purpose of reconstructing the scattering potential from scattering data, in particular scattering problems. A bibliography is given in references (1) and (2). It was noted by the present authors that these various Gelfand-Levitan equations could be considered as special cases of a general formalism, which they gave in reference (1). It was shown that the Gelfand-Levitan equation could be interpreted as an operator equation in Hilbert space for the operator which transforms the unperturbed Hamiltonian into the perturbed Hamiltonian. This operator could be obtained in terms of a positive-definite operator which was called the weight operator. In special cases, the weight operator could be obtained from suitable elements of the scattering operator.

In the present paper we shall use our generalized Gelfand-Levitan technique to obtain the Gelfand-Levitan equation for the three-dimensional scattering problem. Instead of assuming that the scattering potential is simply a real function of the co-ordinates, we shall consider a somewhat more general class of scattering potentials, namely those which, when expressed in terms of polar co-ordinates, are diagonal in terms of the radius variable but are real integral operators in terms of the angular variables. The general Gelfand-Levitan procedure of reference (1) shows that if the Gelfand-Levitan equation has a unique solution, then the potential which gives rise to the scattering will be reproduced by the Gelfand-Levitan procedure. Hence, if the scattering operator was given by a potential which is a multiplicative operator, it will be reproduced.

In the present paper we shall assume familiarity with reference (1). However, we shall change some of the notation in order to simplify various expressions. Also for simplicity, we shall assume there are no point eigenvalues.

(1) I. KAY and H. E. MOSES: *Nuovo Cimento*: part I, **2**, 917 (1955); part II, **3**, 66 (1956); part III, **3**, 276 (1956).

(2) L. D. FADDEEV: *Usp. Math. Nauk*, **14** (88), 57 (1959).

For those familiar with the general Gelfand-Levitan procedure it will be clear how they are to be introduced.

In a later paper, we shall use an elementary procedure to verify that it is the correct one, even though, this proof is not really needed, since the general Gelfand-Levitan procedure automatically gives the correct equation. The advantage of this proof is that one need not have much background in the subject to be persuaded of the validity of the proof.

After the introduction of some notation we shall give the Gelfand-Levitan equation which we shall derive in subsequent sections.

Generally speaking it will be convenient to use polar co-ordinates both for the $\boldsymbol{x}$-representation and the $\boldsymbol{p}$- (*i.e.* momentum) representation. The polar co-ordinates of $\boldsymbol{x}$ will be denoted by $r = |\boldsymbol{x}|$, the angle λ $(0 \leqslant \lambda \leqslant \pi)$ which is the angle made by $\boldsymbol{x}$ with the z-axis and the angle σ $(0 \leqslant \sigma \leqslant 2\pi)$ which is the angle made by the projection of $\boldsymbol{x}$ on the x-y plane with the x-axis. Similarly the vector $\boldsymbol{p}$ will have the polar co-ordinates $p = |\boldsymbol{p}|$, $\theta(0 \leqslant \theta \leqslant \pi)$, and $\varphi(0 \leqslant \varphi \leqslant 2\pi)$. When primes are put on the vectors it is to be understood that the corresponding polar co-ordinates also have primes and conversely.

The unperturbed Hamiltonian H_0 will be taken as

$$H_0 = -\nabla^2 , \tag{1.1}$$

while the perturbed Hamiltonian H is given by

$$H = H_0 + V , \tag{1.2}$$

where V is the scattering potential which in the $\boldsymbol{x}$ representation we assume has the form of an integral operator with the kernel

$$V(\boldsymbol{x}|\boldsymbol{x}') = \delta(r - r')V(r|\lambda, \sigma; \lambda', \sigma') , \tag{1.3}$$

where $V(r|\lambda, \sigma; \lambda', \sigma')$ is real and is the kernel of the integral operator in the angular variables, *i.e.*,

$$\int V(\boldsymbol{x}|\boldsymbol{x}')f(\boldsymbol{x}')\,\mathrm{d}\boldsymbol{x}' = r^2\int_0^{\pi}\sin\lambda'\,\mathrm{d}\lambda'\int_0^{2\pi}\mathrm{d}\sigma'\,V(r|\lambda, \sigma; \lambda', \sigma')f(r, \lambda', \sigma') .$$

Here $f(\boldsymbol{x})$ is a state vector in the $\boldsymbol{x}$-representation and

$$f(\boldsymbol{x}) \equiv f(r, \lambda, \sigma) .$$

A purely local potential is just a special case of the class of potentials which we are considering.

4391

It will be convenient to express the eigenfunctions of H_0 in terms of three different representations. The first representation is the usual momentum representation. These eigenfunctions are denoted by $\Psi_0(\boldsymbol{x}|\boldsymbol{p})$ where

$$\Psi_0(\boldsymbol{x}|\boldsymbol{p}) = \left(\frac{1}{2\pi}\right)^{\frac{3}{2}} \exp\left[i\boldsymbol{p}\cdot\boldsymbol{x}\right]. \tag{1.4}$$

These eigenfunctions satisfy the orthogonality and completeness relations

$$\begin{cases} \int \Psi_0^*(\boldsymbol{x}|\boldsymbol{p})\Psi_0(\boldsymbol{x}|\boldsymbol{p}')\,\mathrm{d}\boldsymbol{x} = \delta(\boldsymbol{p}-\boldsymbol{p}')\,, \\ \int \Psi_0(\boldsymbol{x}|\boldsymbol{p})\Psi_0^*(\boldsymbol{x}'|\boldsymbol{p})\,\mathrm{d}\boldsymbol{p} = \delta(\boldsymbol{x}-\boldsymbol{x}')\,. \end{cases} \tag{1.5}$$

The second way of expressing the eigenfunctions is in terms of polar coordinates of the momentum vector $\boldsymbol{p}$. Let us denote this set by $\Psi_0(\boldsymbol{x}|p, \theta, \varphi)$ where

$$\Psi_0(\boldsymbol{x}|p, \theta, \varphi) = p\sqrt{\sin\theta}\,\Psi_0(\boldsymbol{x}|\boldsymbol{p})\,. \tag{1.6}$$

They satisfy the orthogonality and completeness relations

$$\begin{cases} \int \Psi_0^*(\boldsymbol{x}|p, \theta, \varphi)\Psi_0(\boldsymbol{x}|p', \theta', \varphi')\,\mathrm{d}\boldsymbol{x} = \delta(p-p')\delta(\theta-\theta')\,\delta(\varphi-\varphi')\,, \\ \int_0^\infty \mathrm{d}p\int_0^\pi \mathrm{d}\theta\int_0^{2\pi} \mathrm{d}\varphi\, \Psi_0(\boldsymbol{x}|p, \theta, \varphi)\Psi_0^*(\boldsymbol{x}'|p, \theta, \varphi) = \delta(\boldsymbol{x}-\boldsymbol{x}')\,. \end{cases} \tag{1.7}$$

The third way of expressing the eigenfunctions of H_0 is through the use of the energy and angle variables. We denote this set by $\Psi_0(\boldsymbol{x}|E, \theta, \varphi)$ where

$$\Psi_0(\boldsymbol{x}|E, \theta, \varphi) = \frac{(E)^{\frac{1}{4}}}{\sqrt{2}}\sqrt{\sin\theta}\,\Psi_0(\boldsymbol{x}|\boldsymbol{p})\,, \qquad (E = |\boldsymbol{p}|^2 = p^2), \tag{1.8}$$

and

$$\begin{cases} \int \Psi_0^*(\boldsymbol{x}|E, \theta, \varphi)\Psi_0(\boldsymbol{x}|E', \theta', \varphi')\,\mathrm{d}\boldsymbol{x} = \delta(E-E')\delta(\theta-\theta')\,\mathrm{d}(\varphi-\varphi')\,, \\ \int_0^\infty \mathrm{d}E\int_0^\pi \mathrm{d}\theta\int_0^{2\pi} \mathrm{d}\varphi\,\Psi_0(\boldsymbol{x}|E, \theta, \varphi)\Psi_0^*(\boldsymbol{x}'|E, \theta, \varphi) = \delta(\boldsymbol{x}-\boldsymbol{x}')\,. \end{cases} \tag{1.9}$$

From the viewpoint of scattering problems the third set is most natural though use of the first set is conventional.

We shall assume that the spectrum of H is purely continuous and coincides with that of H_0. Of course, we shall have to specify boundary conditions on the eigenfunctions of H. For each boundary condition we shall introduce three sets of eigenfunctions analogous to those used for H_0.

Let us denote one set of the « outgoing » eigenfunctions of H by $\Psi_-(\boldsymbol{x}|\boldsymbol{p})$ where $\Psi_-(\boldsymbol{x}|\boldsymbol{p})$ satisfies the integral equation

$$\Psi_-(\boldsymbol{x}|\boldsymbol{p}) = \Psi_0(\boldsymbol{x}|\boldsymbol{p}) - \frac{1}{4\pi}\iint \frac{\exp[ip|\boldsymbol{x}-\boldsymbol{x}'|]}{|\boldsymbol{x}-\boldsymbol{x}'|} V(\boldsymbol{x}'|\boldsymbol{x}'')\Psi_-(\boldsymbol{x}''|\boldsymbol{p})\,\mathrm{d}\boldsymbol{x}'\,\mathrm{d}\boldsymbol{x}''. \tag{1.10}$$

We shall define the eigenfunctions $\Psi_-(\boldsymbol{x}|p,\theta,\varphi)$ and $\Psi_-(\boldsymbol{x}|E,\theta,\varphi)$ by

$$\left\{\begin{aligned} &\Psi_-(\boldsymbol{x}|p,\theta,\varphi) = p\sqrt{\sin\theta}\,\Psi_-(\boldsymbol{x}|\boldsymbol{p}), \\ &\Psi_-(\boldsymbol{x}|E,\theta,\varphi) = \frac{(E)^{\frac{1}{4}}}{\sqrt{2}}\sqrt{\sin\theta}\,\Psi_-(\boldsymbol{x}|\boldsymbol{p}). \end{aligned}\right. \tag{1.11}$$

It can be shown that $\Psi_-(\boldsymbol{x}|p,\theta,\varphi)$ and $\Psi_-(\boldsymbol{x}|E,\theta,\varphi)$ satisfy the integral equation (1.10) when $\Psi_0(\boldsymbol{x}|\boldsymbol{p})$ is replaced by $\Psi_0(\boldsymbol{x}|p,\theta,\varphi)$ and $\Psi_0(\boldsymbol{x}|E,\theta,\varphi)$ respectively. The functions are all eigenfunctions of H with the eigenvalue $E=p^2$. They satisfy completeness and orthogonality relations identical to those satisfied by the corresponding eigenfunctions of H_0.

Similarly, we shall introduce « incoming » eigenfunctions of H denoted by

$$\Psi_+(\boldsymbol{x}|\boldsymbol{p}), \qquad \Psi_+(\boldsymbol{x}|p,\theta,\varphi), \qquad \Psi_+(\boldsymbol{x}|E,\theta,\varphi),$$

where

$$\Psi_+(\boldsymbol{x}|\boldsymbol{p}) = \Psi_0(\boldsymbol{x}|\boldsymbol{p}) - \frac{1}{4\pi}\iint \frac{\exp[-ip|\boldsymbol{x}-\boldsymbol{x}'|]}{|\boldsymbol{x}-\boldsymbol{x}'|} V(\boldsymbol{x}'|\boldsymbol{x}'')\Psi_+(\boldsymbol{x}''|\boldsymbol{p})\,\mathrm{d}\boldsymbol{x}'\,\mathrm{d}\boldsymbol{x}''. \tag{1.12}$$

Again $\Psi_+(\boldsymbol{x}|p,\theta,\varphi)$ and $\Psi_+(\boldsymbol{x}|E,\theta,\varphi)$ also satisfy the integral eq. (1.12) with $\Psi_0(\boldsymbol{x}|\boldsymbol{p})$ suitably replaced. The completeness and orthogonality relations satisfied by Ψ_0 are satisfied also by the various eigenfunctions Ψ_+.

Let us define the scattering amplitude $b_p(\lambda,\sigma|\lambda',\sigma')$ by

$$\lim_{r\to\infty}\Psi_-(\boldsymbol{x}|\boldsymbol{p}) = \Psi_0(\boldsymbol{x}|\boldsymbol{p}) - \frac{i}{p\sqrt{2\pi}}\,\frac{1}{\sqrt{\sin\lambda}\sqrt{\sin\theta}}\,\frac{\exp[ipr]}{r}\,b_p(\lambda,\sigma|\theta,\varphi). \tag{1.13}$$

In (1.13) $p=|\boldsymbol{p}|$, $r=|\boldsymbol{x}|$, θ, φ are the polar angles of the vector $\boldsymbol{p}$ while λ, σ are the polar angles of the vector $\boldsymbol{x}$. The somewhat odd normalization of the spherical wave is used to define $b_p(\theta,\varphi|\theta',\varphi')$ because with this normalization the scattering amplitude is essentially the scattering operator.

Although $b_p(\theta,\varphi|\theta',\varphi')$ was originally defined for positive p only we shall show later that the definition can be extended for negative values of p.

In fact

$$b_{-p}(\theta, \varphi \,|\, \theta', \varphi') = [b_p(\theta, \varphi \,|\, \theta' \varphi')]^* \, , \tag{1.14}$$

where the asterisk denotes the complex conjugate. Similarly, though the eigenfunctions $\Psi_0(\boldsymbol{x} \,|\, p, \theta, \varphi)$, $\Psi_{\pm}(\boldsymbol{x} \,|\, p, \theta, \varphi)$ were defined only for $p > 0$, we have by continuing the definition of these functions to negative values of p:

$$\begin{cases} \Psi_0\,(\boldsymbol{x} \,|\, -p, \theta, \varphi) = -\,\Psi_0^*(\boldsymbol{x} \,|\, p, \theta, \varphi) \, , \\ \Psi_{\pm}\,(\boldsymbol{x} \,|\, -p, \theta, \varphi) = -\,\Psi_{\pm}^*(\boldsymbol{x} \,|\, p, \theta, \varphi) \, , \end{cases} \tag{1.15}$$

(The second equation follows from the integral equations for $\Psi_{\pm}$.)

We are now in a position to give the Gelfand-Levitan equation in three dimensions. Let us define

$$\Psi'_0(r, \lambda, \sigma \,|\, p, \theta, \varphi) \quad \Psi_0(\boldsymbol{x} \,|\, p, \theta, \varphi) \, , \tag{1.16}$$

$$\Omega(r, \lambda, \sigma \,|\, r', \lambda', \sigma') = \int_{-\infty}^{+\infty} \mathrm{d}p \int_0^{\pi/2} \mathrm{d}\theta \int_{\pi/2}^{\pi} \mathrm{d}\theta' \cdot \tag{1.17}$$

$$\cdot \int_0^{2\pi} \mathrm{d}\varphi \int_0^{2\pi} \mathrm{d}\varphi' \Psi_0(r, \lambda, \sigma \,|\, p, \theta, \varphi) \Psi_0^*(r', \lambda', \sigma' \,|\, p, \theta', \varphi') \, b_p(\theta, \varphi \,|\, \theta', \varphi') \, .$$

Furthermore let us require the functions of 6 variables $K(r, \lambda, \sigma \,|\, r', \lambda', \sigma')$ be required to satisfy

$$K(r, \lambda, \sigma \,|\, r', \lambda', \sigma') = -\,\Omega(r, \lambda, \sigma \,|\, r', \lambda', \sigma') - \int_r^{\infty} r''^2 \mathrm{d}r'' \int_0^{\pi} \sin \lambda'' \, \mathrm{d}\lambda'' \int_0^{2\pi} \mathrm{d}\sigma'' \cdot \tag{1.18}$$

$$\cdot K(r, \lambda, \sigma \,|\, r'', \lambda'', \sigma'') \Omega(r'', \lambda'', \sigma'' \,|\, r', \lambda', \sigma') \, ,$$

for $r' > r$, then the potential $V(r \,|\, \lambda, \sigma; \lambda', \sigma')$ (see (1.3)) which will reproduce that part of the scattering amplitude which appears in the definition of Ω is given by

$$V(r \,|\, \lambda, \sigma; \lambda', \sigma') = -\,2 \, \frac{1}{r^2} \frac{\partial}{\partial r} \, r^2 K(r, \lambda, \sigma \,|\, r, \lambda', \sigma') . \tag{1.19}$$

Eq. (1.18) together with eq. (1.19) for the potential constitute the Gelfand-Levitan equations for the three-dimensional scattering problem.

As in the case of all Gelfand-Levitan equations for problems which involve degeneracy, the complete scattering operator is not needed to reconstruct the potential. In the case of three-dimensional scattering problem, only that part

of the scattering amplitude which corresponds to scattering in the positive z direction of waves traveling initially in the negative z direction—*i.e.* the back scattering on a hemisphere—is used to reconstruct the potential.
The quantities $K(r, \lambda, \sigma | r', \lambda', \sigma')$ can be used to construct eigenfunctions of H. Let $\Psi(\boldsymbol{x} | p, \theta, \varphi)$ be given by

$$\Psi(\boldsymbol{x} | p, \theta, \varphi) = \Psi(r, \lambda, \sigma | p, \theta, \varphi) = \Psi_0(r, \lambda, \sigma | p, \theta, \varphi) + \\ + \int_r^\infty r'^2 \mathrm{d}r' \int_0^\pi \sin \lambda' \mathrm{d}\lambda' \int_0^{2\pi} \mathrm{d}\sigma' K(r, \lambda, \sigma | r', \lambda', \sigma') \Psi_0(r', \lambda', \sigma' | p, \theta, \varphi) . \tag{1.20}$$

Then $\Psi(\boldsymbol{x} | p, \theta, \varphi)$ will be an eigenfunction of H corresponding to the eigenvalue $E = p^2$. The eigenfunction $\Psi(\boldsymbol{x} | p, \theta, \varphi)$ will satisfy the completeness relation

$$\int_0^\infty \mathrm{d}p \int_0^\pi \mathrm{d}\theta \int_0^{2\pi} \mathrm{d}\varphi \, \Psi(\boldsymbol{x} | p, \theta, \varphi) \Psi^*(\boldsymbol{x}' | p, \theta, \varphi) + \\ + \int_0^\infty \mathrm{d}p \int_0^{\pi/2} \mathrm{d}\theta \int_{\pi/2}^{\pi} \mathrm{d}\theta' \int_0^{2\pi} \mathrm{d}\varphi \int_0^{2\pi} \mathrm{d}\varphi' \Psi(\boldsymbol{x} | p, \theta, \varphi) b_p(\theta, \varphi | \theta', \varphi') \Psi^*(\boldsymbol{x}' | p, \theta', \varphi') + \\ + \int_0^\infty \mathrm{d}p \int_{\pi/2}^{\pi} \mathrm{d}\theta \int_0^{\pi/2} \mathrm{d}\theta' \int_0^{2\pi} \mathrm{d}\varphi \int_0^{2\pi} \mathrm{d}\varphi' \Psi(\boldsymbol{x} | p, \theta, \varphi) b_p^*(\theta', \varphi' | \theta, \varphi) \Psi^*(\boldsymbol{x}' | p, \theta', \varphi') = \delta(\boldsymbol{x} - \boldsymbol{x}') . \tag{1.21}$$

2. – Asymptotic form for bundles of plane waves.

In later sections we shall use the method of reference [1] to construct the weight operator for use in the Gelfand-Levitan equation. But in order to find the weight operator we must indicate how $\Psi_0(\boldsymbol{x} | E, \theta, \varphi)$ behaves as $r = |\boldsymbol{x}| \to \infty$. We must, however, be more explicit as to the nature of the limit we wish to take. Actually we should like to find

$$\lim_{r \to \infty} \iiint \Psi_0(\boldsymbol{x} | E, \theta, \varphi) f(E, \theta, \varphi) \, \mathrm{d}E \, \mathrm{d}\theta \, \mathrm{d}\varphi ,$$

where $f(E, \theta, \varphi)$ is a suitably defined test function, for in this case

$$\iiint \Psi_0(\boldsymbol{x} | E, \theta, \varphi) f(E, \theta, \varphi) \, \mathrm{d}E \, \mathrm{d}\theta \, \mathrm{d}\varphi ,$$

could be considered as a function in Hilbert space. Actually, we need only

4395

use test functions $g(\theta, \varphi)$ for integration over the angular variables θ, φ. Integrals of plane waves over the angular part are called « bundles of plane waves » (see ref. (3)). In evaluating

$$\lim_{r\to\infty}\int_0^{\pi} d\theta \int_0^{2\pi} d\varphi\, \Psi_0(\boldsymbol{x}\,|\,E, \theta, \varphi) g(\theta, \varphi)\,,$$

we shall use saddle-point techniques. For simplicity, however, we shall carry out the saddle-point evaluation in the sense of distributions. Our procedure will be based on the identity,

$$\lim_{\alpha\to\infty} \alpha^{\frac{1}{2}} \exp[\pm i\xi^2\alpha] = \pi^{\frac{1}{2}} \exp[\pm i\pi/4]\,\delta(\xi)\,, \tag{2.1}$$

First, let us write

$$\Psi_0(\boldsymbol{x}\,|\,E, \theta, \varphi) = \frac{(E)^{\frac{1}{4}}}{\sqrt{2}} \sqrt{\sin\theta}\, \exp\left[iprf(\theta, \varphi)\right]\,, \tag{2.2}$$

where

$$f(\theta, \varphi) = \cos\lambda \cos\theta + \sin\lambda \sin\theta \cos(\varphi - \sigma)\,. \tag{2.3}$$

In (2.3) λ and σ are the polar angles of $\boldsymbol{x}$. We shall want to find the values θ_0 and φ_0 such that $f_\theta(\theta_0, \varphi_0) = f_\varphi(\theta_0, \varphi_0) = 0$. Here f_θ and f_φ are the θ derivative and φ derivative of the function f.

Two sets of values θ_0, φ_0 are found. The first set is

$$\theta_0 = \lambda\,, \qquad \varphi_0 = \sigma\,, \tag{2.4}$$

while the second set is

$$\theta_0 = \pi - \lambda\,, \qquad \varphi_0 = \Sigma\,, \tag{2.4a}$$

where

$$\begin{aligned} \Sigma &= \sigma + \pi \qquad && \text{if } 0 < \sigma < \pi\,, \\ &= \sigma - \pi && \text{if } \pi < \sigma < 2\pi\,. \end{aligned}$$

In the vicinity of the first set we have

$$f(\theta, \varphi) = 1 - \frac{(\theta - \lambda)^2}{2} - \sin^2\lambda\, \frac{(\varphi - \sigma)^2}{2} + F(\theta - \lambda, \varphi - \sigma)\,,$$

---

(3) K. O. FRIEDRICHS: *Seminar - Recent Developments in the Theory of Wave Propagation*, Section III-C, Institute for Mathematics and Mechanics of New York University (1949-1950).

4396

where F is a function of its arguments such that $F(0, 0) = 0$. Hence, for values of φ and θ in the vicinity of φ_0 and θ_0 we may write

$$\Psi_0(\boldsymbol{x} \,|\, E, \theta, \varphi) = \frac{(E)^{\frac{1}{4}}}{\sqrt{2}} \frac{\sqrt{\sin\theta}}{(2\pi)^{\frac{3}{2}}} \Big\{ \exp[irp] \cdot \sqrt{\frac{2}{rp}} \sqrt{\frac{rp}{2}} \exp[-irp(\theta-\lambda)^2/2] \cdot$$

$$\sqrt{\frac{2}{rp}} \frac{1}{\sin\lambda} \sin\lambda \sqrt{\frac{rp}{2}} \cdot \exp[-ir(p/2)\sin^2\lambda(\varphi-\sigma)^2] \cdot \exp[irpF(\theta-\lambda, \varphi-\sigma)] \Big\} .$$

On using (2.1) we have

$$\lim_{r\to\infty} \Psi_0(\boldsymbol{x} \,|\, E, \theta, \varphi) = \frac{\sqrt{2}}{(2\pi)^{\frac{3}{2}} (E)^{\frac{1}{4}}} \frac{(-i\pi)}{r} \frac{\exp[irp]}{\sqrt{\sin\theta}} \delta(\theta-\lambda)\delta(\varphi-\sigma) .$$

Similarly in the vicinity of the second set of values of θ_0, φ_0

$$\lim_{r\to\infty} \Psi_0(\boldsymbol{x} \,|\, E, \theta, \varphi) = \frac{\sqrt{2}}{(2\pi)^{\frac{3}{2}} (E)^{\frac{1}{4}}} \frac{(i\pi)}{r} \frac{\exp[-irp]}{\sqrt{\sin\theta}} \delta(\theta-(\pi-\lambda))\,\delta(\varphi-\Sigma) .$$

Or we may write generally for the whole range of integration over the angular variables

$$(2.5) \quad \lim_{r\to\infty} \Psi_0(\boldsymbol{x} \,|\, E, \theta, p) = \frac{\sqrt{2}\, i\pi}{(E)^{\frac{1}{4}} (2\pi)^{\frac{3}{2}} r \sqrt{\sin\theta}} \big[-\exp[irp]\delta(\theta-\lambda)\delta(\varphi-\sigma) + {}$$
$$+ \exp[-irp]\delta(\theta-(\pi-\lambda))\,\delta(\varphi-\Sigma) \big] .$$

Hence, asymptotically a plane wave behaves like a sum of spherical waves.

3. – Triangular form of the K operators. Expression for the potential.

In deriving the Gelfand-Levitan equation for the three-dimensional problem, we shall follow closely the procedure for obtaining the Gelfand-Levitan equation for the one-dimensional problem which was used in Part III of reference [1]. Hence, it will be convenient to give the analogues of the various quantities which appear in the present paper with those of the one-dimensional problem.

The eigenfunctions $\Psi_{\pm}(\boldsymbol{x} \,|\, E, \theta, \varphi)$ and $\Psi(\boldsymbol{x} \,|\, E, \theta, \varphi)$ of the present paper correspond to $\langle x \,|\, E, a \rangle_{\pm}$ and $\langle x \,|\, E, a \rangle$ of the previous paper. The kernel

$$K(\boldsymbol{x} \,|\, \boldsymbol{x}') \equiv K(r, \lambda, \sigma \,|\, r', \lambda', \sigma)$$

corresponds to $\langle x \,|\, K \,|\, x' \rangle$; the kernels $\mu_{-E}^{-1}(\theta, \varphi \,|\, \theta', \varphi')$ and $\mu_{-E}^{*-1}(\theta, \varphi \,|\, \theta', \varphi')$ to $\langle a \,|\, \mu_-^{-1}(E) \,|\, a' \rangle$ and $\langle a \,|\, \mu_-^{*-1}(E) \,|\, a' \rangle$. The weight kernel $\omega_E(\theta, \varphi \,|\, \theta', \varphi')$ corresponds to $\langle a|\omega_c(E)|a'\rangle$; the scattering operator $S_E(\theta, \varphi|\theta', \varphi')$ to $\langle a|S(E)|a'\rangle$; the kernel $\Omega(\boldsymbol{x}|\boldsymbol{x}') \equiv \Omega(r, \lambda, \sigma|r', \lambda', \sigma')$ to $\langle x|\Omega|x'\rangle$.

We shall now follow the procedure of Part III of reference [1] to obtain the expression for the potential and thereby verify that we have given the correct triangularization condition on the operator K. Accordingly we introduce the operators

$$\begin{cases} U = I + K \\ U_0 = I + K_0 \end{cases} \tag{3.1}$$

where K and K_0 are triangular operators in the x-representation; *i.e.*

$$\begin{cases} K(\boldsymbol{x}|\boldsymbol{x}') \equiv K(r, \lambda, \sigma|r', \lambda', \sigma') = 0 \\ K_0(\boldsymbol{x}|\boldsymbol{x}') \equiv K_0(r, \lambda, \sigma|r', \lambda', \sigma') = 0 \end{cases} \qquad \text{for } r > r'. \tag{3.2}$$

The operator which corresponds to the scattering potential is given by (cf. eq. (5.7) Part III, ref. [1]).

$$V = H - H_0 = UH_0U_0 - H_0UU_0 = (KH_0 - H_0K) + (KH_0 - H_0K)K_0 \tag{3.3}$$

or equivalently

$$V(\boldsymbol{x}|\boldsymbol{x}') = -\nabla'^2 K(\boldsymbol{x}|\boldsymbol{x}') + \nabla^2 K(\boldsymbol{x}|\boldsymbol{x}') - \int \nabla''^2 K(\boldsymbol{x}|\boldsymbol{x}'')\cdot K_0(\boldsymbol{x}''|\boldsymbol{x}')\,\mathrm{d}\boldsymbol{x}'' + \int \nabla^2 K(\boldsymbol{x}|\boldsymbol{x}'')\,K_0(\boldsymbol{x}''|\boldsymbol{x}')\,\mathrm{d}\boldsymbol{x}''\,. \tag{3.4}$$

In (3.4) the primes on the Laplacian operators refer to the variables on which they act. A dot has been inserted in the first integral to indicate that the Laplacian acts on the first factor of the integral only. We shall have occasion to use this convention later. Also, we have used the Hermitian nature of the Laplacian operator. It will be convenient to indicate the triangular nature of K and K_0 operators explicitly by writing

$$\begin{cases} K(r, \lambda, \sigma|r', \lambda', \sigma') = \eta(r' - r)\, K(r, \lambda, \sigma|r', \lambda', \sigma') \\ K_0(r, \lambda, \sigma|r', \lambda', \sigma') = \eta(r' - r)\, K_0(r, \lambda, \sigma|r', \lambda', \sigma') \end{cases} \tag{3.5}$$

where

$$\begin{cases} \eta(x) = 0 \quad \text{if } x < 0. \\ \eta(x) = 1 \quad \text{if } x > 0. \end{cases} \tag{3.6}$$

On expressing (3.4) in terms of spherical co-ordinates and using (3.5) we find that $V(\boldsymbol{x}|\boldsymbol{x}')$ is the sum of two terms one of which has $\delta(r-r')$ as a factor and the other of which has $\eta(r'-r)$ as a factor. From the general theory, the scattering potential V *must* be Hermitian. Hence the coefficient of the $\eta(r'-r)$ is zero, since the term which has this factor cannot possibly be Hermitian. We are left with the terms that have the δ-function factor

$$V(\boldsymbol{x}|\boldsymbol{x}') = \delta(r-r')V(r|\lambda,\sigma;\lambda',\sigma') , \tag{3.7}$$

where

$$V(r|\lambda,\sigma;\lambda',\sigma') = -\frac{2}{r^2}\frac{\partial}{\partial r}r^2K(r,\lambda,\sigma|r,\lambda',\sigma') . \tag{3.8}$$

Hence our triangularity conditions assure us that the scattering potential is diagonal in the radial co-ordinate.

4. – The weight operator; the scattering operator; the Gelfan-Levitan equation.

In accordance with the general procedure we introduce the eigenfunctions of H, $\Psi(\boldsymbol{x}|E,\theta,\varphi)\equiv\Psi(r,\lambda,\sigma|E,\theta,\varphi)$ given by

$$\begin{aligned}\Psi(\boldsymbol{r},\lambda,\sigma|E,\theta,\varphi) &= \Psi_0(r,\lambda,\sigma|E,\theta,\varphi) + \\ &+\int_r^\infty r'^2\,\mathrm{d}r'\int_0^\pi \sin\lambda'\,\mathrm{d}\lambda'\int_0^{2\pi}\mathrm{d}\sigma'\,K(r,\lambda,\sigma|r',\lambda',\sigma')\Psi_0(r',\lambda',\sigma'|E,\theta,\varphi) .\end{aligned} \tag{4.1}$$

These eigenfunctions will enable us to find the weight operator needed for the completeness relationship in terms of the scattering operator. Let us write

$$\int_0^\pi \mathrm{d}\theta\int_0^{2\pi}\mathrm{d}\varphi\,\Psi(\boldsymbol{x}|E,\theta,\varphi)\mu_{-E}^{-1}(\theta,\varphi|\theta',\varphi') = \Psi_-(\boldsymbol{x}|E,\theta',\varphi') . \tag{4.2}$$

We are required to find the functions $\mu_{-E}^{-1}(\theta,\varphi|\theta',\varphi')$ and $\mu_{-E}^{*-1}(\theta,\varphi|\theta',\varphi')$ where

$$\mu_{-E}^{*-1}(\theta,\varphi|\theta',\varphi') = [\mu_{-E}^{-1}(\theta',\varphi'|\theta,\varphi)]^* . \tag{4.3}$$

(The asterisk on the left-hand side arose from the notation of Hermitian adjoint, but that on the right-hand side is simply complex conjugate.) The weight operator is given by

$$\omega_E(\theta,\varphi|\theta',\varphi') = \int_0^\pi \mathrm{d}\theta''\int_0^{2\pi}\mathrm{d}\varphi''\mu_{-E}^{-1}(\theta,\varphi|\theta'',\varphi'')\mu_{-E}^{*-1}(\theta'',\varphi''|\theta',\varphi') . \tag{4.4}$$

From the triangularity conditions on K it is clear that

$$\lim_{r\to\infty} \Psi(\boldsymbol{x}\,|\,E,\theta,\varphi) - \Psi_0(\boldsymbol{x}\,|\,E,\theta,\varphi) = 0\,. \tag{4.5}$$

It is also possible to show

$$\lim_{r\to\infty} \Psi_\pm(\boldsymbol{x}\,|\,E,\theta,\varphi) = \Psi_0(\boldsymbol{x}\,|\,E,\theta,\varphi) - \\ - \frac{1}{4\pi}\,\frac{(2\pi)^{\frac{3}{2}}\sqrt{2}}{\sqrt{\sin\lambda}\,(E)^{\frac{1}{4}}}\,\frac{\exp[\pm\,ipr]}{r}\,T_{\pm E}(\lambda,\sigma\,|\,\theta,\varphi)\,, \tag{4.6}$$

where

$$T_{\pm E}(\theta,\varphi\,|\,\theta',\varphi') = \iint \Psi_0^*(\boldsymbol{x}\,|\,E,\theta,\varphi)\,V(\boldsymbol{x}\,|\,\boldsymbol{x}')\,\Psi_\pm(\boldsymbol{x}'\,|\,E,\theta,\varphi)\,\mathrm{d}\boldsymbol{x}\,\mathrm{d}\boldsymbol{x}'\,. \tag{4.7}$$

The importance of the function $T_{-E}(\theta,\varphi\,|\,\theta',\varphi')$ is that the scattering operator $S_E(\theta,\varphi\,|\,\theta',\varphi')$ is simply expressed in terms of it:

$$S_E(\theta,\varphi\,|\,\theta',\varphi') = \delta(\theta-\theta')\,\delta(\varphi-\varphi') - 2\pi i\,T_{-E}(\theta,\varphi\,|\,\theta',\varphi')\,. \tag{4.8}$$

The scattering operator $S_E(\theta,\varphi\,|\,\theta',\varphi')$ of course is the scattering amplitude for scattering in the direction θ, φ when the incident wave is in the direction θ', φ'. It is unitary in the sense

$$\left\{\begin{aligned} &\int_0^\pi \mathrm{d}\theta''\int_0^{2\pi}\mathrm{d}\varphi''\,S_E(\theta,\varphi\,|\,\theta'',\varphi'')[S_E(\theta',\varphi'\,|\,\theta'',\varphi'')]^* = \delta(\theta-\theta')\,\delta(\varphi-\varphi')\,,\\ &\int_0^\pi \mathrm{d}\theta''\int_0^{2\pi}\mathrm{d}\varphi''\,[S_E(\theta'',\varphi''\,|\,\theta,\varphi)]^*\,S_E(\theta'',\varphi''\,|\,\theta',\varphi') = \delta(\theta-\theta')\,\delta(\varphi-\varphi')\,. \end{aligned}\right. \tag{4.9}$$

To find $\mu_{-E}^{-1}(\theta,\varphi\,|\,\theta',\varphi')$ we take the limit of (4.2) as $r\to\infty$. On using (2.5) and (4.6) we find as our equation for $\mu_{-E}^{-1}(\theta,\varphi\,|\,\theta',\varphi')$ after some cancellation of common factors

$$\begin{aligned} -\exp[irp]\mu_{-E}^{-1}(\lambda,\sigma\,|\,\theta',\varphi') &+ \exp[irp]\mu_{-E}^{-1}(\pi-\lambda,\Sigma\,|\,\theta',\varphi') = \\ &= -\exp[irp]\,\delta(\theta'-\lambda)\,\delta(\varphi'-\sigma) + \exp[-irp]\,\delta(\theta'-(\pi-\lambda))\,\delta(\varphi'-\Sigma) + \\ &+ 2\pi i\exp[ipr]\,T_{-E}(\lambda,\sigma\,|\,\theta',\varphi')\,, \end{aligned}$$

where Σ is defined below eq. (2.4a).

4100

On equating coefficients of like exponentials we have

$$(4.10)\quad \begin{cases} \mu^{-1}_{-E}(\lambda, \sigma\,|\,\theta', \varphi') = \delta(\theta' - \lambda)\,\delta(\theta - \sigma) - 2\pi i\, T_{-E}(\lambda, \sigma\,|\,\theta', \varphi') = S_E(\lambda, \sigma\,|\,\theta', \varphi')\,, \\ \mu^{-1}_{-E}(\pi - \lambda, \Sigma\,|\,\theta', \varphi') = \delta(\theta' - (\pi - \lambda))\,\delta(\varphi' - \Sigma)\,. \end{cases}$$

Eq. (4.10) lead to

$$(4.11)\quad \begin{cases} \mu^{-1}_{-E}(\lambda, \sigma\,|\,\theta', \varphi') = S_E(\lambda, \sigma\,|\,\theta', \varphi')\,, & \left(0 \leqslant \lambda < \dfrac{\pi}{2}\right), \\ \mu^{-1}_{-E}(\lambda, \sigma\,|\,\theta', \varphi') = \delta(\lambda - \theta')\,\delta(\sigma - \varphi') & \left(\dfrac{\pi}{2} \leqslant \lambda \leqslant \pi\right), \end{cases}$$

or

$$(4.12)\quad \mu^{-1}_{-E}(\theta, \varphi\,|\,\theta', \varphi') = \eta\left(\frac{\pi}{2} - \theta\right) S_E(\theta, \varphi\,|\,\theta', \varphi') + \eta\left(\theta - \frac{\pi}{2}\right)\delta(\theta - \theta')\,\delta(\varphi - \varphi')\,.$$

Also

$$(4.13)\quad \mu^{*-1}_{-E}(\theta, \varphi\,|\,\theta', \varphi') = \eta\left(\frac{\pi}{2} - \theta'\right)[S_E(\theta', \varphi'\,|\,\theta, \varphi)]^* + \\ + \eta\left(\theta - \frac{\pi}{2}\right)\delta(\theta - \theta')\,\delta(\varphi - \varphi')\,.$$

The weight operator is given by

$$(4.14)\quad \omega_E(\theta, \varphi\,|\,\theta', \varphi') = \int_0^{\pi} d\theta'' \int_0^{2\pi} d\varphi''\, \mu^{-1}_{-E}(\theta, \varphi\,|\,\theta'', \varphi'')\,\mu^{*-1}_{-E}(\theta'', \varphi''\,|\,\theta', \varphi') = \\ = \delta(\theta - \theta') + \eta\left(\theta - \frac{\pi}{2}\right)\eta\left(\frac{\pi}{2} - \theta'\right)[S_E(\theta', \varphi'\,|\,\theta, \varphi)]^* + \\ + \eta\left(\frac{\pi}{2} - \theta\right)\eta\left(\theta' - \frac{\pi}{2}\right) S_E(\theta, \varphi\,|\,\theta', \varphi')\,.$$

In (4.14) we have used the unitary property of $S_E(\theta, \varphi\,|\,\theta', \varphi')$ (eq. (4.9)). Hence the kernel of the Gelfand-Levitan equation is

$$(4.15)\quad \Omega(\boldsymbol{x}\,|\,\boldsymbol{x}') = \int_0^{\infty} dE \int_0^{\pi/2} d\theta \int_{\pi/2}^{\pi} d\theta' \int_0^{2\pi} d\varphi \int_0^{2\pi} d\varphi'\, \Psi_0(\boldsymbol{x}\,|\,E, \theta, \varphi)\, S_E(\theta, \varphi\,|\,\theta', \varphi')\cdot \\ \cdot \Psi_0^*(\boldsymbol{x}'\,|\,E, \theta', \varphi') + \int_0^{\infty} dE \int_{\pi/2}^{\pi} d\theta \int_0^{\pi/2} d\theta' \int_0^{2\pi} d\varphi \int_0^{2\pi} d\varphi' \cdot \\ \cdot \Psi_0(\boldsymbol{x}\,|\,E, \theta, \varphi)[S_E(\theta', \varphi'\,|\,\theta, \varphi)]^*\, \Psi_0^*(\boldsymbol{x}'\,|\,E, \theta, \varphi)\,.$$

In principle, the expression (4.15) is all that is required to set up the Gelfand-Levitan equation (1.18). Eq. (1.18) follows from

$$K(\boldsymbol{x}\,|\,\boldsymbol{x}') = -\,\Omega(\boldsymbol{x}\,|\,\boldsymbol{x}') - \int K(\boldsymbol{x}\,|\,\boldsymbol{x}'')\,\Omega(\boldsymbol{x}''\,|\,\boldsymbol{x}')\,\mathrm{d}\boldsymbol{x}'' ,$$

when we use polar co-ordinates and the triangularity condition (3.2).

However, we wish to simplify the expression (4.15) for Ω. Toward this end we shall need more properties of the scattering operator and the eigenfunctions of H and H_0 introduced in Section **1**.

First of all we note

$$\Psi_0(\boldsymbol{x}\,|\,E,\,\pi-\theta,\,\Phi) = \Psi_0^*(\boldsymbol{x}\,|\,E,\,\theta,\,\varphi) , \tag{4.16}$$

where

$$\begin{cases} \Phi = \varphi + \pi & \text{if } \; 0 \leqslant \varphi < \pi , \\ \Phi = \varphi - \pi & \text{if } \; \pi \leqslant \varphi < 2\pi . \end{cases} \tag{4.16a}$$

Secondly we have

$$\begin{cases} \Psi_-(\boldsymbol{x}\,|\,E,\,\pi-\theta,\,\Phi) = \Psi_+^*(\boldsymbol{x}\,|\,E,\,\theta,\,\varphi) , \\ \Psi_+(\boldsymbol{x}\,|\,E,\,\pi-\theta,\,\Phi) = \Psi_-^*(\boldsymbol{x}\,|\,E,\,\theta,\,\varphi) . \end{cases} \tag{4.17}$$

These equations follow from the integral equations for $\Psi_\pm(\boldsymbol{x}\,|\,E,\,\theta,\,\varphi)$ and the reality condition on $V(\boldsymbol{x}\,|\,\boldsymbol{x}')$. Likewise

$$\begin{cases} \Psi_0\,(\boldsymbol{x}\,|\,p,\,\pi-\theta,\,\Phi) = \Psi_0^*(\boldsymbol{x}\,|\,p,\,\theta,\,\varphi) , \\ \Psi_\pm\,(\boldsymbol{x}\,|\,p,\,\pi-\theta,\,\Phi) = \Psi_\mp^*(\boldsymbol{x}\,|\,p,\,\theta,\,\varphi) . \end{cases} \tag{4.18}$$

We wish now to prove

$$S_E(\theta,\,\varphi\,|\,\theta',\,\varphi') = S_E(\pi-\theta',\,\Phi'\,|\,\pi-\theta,\,\Phi) , \tag{4.19}$$

where Φ' is the related to φ' as Φ is to φ (eq. (4.16a)).

Eq. (4.19) follows in part from the reciprocity theorem in the theory of the scattering operator which states

$$T_{-E}(\theta,\,\varphi\,|\,\theta',\,\varphi') = [T_{+E}(\theta',\,\varphi'\,|\,\theta,\,\varphi)]^* \tag{4.20}$$

(see *e.g.* ref. [4]).

The quantities $T_{\pm E}(\theta,\,\varphi\,|\,\theta',\,\varphi')$ are defined by eq. (4.6) and (4.7).

[4] H. E. MOSES: *Nuovo Cimento*, **1**, 103 (1955).

4402

But on using (4.16) and (4.17)

$$(4.21)\qquad \begin{aligned} T_{-E}(\theta, \varphi\,|\,\theta', \varphi') &= [T_{+}(\theta', \varphi'\,|\,\theta, \varphi)]^{*} = \\ &= \iint \Psi_0(\boldsymbol{x}\,|\,E, \theta', \varphi')\, V(\boldsymbol{x}\,|\,\boldsymbol{x}')\Psi_{+}^{*}(\boldsymbol{x}'\,|\,E, \theta, \varphi)\,\mathrm{d}\boldsymbol{x}\,\mathrm{d}\boldsymbol{x}' = \\ &= \iint \Psi_0^{*}(\boldsymbol{x}\,|\,E, \pi-\theta', \Phi')\, V(\boldsymbol{x}\,|\,\boldsymbol{x}')\Psi_{-}(\boldsymbol{x}'\,|\,E, \pi-\theta, \Phi)\,\mathrm{d}\boldsymbol{x}\,\mathrm{d}\boldsymbol{x}' = \\ &= T_{-E}(\pi-\theta', \Phi'\,|\,\pi-\theta, \Phi)\,, \end{aligned}$$

and finally the use of (4.8) leads to the desired result (4.19).

On using (4.19) and (4.16) we find that our expression for $\Omega(\boldsymbol{x}\,|\,\boldsymbol{x}')$ becomes:

$$(4.22)\qquad \begin{aligned} \Omega(\boldsymbol{x}\,|\,\boldsymbol{x}') &= \int\limits_0^{\infty}\mathrm{d}E\int\limits_0^{\pi/2}\mathrm{d}\theta\int\limits_{\pi/2}^{\pi}\mathrm{d}\theta'\int\limits_0^{2\pi}\mathrm{d}\varphi\int\limits_0^{2\pi}\mathrm{d}\varphi'\,\Psi_0(\boldsymbol{x}\,|\,E, \theta, \varphi)\cdot \\ &\cdot S_E(\theta, \varphi\,|\,\theta', \varphi')\Psi_0^{*}(\boldsymbol{x}'\,|\,E, \theta', \varphi') + \int\limits_0^{\infty}\mathrm{d}E\int\limits_0^{\pi/2}\mathrm{d}\theta\int\limits_{\pi/2}^{\pi}\mathrm{d}\theta'\int\limits_0^{2\pi}\mathrm{d}\varphi\int\limits_0^{2\pi}\mathrm{d}\varphi'\,\Psi_0^{*}(\boldsymbol{x}\,|\,E, \theta, \varphi)\cdot \\ &\cdot[S_E(\theta, \varphi\,|\,\theta', \varphi')]^{*}\Psi_0(\boldsymbol{x}'\,|\,E, \theta', \varphi')\,. \end{aligned}$$

Clearly $\Omega(\boldsymbol{x}\,|\,\boldsymbol{x}')$ is real. Hence the Gelfand-Levitan eq. (1.18) set up with this kernel has a real solution if the solution is unique. To get an even simpler representation of Ω we use the eigenfunctions $\Psi_0(\boldsymbol{x}\,|\,p, \theta, \varphi)$, $\Psi_{\pm}(\boldsymbol{x}\,|\,p, \theta, \varphi)$. First of all, it is clear that though Ψ_0 has been defined for positive p only, we have by continuing p to negative values

$$(4.23)\qquad \Psi_0(\boldsymbol{x}\,|-p, \theta, \varphi) = -\Psi_0^{*}(\boldsymbol{x}\,|\,p, \theta, \varphi)\,.$$

Similarly, using the integral equation for $\Psi_{\pm}(\boldsymbol{x}\,|\,p, \theta, \varphi)$, we find

$$(4.24)\qquad \Psi_{\pm}(\boldsymbol{x}\,|-p, \theta, \varphi) = -\Psi_{\pm}^{*}(\boldsymbol{x}\,|\,\boldsymbol{p}, \theta, \varphi)\,.$$

Let us now introduce the function $b_p(\theta, \varphi\,|\,\theta', \varphi')$ defined by

$$(4.15)\qquad b_p(\theta, \varphi\,|\,\theta', \varphi') = S_E(\theta, \varphi\,|\,\theta', \varphi') \qquad (p = \sqrt{E}).$$

Now

$$(4.26)\qquad \left\{\begin{aligned} \Psi_0(\boldsymbol{x}\,|\,E, \theta, \varphi) &= \frac{1}{\sqrt{2p}}\,\Psi_0(\boldsymbol{x}\,|\,p, \theta, \varphi)\,, \\ \Psi_{\pm}(\boldsymbol{x}\,|\,E, \theta, \varphi) &= \frac{1}{\sqrt{2p}}\,\Psi_{\pm}(\boldsymbol{x}\,|\,p, \theta, \varphi)\,, \end{aligned}\right. \qquad (p > 0);$$

(see eq. (1.11)).

On using (4.7) and (4.8) we can express $b_p(\theta, \varphi|\theta', \varphi')$ in terms of $\Psi_0(\boldsymbol{x}|p, \theta, \varphi)$ and $\Psi_-(\boldsymbol{x}|p, \theta, \varphi)$

$$b_p(\theta, \varphi|\theta', \varphi') = \delta(\theta - \theta')\delta(\varphi - \varphi') - \\ - \frac{2\pi i}{2p}\iint \Psi_0^*(\boldsymbol{x}|p, \theta, \varphi)V(\boldsymbol{x}|\boldsymbol{x}')\Psi_-(\boldsymbol{x}'|p, \theta', \varphi')\,\mathrm{d}\boldsymbol{x}\mathrm{d}\boldsymbol{x}' . \tag{4.27}$$

If we continue to negative values of p we find (see eq. (1.15)),

$$b_{-p}(\theta, \varphi|\theta', \varphi') = [b_p(\theta, \varphi|\theta', \varphi')]^* . \tag{4.28}$$

We can now obtain our simple expression for $\Omega(\boldsymbol{x}|\boldsymbol{x}')$. We obtain

$$\Omega(\boldsymbol{x}|\boldsymbol{x}') = \int_{-\infty}^{+\infty}\mathrm{d}p\int_0^{\pi/2}\mathrm{d}\theta\int_{\pi/2}^{\pi}\mathrm{d}\theta'\int_0^{2\pi}\mathrm{d}\varphi\int_0^{2\pi}\mathrm{d}\varphi' \cdot \\ \cdot \Psi_0(\boldsymbol{x}|p, \theta, \varphi)b_p(\theta, \varphi|\theta', \varphi')\Psi_0^*(\boldsymbol{x}'|p, \theta', \varphi'). \tag{4.29}$$

5. – The completeness relationship.

In accordance with the general formalism of reference (1), the eigenfunctions of $H, \Psi(\boldsymbol{x}|E, \theta, \varphi)$ given by (4.1) must satisfy the completeness relation

$$\int_0^{\infty}\mathrm{d}E\int_0^{\pi}\mathrm{d}\theta\int_0^{\pi}\mathrm{d}\theta'\int_0^{2\pi}\mathrm{d}\varphi\int_0^{2\pi}\mathrm{d}\varphi'\Psi(\boldsymbol{x}|E, \theta, \varphi)\omega_E(\theta, \varphi|\theta', \varphi')\Psi^*(\boldsymbol{x}'|E, \theta', \varphi') = \delta(\boldsymbol{x} - \boldsymbol{x}') . \tag{5.1}$$

In terms of the eigenfunctions of H given by

$$\Psi(\boldsymbol{x}|p, \theta, \varphi) = \Psi_0(\boldsymbol{x}|p, \theta, \varphi) + \iint K(\boldsymbol{x}|\boldsymbol{x}')\Psi_0(\boldsymbol{x}'|p, \theta, \varphi)\,\mathrm{d}\boldsymbol{x}' = \\ = \sqrt{2p}\,\Psi(\boldsymbol{x}|E, \theta, \varphi) , \tag{5.2}$$

the completeness relation is:

$$\int_0^{\infty}\mathrm{d}p\int_0^{\pi}\mathrm{d}\theta\int_0^{2\pi}\mathrm{d}\varphi\, \Psi(\boldsymbol{x}|p, \theta, \varphi)\Psi^*(\boldsymbol{x}'|p, \theta, \varphi) + \\ + \int_0^{\infty}\mathrm{d}p\int_0^{\pi/2}\mathrm{d}\theta\int_{\pi/2}^{\pi}\mathrm{d}\theta'\int_0^{2\pi}\mathrm{d}\varphi\int_0^{2\pi}\mathrm{d}\varphi'\Psi(\boldsymbol{x}|p, \theta, \varphi)b_p(\theta, \varphi|\theta', \varphi')\Psi^*(\boldsymbol{x}'|p, \theta', \varphi') + \\ + \int_0^{\infty}\mathrm{d}p\int_{\pi/2}^{\pi}\mathrm{d}\theta\int_0^{\pi/2}\mathrm{d}\theta'\int_0^{2\pi}\mathrm{d}\varphi\int_0^{2\pi}\mathrm{d}\varphi'\Psi(\boldsymbol{x}|p, \theta, \varphi)b_p(\theta', \varphi'|\theta, \varphi)]^*\Psi^*(\boldsymbol{x}'|\theta', \varphi') = \delta(\boldsymbol{x} - \boldsymbol{x}') . \tag{5.3}$$

Since $K(\boldsymbol{x}|\boldsymbol{x}')$ is real, we obtain on using (4.18), (4.23) and (5.2)

$$\left\{\begin{aligned} &\Psi(\boldsymbol{x}|-p,\theta,\varphi) = -\Psi^*(\boldsymbol{x}|p,\theta,\varphi)\,,\\ &\Psi(\boldsymbol{x}|p,\pi-\theta,\Phi) = \Psi^*(\boldsymbol{x}|p,\theta,\varphi)\,,\end{aligned}\right. \tag{5.4}$$

(Φ defined previously).

Using this result together with the relations for $b_p(\theta,\varphi|\theta',\varphi')$ obtainable from (4.19), (4.25), and (4.28) we have the somewhat simpler form of the completeness relation

$$\int\limits_0^\infty \mathrm{d}p\int\limits_0^\pi \mathrm{d}\theta\int\limits_0^{2\pi}\mathrm{d}\varphi\,\Psi(\boldsymbol{x}|p,\theta,\varphi)\Psi^*(\boldsymbol{x}'|p,\theta,\varphi)\,+ \tag{5.5}$$

$$+\int\limits_{-\infty}^{+\infty}\mathrm{d}p\int\limits_0^{\pi/2}\mathrm{d}\theta\int\limits_{\pi/2}^{\pi}\mathrm{d}\theta'\int\limits_0^{2\pi}\mathrm{d}\varphi\int\limits_0^{2\pi}\mathrm{d}\varphi'\,\Psi(\boldsymbol{x}|p,\theta,\varphi)b_p(\theta,\varphi|\theta',\varphi')\Psi^*(\boldsymbol{x}'|p,\theta',\varphi') = \delta(\boldsymbol{x}-\boldsymbol{x}')\,.$$

RIASSUNTO (*)

Introduciamo una equazione di Gelfand-Levitan che permette di calcolare i potenziali di scattering da una parte dell'ampiezza di scattering nei problemi di scattering tridimensionale. L'equazione di Gelfand-Levitan introdotta, è una generalizzazione della equazione di Gelfand-Levitan unidimensionale di Kay e Moses. La parte richiesta dell'ampiezza di scattering è l'analogo del coefficiente di riflessione del caso unidimensionale. Il potenziale ottenuto è più generale di quello generalmente assunto nel problema di scattering tridimensionale, in quanto è diagonale nella variabile radiale ma è un operatore integrale reale nelle variabili angolari. Il potenziale puramente locale (cioè, il potenziale diagonale sia nella variabile radiale che in quelle angolari) risulta essere un caso speciale. In questo lavoro noi supponiamo che l'Hamiltoniano perturbato abbia uno spettro puramente continuo che coincide con quello dell'Hamiltoniano non perturbato. Il caso in cui vi siano anche autovalori puntuali si omette allo scopo di presentare le idee principali senza inutili complicazioni. Deriveremo l'equazione di Gelfand-Levitan dal metodo di Gelfand-Levitan più generalizzato da Kay e Moses. Questo formalismo garantisce già che il potenziale ottenuto riprodurrà il coefficiente di riflessione e che l'equazione è corretta. In un lavoro successivo, comunque, dimostreremo, con un formalismo elementare che non fa alcun uso della tecnica generale di Gelfand-Levitan, che l'equazione di Gelfand-Levitan, che deriviamo, è corretta; questa procedura ha il vantaggio di convincere immediatamente che l'equazione è quella giusta.

(*) *Traduzione a cura della Redazione.*

4405

PHYSICAL REVIEW VOLUME 102, NUMBER 2 APRIL 15, 1956

Calculation of the Scattering Potential from Reflection Coefficients*

H. E. Moses
New York University, New York, New York
(Received November 17, 1955)

It is shown how the scattering potential can be calculated from suitably defined reflection coefficients in the case of the one- and three-dimensional reduced wave equations by means of a formal series expansion. The more general problem of calculating scattering potentials from elements of the scattering operator is also discussed and it is shown that to calculate the scattering potential it is often sufficient to prescribe the representation in which it is to be diagonal.

1. INTRODUCTION AND SUMMARY

JOST and Kohn[1] have developed a procedure for finding spherically symmetric potentials from scattering phases. This problem is simplified by the fact that the solutions of the radial equation are not degenerate.

It has been found possible to generalize their procedure to cases where the outgoing eigenfunctions are degenerate. One is able to show that in many cases the scattering potential can be obtained from certain elements of the scattering operator, provided one specifies the representation in which V is to be diagonal.

In the present paper the method is applied to the one- and three-dimensional scattering problems, where the scattering potential V is assumed to be a function (not necessarily symmetric) of the space variables. It is shown that in the one-dimensional case, the potential can be obtained from the reflection coefficient at one end. The potential is calculated explicitly to the first two orders in the reflection coefficient where it is shown that the results are the same as those obtained using the Gelfand-Levitan procedure.[2,3]

In the three-dimensional case it is shown that the potential can be obtained from the amplitudes of the spherical waves reflected back along the rays on which the incident plane waves are sent, the totality of such rays being those pointing out at right angles to a hemisphere whose center is at the origin.

The general procedure for obtaining the scattering potential from the scattering operator is also discussed. It is shown that one must specify the representation in which V is diagonal to get a unique answer.

We restrict our discussion to cases in which the unperturbed and perturbed Hamiltonians have purely continuous spectra which coincide. However, it is possible to generalize the results to cases where the

* The research reported in this article was done at the Institute of Mathematical Sciences, New York University, and has been made possible through support and sponsorship extended by Geophysics Research Directorate, Air Force Cambridge Research Center.

[1] R. Jost and W. Kohn, Phys. Rev. 87, 977 (1952).

[2] I. Kay, "On the Determination of a Linear System from the Reflection Coefficient," Research Report No. EM-74, Institute of Mathematical Science, Division of Electromagnetic Research, New York University, 1955 (unpublished).

[3] I. Kay and H. E. Moses, "The Determination of the Scattering Potential from the Spectral Measure Function. Part III. Calculation of the Scattering Potential from the Scattering Operator for the One-Dimensional Schrödinger Equation," Nuovo cimento (to be published).

total Hamiltonian has point eigenvalues in addition to the continuous spectrum.

2. REFLECTION COEFFICIENT AND OUTGOING WAVE SOLUTION: ONE-DIMENSIONAL CASE

Let us consider the one-dimensional reduced equation:

$$\left[-\frac{d^2}{dx^2}+V(x)\right]\Psi(x|k)=k^2\Psi(x|k), \tag{2.1}$$

where the scattering potential $V(x)$ vanishes as $|x|\rightarrow\infty$.

For $k>0$, let $\Psi(x|k)$ have the asymptotic form:

$$\begin{aligned}\Psi(x|k)&\rightarrow\frac{e^{ikx}}{(2\pi)^{\frac{1}{2}}}+b(k)\frac{e^{-ikx}}{(2\pi)^{\frac{1}{2}}}, \quad \text{as} \quad x\rightarrow-\infty;\\ \Psi(x|k)&\rightarrow c(k)e^{ikx}, \quad \text{as} \quad x\rightarrow+\infty.\end{aligned} \tag{2.2}$$

Then we define $b(k)$ to be the reflection coefficient. The reflection coefficient gives the amplitude of the reflected wave moving toward the left when the incident waves moves from $-\infty$ toward the origin.

It is our objective to show how the potential $V(x)$ can be calculated from the reflection coefficient $b(k)$.

The "outgoing wave solution" $\Psi_-(x|k)$ is the solution of (2.1) which is the sum of an incident wave and an outgoing scattered wave. It satisfies the integral equation

$$\Psi_-(x|k)=\frac{1}{(2\pi)^{\frac{1}{2}}}e^{ikx} -\frac{i}{2|k|}\int_{-\infty}^{+\infty}e^{i|k|\,|x-x'|}V(x')\Psi_-(x'|k)dx'. \tag{2.3}$$

If we take $k>0$ and let x approach $-\infty$, we see that $\Psi_-(x|k)$ takes the form (2.2), where

$$b(k)=-\frac{i\pi}{k}\int_{-\infty}^{+\infty}\frac{1}{(2\pi)^{\frac{1}{2}}}e^{ikx}V(x)\Psi_-(x|k)dx, \quad k>0. \tag{2.4}$$

Though $b(k)$ has been defined for $k>0$ only, we can continue this function analytically to negative values of k. From (2.3) and (2.4) it can be shown that

$$b(-k)=b^*(k), \quad k>0. \tag{2.5}$$

The asterisk means complex conjugate.

It will be useful to express $\Psi_-(x|k)$ in terms of the representation of the operator $-d^2/dx^2$. If we write

$$\frac{1}{(2\pi)^{\frac{1}{2}}}\int_{-\infty}^{+\infty}e^{-ik'x}\Psi_-(x|k)dx=u_-(k'|k), \tag{2.6}$$

it can be shown that the integral equation (2.3) can be written

$$u_-(k'|k)=\delta(k'-k)+\gamma_-(k^2-k'^2)T_-(k'|k), \tag{2.7}$$

where

$$T_-(k'|k)=\frac{1}{(2\pi)^{\frac{1}{2}}}\int_{-\infty}^{+\infty}e^{-ik'x}V(x)\Psi_-(x|k)dx =\int_{-\infty}^{+\infty}V(k'|k'')u_-(k''|k)dk''. \tag{2.8}$$

In (2.8), $V(k|k')$ is just the operator $V(x)$ in the representation of the operator $-d^2/dx^2$, i.e.,

$$V(k|k')=\frac{1}{2\pi}\int_{-\infty}^{+\infty}e^{-ikx}V(x)e^{ik'x}dx =\frac{1}{2\pi}\int_{-\infty}^{+\infty}V(x)e^{-i(k-k')x}dx. \tag{2.9}$$

The function $\gamma_-(\xi)$ is defined by

$$\gamma_-(\xi)=-i\pi\delta(\xi)+\frac{\mathrm{P}}{\xi}=\lim_{\epsilon\rightarrow+0}\frac{1}{\xi+i\epsilon}, \tag{2.10}$$

where the symbol P means that in integrations over ξ, the principal part of the integral is to be used.

From (2.4) and (2.8), we see that

$$b(k)=\frac{-i\pi}{k}T_-(-k|k), \quad k>0. \tag{2.11}$$

The eigenfunctions $\Psi_-(x|k)$ or equivalently $u_-(k'|k)$ can be shown to form an orthonormal set. They satisfy

$$\int_{-\infty}^{+\infty}\Psi_-^*(x|k)\Psi_-(x|k')dx=\delta(k-k'), \tag{2.12}$$

$$\int_{-\infty}^{+\infty}u_-^*(k''|k)u_-(k''|k')dk''=\delta(k-k'). \tag{2.12a}$$

We shall assume that there are no bound states. Thus the following completeness relations are valid:

$$\int_{-\infty}^{+\infty}\Psi^*(x|k)\Psi(x'|k)dk=\delta(x-x'), \tag{2.13}$$

$$\int_{-\infty}^{+\infty}u_-^*(k|k'')u_-(k'|k'')dk''=\delta(k-k'). \tag{2.13a}$$

We shall now show how $V(x)$ and $V(k|k')$ can be obtained in terms of $V(-k|k)$. Let us write

$$W(k)=V(-k|k)=\frac{1}{2\pi}\int_{-\infty}^{+\infty}V(x)e^{2ikx}dx. \tag{2.14}$$

From (2.14), a relation which will later prove useful is obtained, namely,

$$W(-k)=W^*(k). \tag{2.15}$$

From (2.14), it is seen that

$$V(x)=2\int_{-\infty}^{+\infty} W(k)e^{-2ikx}dk. \tag{2.16}$$

On using (2.9) and (2.16), we find

$$V(k|k')=W\left(\frac{k'-k}{2}\right). \tag{2.17}$$

Equation (2.17) shows that any element of $V(k|k')$ can be obtained from the knowledge of the elements $V(-k|k)=W(k)$. This observation is, in fact, the principal departure from the original Jost-Kohn treatment.

3. BASIC EQUATIONS

We shall now set up two equations which, with Eq. (2.17), will permit us to solve for the three unknowns $W(k)$, $V(k|k')$, and $T_-(k|k')$ in terms of the reflection coefficient $b(k)$. Having obtained $W(k)$, we can find $V(x)$ from (2.16).

Our first equation is that for $T_-(k|k')$. From (2.8) and (2.7), we have

$$T_-(k|k')=V(k|k')+\int_{-\infty}^{+\infty} V(k|k'')\gamma_-(k'^2-k''^2)T_-(k''|k')dk''. \tag{3.1}$$

We now want to find $V(k|k'')$ in terms of $T_-(k|k')$. In reference 1 this task is accomplished by solving (3.1) for $T_-(k|k')$ in terms of $V(k|k')$ through the use of a series. The series is then inverted to get $V(k|k')$ in terms of $T_-(k|k')$. Rather than go through this complicated procedure, we shall use the completeness relation to find $V(k|k')$ in terms of $T_-(k|k')$. From (2.8) and (2.13a), we have

$$V(k|k')=\int_{-\infty}^{+\infty} T_-(k|k'')u_-^*(k'|k'')dk''. \tag{3.2}$$

However, on using (2.7), we see that

$$V(k|k')=T_-(k|k')+\int_{-\infty}^{+\infty} T_-(k|k'')\gamma_-^*(k''^2-k'^2)T_-^*(k'|k'')dk''. \tag{3.3}$$

In particular,

$$V(-k|k)=T_-(-k|k)+\int_{-\infty}^{+\infty} T_-(-k|k'')\gamma_-^*(k''^2-k^2)T_-^*(k|k'')dk''. \tag{3.4}$$

If $k>0$ we have, on using (2.14) and (2.11),

$$W(k)=\frac{ik}{\pi}b(k)+\int_{-\infty}^{+\infty} T_-(-k|k'')\gamma_-^*(k''^2-k^2)\times T_-^*(k|k'')dk'', \quad k>0. \tag{3.5}$$

To extend this result for $k<0$, let $k=-p$, where $p>0$. Then from (2.15) and (2.5),

$$W(k)=W^*(p)=-\frac{ip}{\pi}b^*(p)+\int_{-\infty}^{+\infty} T_-^*(-p|k'')\gamma_-(k''^2-p^2)T_-(p|k'')dk'', \tag{3.6}$$

or

$$W(k)=\frac{ikb(k)}{\pi}+\int_{-\infty}^{+\infty} T_-(-k|k'')\gamma_-(k''^2-k^2)\times T_-^*(k|k'')dk'', \quad k<0. \tag{3.7}$$

We can combine (3.5) and (3.7) as

$$W(k)=\frac{ikb(k)}{\pi}+\int_{-\infty}^{+\infty} T_-(-k|k')[\eta(k)\gamma_-^*(k'^2-k^2)+\eta(-k)\gamma_-(k'^2-k^2)]T_-^*(k|k')dk', \tag{3.8}$$

where $\eta(k)$ is the Heaviside step function defined by

$$\eta(k)=1, \quad k>0; \qquad \eta(k)=0, \quad k<0. \tag{3.9}$$

Equations (3.1), (3.8), and (2.17) are our basic equations for $T_-(k|k')$, $W(k)$, and $V(k|k')$. To solve these equations we replace $b(k)$ in (3.8) by $\epsilon b(k)$, where ϵ is a smallness parameter. Furthermore, we write

$$W(k)=\sum_{n=1}^{\infty}\epsilon^n W^{(n)}(k),$$
$$T_-(k|k')=\sum_{n=1}^{\infty}\epsilon^n T_-^{(n)}(k|k'), \tag{3.10}$$
$$V(k|k')=\sum_{n=1}^{\infty}\epsilon^n V^{(n)}(k|k'),$$

substitute into (3.1), (3.8), and (2.17), and equate coefficients of the same power of ϵ. The lowest orders can be expressed quite simply in terms of $b(k)$. It is seen that

$$W^{(1)}(k)=ikb(k)/\pi,$$
$$V^{(1)}(k|k')=\frac{i(k'-k)}{2\pi}b\left(\frac{k'-k}{2}\right),$$
$$T_-^{(1)}(k|k')=V^{(1)}(k|k')=\frac{i(k'-k)}{2\pi}b\left(\frac{k'-k}{2}\right), \tag{3.11}$$
$$W^{(2)}(k)=\frac{1}{(2\pi)^2}\int_{-\infty}^{+\infty} b\left(\frac{k''+k}{2}\right)b^*\left(\frac{k''-k}{2}\right)dk''=\frac{2}{(2\pi)^2}\int_{-\infty}^{+\infty} b(k'')b(k-k'')dk'',$$

etc. It is easily seen by induction that all orders of the unknown can be obtained in terms of a knowledge of $b(k)$.

We can obtain $V(x)$ in terms of the lowest orders of $b(k)$ quite simply. Writing

$$V(x)=\sum_{n=1}^{\infty}\epsilon^n V^{(n)}(x), \tag{3.12}$$

we see, from (2.17), that

$$V^{(n)}(x)=2\int_{-\infty}^{+\infty}W^{(n)}(k)e^{-2ikx}dk. \tag{3.13}$$

On using (3.11), we obtain

$$V^{(1)}(x)=\frac{2i}{\pi}\int_{-\infty}^{+\infty}kb(k)e^{-2ikx}dk=-2\frac{d}{dx}F(2x), \tag{3.14}$$

where $F(x)$ is the Fourier transform of $b(k)$,

$$F(x)=\frac{1}{2\pi}\int_{-\infty}^{+\infty}b(k)e^{-ikx}dx. \tag{3.15}$$

[It is to be noted, that as a consequence of (2.5), $F(x)$ is real.] Also,

$$\begin{aligned}V^{(2)}(x)&=\frac{4}{(2\pi)^2}\int_{-\infty}^{+\infty}dke^{-2ikx}\int_{-\infty}^{+\infty}b(k'')b(k-k'')dk''\\&=\frac{4}{(2\pi)^2}\int_{-\infty}^{+\infty}e^{-2ik''x}b(k'')dk''\\&\qquad\times\int_{-\infty}^{+\infty}e^{-2ix(k-k'')}b(k-k'')dk\\&=4[F(2x)]^2.\end{aligned} \tag{3.16}$$

The expressions for $V^{(1)}(x)$ and $V^{(2)}(x)$ are precisely those which one would obtain using the Gelfand-Levitan procedure of references 2 and 3.

An alternative, but entirely equivalent procedure for obtaining the potential $V(x)$, is obtained when one substitutes $W(k)$, as given by (3.8), into (2.16). On using (2.8) and the fact that

$$\frac{1}{2\pi}\int_{-\infty}^{+\infty}e^{ikx}[\eta(k)\gamma_-^*(k'^2-k^2)+\eta(-k)\gamma_-(k'^2-k^2)]dk=-\eta(-x)\frac{\sin k'x}{k'}, \tag{3.17}$$

it is seen that

$$\begin{aligned}V(x)&=\frac{2i}{\pi}\int_{-\infty}^{+\infty}kb(k)e^{-2ikx}dk\\&\quad-2\int_{-\infty}^{+\infty}\frac{dk}{k}\int_{-\infty}^{+\infty}V(x'')\Psi_-^*(x''|k)dx''\\&\quad\times\int_{-\infty}^{2x-x''}V(x')\Psi_-(x'|k)\sin k(x'+x''-2x)dx'.\end{aligned} \tag{3.18}$$

Equation (3.18) together with (2.3), which we rewrite as

$$\Psi_-(x|k)=\frac{1}{(2\pi)^{\frac{1}{2}}}e^{ikx}-\frac{i}{2|k|}\int_{-\infty}^{+\infty}e^{i|k|\,|x-x'|}V(x')\Psi_-(x'|k)dx', \tag{3.19}$$

are sufficient to solve for $\Psi_-(x|k)$ and $V(x)$. As before, one replaces $b(k)$ by $\epsilon b(k)$ and writes

$$V(x)=\sum_{n=1}^{\infty}\epsilon^n V^{(n)}(x), \tag{3.20}$$

$$\Psi_-(x|k)=\frac{1}{(2\pi)^{\frac{1}{2}}}e^{ikx}+\sum_{n=1}^{\infty}\epsilon^n\Psi_-^{(n)}(x|k). \tag{3.21}$$

On substituting (3.21) and (3.20) into (3.18) and (3.19) and equating coefficients of ϵ, one again sees how one may obtain $V^{(n)}(x)$, $\Psi_-^{(n)}(x|k)$ from a knowledge of $b(k)$ alone. On looking at (3.18), it is seen that as $x\to-\infty$, the second term on the right becomes small compared to first term. Hence, when $x\to-\infty$, $V(x)$ is approximated by the first term which is just $V^{(1)}(x)$, i.e.,

$$V(x)\cong V^{(1)}(x)=\frac{2i}{\pi}\int_{-\infty}^{+\infty}kb(k)e^{-2ikx}dk,\qquad\text{for } x\to-\infty. \tag{3.22}$$

It is also useful to note that even to the lowest order approximation,

$$\Psi_-(x|k)\cong\frac{1}{(2\pi)^{\frac{1}{2}}}e^{ikx}+\Psi_-^{(1)}(x|k)$$

will reproduce the *exact* reflection coefficient $b(k)$. For

$$\begin{aligned}\Psi_-^{(1)}(x|k)&\cong-\frac{i}{2k}\frac{e^{-ikx}}{(2\pi)^{\frac{1}{2}}}\int_{-\infty}^{+\infty}V^{(1)}(x')e^{2ikx'},\qquad\text{when } k>0 \text{ and } x\to-\infty,\\&=\frac{e^{-ikx}}{(2\pi)^{\frac{1}{2}}}b(k).\end{aligned} \tag{3.23}$$

It can further be shown that $\Psi^{(n)}(x|k)\to0$ when $x\to-\infty$, $k>0$. Hence, when $V(x)$ and $\Psi_-(x|k)$ are solved to *any* order in ϵ, the *exact* reflection coefficient $b(k)$ will be reproduced.

4. SIMPLE EXAMPLE

Let us consider the case where the scattering potential is

$$V(x)=A\delta(x). \tag{4.1}$$

From the direct problem, it can be shown that

$$\Psi_-(x|k)=\frac{1}{(2\pi)^{\frac{1}{2}}}e^{ikx}+\frac{1}{(2\pi)^{\frac{1}{2}}}e^{i|k|\,|x|}C(|k|), \quad (4.2)$$

where

$$C(|k|)=-\tfrac{1}{2}iA\left(\frac{1}{|k|+\frac{1}{2}A}\right). \quad (4.3)$$

From (2.2), it is clear that the reflection coefficient is given by

$$b(k)=-\tfrac{1}{2}iA\left(\frac{1}{k+\frac{1}{2}iA}\right)=C(k). \quad (4.4)$$

From the inverse point of view, we must show that with $b(k)$ given by (4.4), Eqs. (3.18) and (3.19) for $V(x)$ and $\Psi_-(x|k)$ are satisfied by (4.1) and (4.2). That this is the case is easily proved by substitution.

It is, perhaps, interesting to evaluate the lowest two orders for $V(x)$ using the expansion scheme. We have

$$F(x)=\frac{1}{2\pi}\int_{-\infty}^{+\infty}b(k)e^{-ikx}dk=-\frac{A}{2}\eta(x)e^{-Ax/2}. \quad (4.5)$$

Hence, from (3.14),

$$V^{(1)}(x)=-2\frac{d}{dx}F(2x)=A\delta(x)-A^2\eta(x)e^{-Ax}; \quad (4.6)$$

while from (3.16),

$$V^{(2)}(x)=4[F(2x)]^2=A^2\eta(x)e^{-2Ax}. \quad (4.7)$$

Hence to the two lowest orders in the smallness parameter, after setting $\epsilon=1$.

$$V(x)=A\delta(x)-A^2\eta(x)[e^{-Ax}-e^{-2Ax}]. \quad (4.8)$$

The second term on the right of (4.8) represents the error arising from the fact that we are working only to the second approximation.

5. REFLECTION COEFFICIENT AND OUTGOING WAVE SOLUTION: THREE-DIMENSIONAL CASE

In order to define the reflection coefficient for the three-dimensional wave equation, it is useful to introduce the "ray" or "optical" description of vectors which pass through the origin of coordinates.

Let us introduce a rectangular coordinate system. Let us also introduce a system of directed rays. These rays are described by unit vectors whose origin is the origin of coordinates. These vectors may point in any directions so long as the angle θ which this vector makes with the positive z-axis is less than $\pi/2$. An arbitrary vector, say $\mathbf{k}$, which passes through the origin may then be written

$$\mathbf{k}=k\mathbf{r}, \quad (5.1)$$

where $\mathbf{r}$ is the unit vector describing the direction of the ray on which the vector $\mathbf{k}$ lies and

$$k=|\mathbf{k}|\ \mathrm{sign}\ k_z, \quad (5.2)$$

where k_z is the z component of $\mathbf{k}$.

Any function of the vector $\mathbf{k}$, $f(\mathbf{k})$ may be written

$$f(\mathbf{k})=f(k,\theta,\phi),$$
$$0\leq\phi<2\pi,\quad 0\leq\theta<\pi/2,\quad -\infty<k<\infty, \quad (5.3)$$

if one wishes to show explicitly the dependence of the function $f(\mathbf{k})$ on k, and the angles θ (which is the angle made by the ray with the positive z axis) and ϕ (which is the angle which the projection of the ray on the x-y plane makes with the positive x axis). Vector integrals, written in terms of ray coordinates, have the following form:

$$\int f(\mathbf{x})d\mathbf{x}=\int_{-\infty}^{+\infty}dkk^2\int_0^{2\pi}d\phi\int_0^{\pi/2}\sin\theta d\theta f(k,\theta,\phi). \quad (5.4)$$

The integral on the right of (5.4) is the same as that obtained using the usual spherical coordinates except the limits of the integral in the variable θ now range from 0 to $\pi/2$, and the integral in the variable k ranges from $-\infty$ to $+\infty$.

In terms of the ray description, we can now define a reflection coefficient along a ray in three dimensions analogous to the reflection coefficient for one dimension.

Let us consider the solutions $\Psi_-(\mathbf{x}|\mathbf{k})$ of the equation

$$[-\nabla^2+V(\mathbf{x})]\Psi_-(\mathbf{x}|\mathbf{k})=|\mathbf{k}|^2\Psi_-(\mathbf{x}|\mathbf{k}), \quad (5.5)$$

which satisfy the condition that they are the sum of an incident wave and an outgoing spherical wave. Such solutions satisfy the integral equation:

$$\Psi_-(\mathbf{x}|\mathbf{k})=\frac{1}{(2\pi)^{\frac{3}{2}}}e^{i\mathbf{k}\cdot\mathbf{x}}-\frac{1}{4\pi}\int\frac{e^{i|\mathbf{k}|\,|\mathbf{x}-\mathbf{x}'|}}{|\mathbf{x}-\mathbf{x}'|}V(\mathbf{x}')\Psi_-(\mathbf{x}'|\mathbf{k})d\mathbf{x}'. \quad (5.6)$$

We shall define the reflection coefficient $b(\mathbf{k})=b(k,\theta,\phi)$ as being the amplitude of the spherical wave which one obtains from (5.6) when the optical coordinate k is taken positive and when $|\mathbf{x}|$ approaches infinity in the negative direction on the ray on which $\mathbf{k}$ lies. Generally, when $|\mathbf{x}|\to\infty$ we have, from (5.6),

$$\Psi_-(\mathbf{x}|\mathbf{k})=\frac{1}{(2\pi)^{\frac{3}{2}}}e^{i\mathbf{k}\cdot\mathbf{x}}-\left(\frac{\pi}{2}\right)^{\frac{1}{2}}\frac{e^{i|\mathbf{k}|\,|\mathbf{x}|}}{|\mathbf{x}|}\times\int\frac{1}{(2\pi)^{\frac{3}{2}}}e^{-i\mathbf{k}'\cdot\mathbf{x}'}V(\mathbf{x}')\Psi_-(\mathbf{x}'|\mathbf{k})d\mathbf{x}', \quad (5.7)$$

where

$$\mathbf{k}'=|\mathbf{k}|\times\frac{\mathbf{x}}{|\mathbf{x}|}.$$

On setting $k>0$ and $\mathbf{k}'=-\mathbf{k}$ we have, by definition of $b(\mathbf{k})$,

$$\Psi_-(\mathbf{x}|\mathbf{k})=\frac{e^{i\mathbf{k}\cdot\mathbf{x}}}{(2\pi)^{\frac{3}{2}}}+\frac{e^{i|\mathbf{k}|\,|\mathbf{x}|}}{|\mathbf{x}|}b(\mathbf{k}), \tag{5.8a}$$

where

$$b(\mathbf{k})=b(k,\theta,\phi)=-\left(\frac{\pi}{2}\right)^{\frac{1}{2}}\int\frac{1}{(2\pi)^{\frac{3}{2}}}e^{ik\cdot x'}$$
$$\times V(\mathbf{x}')\Psi_-(x'|\mathbf{k})dx',\quad k>0. \tag{5.8b}$$

It is our objective to show how the scattering potential can be obtained from $b(\mathbf{k})$, where $\mathbf{k}$ is such that it makes an angle less than $\pi/2$ radians with the positive z-axis. The development is almost identical with the one-dimensional case.

We note first that though $b(\mathbf{k})=b(k,\theta,\phi)$ is defined for $k>0$ only, we can analytically continue $b(k,\theta,\phi)$ (for fixed θ,ϕ) to negative values of k. It can be shown that

$$b(-\mathbf{k})=b^*(\mathbf{k}),\quad k>0 \tag{5.9}$$

or

$$b(-k,\theta,\phi)=b^*(k,\theta,\phi),\quad k>0. \tag{5.9a}$$

It is convenient to express $\Psi_-(\mathbf{x}|\mathbf{k})$ in terms of the representation of the operator $-\nabla^2$. If we write

$$u_-(\mathbf{k}'|\mathbf{k})=\frac{1}{(2\pi)^{\frac{3}{2}}}\int e^{-i(\mathbf{k}'\cdot\mathbf{x})}\Psi_-(\mathbf{x}|\mathbf{k})d\mathbf{x}, \tag{5.10}$$

it can be shown that the integral equation (5.6) yields the following equation for $u_-(\mathbf{k}|\mathbf{k}')$:

$$u_-(\mathbf{k}|\mathbf{k}')=\delta(\mathbf{k}-\mathbf{k}')+\gamma_-(k'^2-k^2)T_-(\mathbf{k}|\mathbf{k}') \tag{5.11}$$

where

$$T_-(\mathbf{k}|\mathbf{k}')=\frac{1}{(2\pi)^{\frac{3}{2}}}\int e^{-i\mathbf{k}\cdot\mathbf{x}}V(\mathbf{x})\Psi_-(\mathbf{x}|\mathbf{k}')dx$$
$$=\int V(\mathbf{k}|\mathbf{k}'')u_-(\mathbf{k}''|\mathbf{k}')d\mathbf{k}'' \tag{5.12}$$

and

$$V(\mathbf{k}|\mathbf{k}')=\frac{1}{(2\pi)^3}\int e^{i(\mathbf{k}'-\mathbf{k})\cdot\mathbf{x}}V(\mathbf{x})dx. \tag{5.13}$$

If we assume that there are no bound states, we have the equations

$$\int u_-^*(k|k')u_-(k|k'')dk=\delta(k'-k''),$$
$$\int u_-^*(k'|k)u_-(k''|k)dk=\delta(k'-k''). \tag{5.14}$$

From (5.8b) it is seen that

$$b(\mathbf{k})=-(\pi/2)^{\frac{1}{2}}T_-(-\mathbf{k}|\mathbf{k}),\quad k>0. \tag{5.15}$$

6. BASIC EQUATIONS IN THREE DIMENSIONS

Since the procedure for finding the basic equations is almost identical to that for the one-dimensional case, we shall merely summarize our results below.

The first basic equation is

$$T_-(\mathbf{k}|\mathbf{k}')=V(\mathbf{k}|\mathbf{k}')$$
$$+\int V(\mathbf{k}|\mathbf{k}'')\gamma_-(k'^2-k''^2)T_-(\mathbf{k}''|\mathbf{k}')d\mathbf{k}''. \tag{6.1}$$

The second is

$$W(\mathbf{k})=-\left(\frac{2}{\pi}\right)^{\frac{1}{2}}b(\mathbf{k})+\int T_-(-\mathbf{k}|\mathbf{k}')[\eta(k)\gamma_-^*(k'^2-k^2)$$
$$+\eta(-k)\gamma_-(k'^2-k^2)]T_-^*(\mathbf{k}|\mathbf{k}')d\mathbf{k}', \tag{6.2}$$

while the third is

$$V(\mathbf{k}|\mathbf{k}')=W\left(\frac{\mathbf{k}'-\mathbf{k}}{2}\right). \tag{6.3}$$

One solves these three equations as in the one-dimensional case. One replaces $b(\mathbf{k})$ by $\epsilon b(\mathbf{k})$ and writes

$$T_-(\mathbf{k}|\mathbf{k}')=\sum_{n=1}^{\infty}\epsilon^n T_-^{(n)}(\mathbf{k}|\mathbf{k}'),$$
$$W(\mathbf{k})=\sum_{n=1}^{\infty}\epsilon^n W^{(n)}(\mathbf{k}), \tag{6.4}$$
$$V(\mathbf{k}|\mathbf{k}')=\sum_{n=1}^{\infty}\epsilon^n W^{(n)}\left(\frac{\mathbf{k}'-\mathbf{k}}{2}\right).$$

Upon substituting (6.4) into (6.1)–(6.3), one sees that $T^{(n)}$ and $W^{(n)}$ can be obtained from a knowledge of $b(\mathbf{k})$ alone. To obtain $V(\mathbf{x})$ from $W(\mathbf{k})$, one uses

$$V(\mathbf{x})=8\int W(\mathbf{k})e^{-2i\mathbf{k}\cdot\mathbf{x}}d\mathbf{k}''. \tag{6.5}$$

In the special case where $V(\mathbf{x})$ is a function of $|\mathbf{x}|$ alone, i.e., where $V(\mathbf{x})$ is a spherically symmetric potential, $b(\mathbf{k})$ will be independent of the direction of the ray. In principle, one could calculate such a potential by observing the amplitude for all energies of the spherical wave reflected backward along a single ray upon which an incident plane wave was sent. This datum is quite different from that of reference 1, where one requires knowledge of the value of a scattering phase for all energies for a single angular momentum. It also differs from the data required in the work of Wheeler.[4] Here knowledge of all phases is required for a single energy. The classical analog of our result for spherically symmetric potentials is given by Keller *et al.*[5]

[4] J. A. Wheeler, Phys. Rev. **99**, 630 (1955).
[5] Keller, Kay, and Shmoys, Phys. Rev. **102**, 557 (1956).

One can write the basic equations in the $\mathbf{x}$ representation in analogy to the one-dimensional case. Our basic equations are then

$$V(\mathbf{x})=-\left(\frac{128}{\pi}\right)^{\frac{1}{2}}\int b(\mathbf{k})e^{-2i\mathbf{k}\cdot\mathbf{x}}d\mathbf{k}$$
$$+8\int d\mathbf{k}\int V(\mathbf{x}')\Psi_-(\mathbf{x}'|\mathbf{k})d\mathbf{x}',$$
$$\times\int V(\mathbf{x}'')\Psi_-^*(\mathbf{x}''|\mathbf{k})d\mathbf{x}''g_{\mathbf{k}}(\mathbf{x}'+\mathbf{x}''-2\mathbf{x}), \quad (6.6)$$

and

$$\Psi_-(\mathbf{x}|\mathbf{k})=\frac{e^{i\mathbf{k}\cdot\mathbf{x}}}{(2\pi)^{\frac{3}{2}}}$$
$$-\frac{1}{4\pi}\int e^{i|\mathbf{k}|\,|\mathbf{x}-\mathbf{x}'|}V(\mathbf{x}')\Psi_-(\mathbf{x}'|\mathbf{k})d\mathbf{x}'. \quad (6.7)$$

In (6.6), the function $g_{\mathbf{k}}(\mathbf{x})$ is a Green's function for the operator $(\nabla^2+\mathbf{k}^2)$ given by

$$g_{\mathbf{k}}(\mathbf{x})=\frac{1}{(2\pi)^3}\int e^{i\mathbf{p}\cdot\mathbf{x}}[\eta(p)\gamma_-^*(k^2-p^2)$$
$$+\eta(-p)\gamma_-(k^2-p^2)]d\mathbf{p}$$
$$=\frac{-\eta(-z)}{(2\pi)^2}\int_{-\infty}^{+\infty}d\lambda\int_{-\infty}^{+\infty}d\nu e^{i\lambda x}e^{i\nu y}$$
$$\times\frac{\sin[(k^2-\lambda^2-\nu^2)^{\frac{1}{2}}z]}{(k^2-\lambda^2-\nu^2)^{\frac{1}{2}}}. \quad (6.8)$$

In (6.8), k and p are the optical coordinates of $\mathbf{k}$ and $\mathbf{p}$; x, y, and z are the Cartesian coordinates of the vector $\mathbf{x}$.

To solve (6.6) and (6.8) simultaneously, one replaces $b(\mathbf{k})$ by $\epsilon b(\mathbf{k})$ and writes

$$V(\mathbf{x})=\sum_{n=1}^{\infty}\epsilon^n V^{(n)}(\mathbf{x}),$$
$$\Psi_-(\mathbf{x}|\mathbf{k})=\frac{1}{(2\pi)^{\frac{3}{2}}}e^{i\mathbf{k}\cdot\mathbf{x}}+\sum_{n=1}^{\infty}\epsilon^n\Psi_-^{(n)}(\mathbf{x}|\mathbf{k}). \quad (6.9)$$

Upon substituting into (6.6) and (6.8), one obtains an iteration procedure for finding $V^{(n)}(\mathbf{x})$ and $\Psi_-^{(n)}(\mathbf{x})$. In analogy to the one-dimensional case, it can be shown that, to any approximation, $\Psi_-(\mathbf{x}|\mathbf{k})$ reproduces the reflection coefficient $b(\mathbf{k})$.

7. GENERAL PROCEDURE

There is a more general procedure which may often be used to obtain the scattering potential from appropriate elements of the scattering operator provided one has specified the representation in which the scattering potential is diagonal. In fact, the writer originally obtained the results for the one- and three-dimensional case in terms of this procedure.

We shall now discuss the problem in detail. We shall use the notation and equations of a previous paper.[6]

Let H and H_0 be the total and unperturbed Hamiltonians respectively, and let V denote the perturbation. Then we have

$$H=H_0+V. \quad (7.1)$$

Let the eigenfunctions of H_0 be denoted by $\omega_0(E,\alpha)$, where E is the quantum number associated with the operator H_0 $(0\leq E<\infty)$ and α labels the degeneracy. For example, α might denote the direction or angular momentum in the case of a single particle. We shall assume that H has the same spectrum as H_0 (more general Hamiltonians can be treated similarly). There is one class of eigenfunctions of H which is of interest, namely the "outgoing" eigenfunctions. We shall call them $\omega_-(F,\beta)$. The variable F denotes the eigenvalues of H $(0\leq F<\infty)$. The variable β is the degeneracy label for H. Since we have assumed the spectrum of H to be the same as that of H_0, β will have the same range as α.

It is convenient to work in the H_0 representation. The variables corresponding to ω_- in the H_0 representation, denoted by $u_-(E,\alpha|F,\beta)$, are

$$u_-(E,\alpha|F,\beta)=(\omega_0(E,\alpha),\omega_-(F,\beta)); \quad (7.2)$$

the right-hand side of (7.2) indicates the inner product of the eigenvectors involved. The eigenfunctions ω_- are characterized by the fact that the corresponding variables in the H_0 representation satisfy the integral equations

$$u_-(E,\alpha|F,\beta)=\delta(F-E)\delta(\beta,\alpha)$$
$$+\gamma_-(F-E)T_-(E,\alpha|F,\beta), \quad (7.3)$$

where $\delta(\alpha,\beta)$ is a suitable generalization of the Dirac or Kronecker δ function, and where

$$\gamma_-(x)=-i\pi\delta(x)+\mathrm{P}/x \quad (7.4)$$

(P means Cauchy principal value), and

$$T_-(E,\alpha|F,\beta)$$
$$=\int\int V(E,\alpha|E',\alpha')u_-(E',\alpha'|F,\beta)d\alpha'dE'. \quad (7.5)$$

In (7.5) $V(E,\alpha|E',\alpha')$ is the "matrix" of V in the H_0-representation. Explicitly,

$$V(E,\alpha|E',\alpha')=(\omega_0(E,\alpha),V\omega_0(E',\alpha')). \quad (7.6)$$

It can be shown that the scattering operator in the H_0-representation is given by

$$S(E,\alpha|E',\alpha')=\delta(E-E')\delta(\alpha,\alpha')$$
$$-2\pi i\delta(E-E')T_-(E',\alpha|E',\alpha'). \quad (7.7)$$

---

[6] H. E. Moses, Nuovo cimento 1, 103 (1955).

Now, our problem is to find $V(E,\alpha|E',\alpha')$ in terms of a given $S(E,\alpha|E',\alpha')$, or equivalently $T_-(E',\alpha|E',\alpha')$.

From (7.3) and (7.5) it is clear that

$$T_-(E,\alpha|F,\beta)=V(E,\alpha|F,\beta)+\int\int V(E,\alpha|E',\alpha')\times\gamma_-(F-E')T_-(E',\alpha'|F,\beta)dE'd\alpha'. \quad (7.8)$$

Since we have assumed that the spectrum of H is the same as that of H_0, the completeness condition

$$\int\int u_-^*(E,\alpha|F,\beta)u_-(E',\alpha'|F,\beta)dFd\beta=\delta(E-E')\delta(\alpha,\alpha') \quad (7.9)$$

must be satisfied. Using (7.9) and (7.5), we have

$$\int\int T_-(E,\alpha|F,\beta)u_-^*(E'',\alpha''|F,\beta)dFd\beta$$
$$=\int\int V(E,\alpha|E',\alpha')d\alpha'dE'\times\int\int u_-(E'',\alpha''|F,\beta)u_-^*(E',\alpha'|F,\beta)dFd\beta$$
$$=V(E,\alpha|E'',\alpha'').$$

Using (7.3) we have finally

$$V(E,\alpha|E',\alpha')=T_-(E,\alpha|E',\alpha')+\int\int T_-(E,\alpha|F,\beta)\times\gamma_-^*(F-E')T_-^*(E',\alpha'|F,\beta)dFd\beta. \quad (7.10)$$

On first thought it might seem that Eqs. (7.8) and (7.10) can be considered as two equations in two unknowns, namely $V(E,\alpha|E',\alpha')$ and $T_-(E,\alpha|F,\beta)$. However, this conjecture is false, for if $V(E,\alpha|E',\alpha')$ as given by (7.10) is substituted in (7.8) so as to obtain an equation for T_- only, and if the orthogonality relations

$$\int\int u_-^*(E,\alpha|F,\beta)u_-(E,\alpha|F',\beta')dEd\alpha=\delta(F-F')\delta(\beta,\beta')$$

are used, the result is a trivial identity. Hence (7.8) and (7.10) do not contain enough information to solve for V. The additional information needed is obtained, for example, by specifying the representation in which V is to be diagonal. It will be shown that with such a specification it is possible to solve for $T_-(E,\alpha|E',\alpha')$ in terms of $T_-(E',\alpha|E',\alpha')$ (i.e., in terms of the scattering operator). Then we can find $V(E,\alpha|E',\alpha')$ from (7.10).

Let q denote collectively the eigenvalues of the complete set of commuting variables in which we wish V to be diagonal. Furthermore, let $\Psi(q|E,\alpha)$ (assumed known) be the eigenfunctions $\omega_0(E,\alpha)$ of H_0 as given in the q-representation. Then we have

$$V(E,\alpha|E',\alpha')=\int\Psi^*(q|E,\alpha)V(q)\Psi(q|E',\alpha')dq, \quad (7.11)$$

where $V(q)$ is the potential as expressed in the q-representation. In particular,

$$V(E,\alpha|E,\alpha')=\int\Psi^*(q|E,\alpha)\Psi(q|E,\alpha')V(q)dq$$
$$=\int L(E,\alpha,\alpha'|q)V(q)dq. \quad (7.11a)$$

The kernel L is defined by the second equation (7.11a). Equation (7.11a) is to be considered as an integral equation for $V(q)$ which is to be solved in terms of $V(E,\alpha|E,\alpha')$.

$$V(q)=\int\int\int L^{-1}(E,\alpha,\alpha'|q)\times V(E,\alpha|E,\alpha')dEd\alpha d\alpha', \quad (7.12)$$

where the operator L^{-1} is the inverse of L. The existence of this operator will depend on the choice of q. From (7.11) we see that we can write $V(E,\alpha|E',\alpha')$ in terms of $V(E,\alpha|E,\alpha')$ as follows:

$$V(E,\alpha|E',\alpha')$$
$$=\int\int\int\int\Psi^*(E,\alpha|q)\Psi(E',\alpha'|q)\times L^{-1}(F',\beta,\beta'|q)V(F',\beta|F',\beta')dqdF'd\beta d\beta'$$
$$=\int\int\int K(E,\alpha|E',\alpha'||F',\beta|F',\beta')\times V(F',\beta|F',\beta')dF'd\beta d\beta', \quad (7.13)$$

where $K(E,\alpha|E',\alpha'||F',\beta|F',\beta')$ is a known function of its arguments and depends on the representation to be chosen for the diagonalization of V. Equation (7.13) provides the additional information needed to solve for $T_-(E,\alpha|F,\beta)$ in terms of $T_-(E,\alpha|E,\alpha')$. Equation (7.13) shows how all of the matrix elements $V(E,\alpha|E',\alpha')$ may be obtained from a knowledge of some of the elements $V(E,\alpha|E,\alpha')$. Equations (2.17) and (6.3) are the forms that Eq. (7.13) takes in the one- and three-dimensional cases. As will be shown below, those elements of $V(E,\alpha|E,\alpha')$ which are used to determine the remainder of the elements $V(E,\alpha|E',\alpha')$ determine also which elements of the scattering operator $T_-(E,\alpha|E,\alpha')$ are needed to find V.

Using (7.13) and (7.10), we have

$$V(E,\alpha|E',\alpha')=\iiint K(E,\alpha|E',\alpha'||F,\beta|F,\beta')$$
$$\times T_-(F,\beta|F,\beta')dFd\beta d\beta'$$
$$+\iiiii K(E,\alpha|E',\alpha'||F,\beta|F,\beta')$$
$$\times T_-(F,\beta|G,\gamma)\gamma_-^*(G-F)$$
$$\times T_-^*(F,\beta'|G,\gamma)dFd\beta d\beta' dGd\gamma, \quad (7.14)$$

which we substitute into (7.18) to obtain

$$T_-(E,\alpha|E',\alpha')=\iiint K(E,\alpha|E',\alpha'||F,\beta|F,\beta')$$
$$\times T_-(F,\beta|F,\beta')dFd\beta d\beta'$$
$$+\iiiii K(E,\alpha|E',\alpha'||F,\beta|F,\beta')$$
$$\times T_-(F,\beta|G,\gamma)\gamma_-^*(G-F)$$
$$\times T_-^*(F,\beta'|G,\gamma)dFd\beta d\beta' dGd\gamma$$
$$+\iiiii K(E,\alpha|D,\delta||F,\beta|F,\beta')$$
$$\times T_-(F,\beta|F,\beta')\gamma_-(E'-D)$$
$$\times T_-(D,\delta|E',\alpha')dFd\beta d\beta' dDd\delta$$
$$+\iiiiiii K(E,\alpha|D,\delta||F,\beta|F,\beta')$$
$$\times T_-(F,\beta|G,\gamma)\gamma_-^*(G-F)$$
$$\times T_-^*(F,\beta'|G,\gamma)\gamma_-(E'-D)T_-(D,\delta|E',\alpha')$$
$$\times dFd\beta d\beta' dGd\gamma dDd\delta. \quad (7.15)$$

When this equation is solved for $T_-(E,\alpha|E',\alpha')$, one can substitute in (7.10) to obtain $V(E,\alpha|F,\beta)$. The Jost-Kohn procedure for solving (7.15) is to replace $T_-(F,\beta|F,\beta')$ by $\epsilon T(F,\beta|F,\beta')$, where ϵ is a smallness parameter. One then expands $T_-(E,\alpha|E',\alpha')$ in an infinite series in ϵ as follows:

$$T_-(E,\alpha|F,\beta)=\sum_{n=1}^{\infty}\epsilon^n T_-^{(n)}(E,\alpha|F,\beta), \quad (7.16)$$

and, substituting (7.16) into (7.15) and equating coefficients of equal powers ϵ, one obtains a series of expressions in which the $T_-^{(n)}(E,\alpha|F,\beta)$ are given in terms of $T_-^{(m)}(E,\alpha|F,\beta)$, $(m<n)$. One then can solve for $T_-^{(n)}$ by induction and substitute into (7.16) to obtain $T_-(E,\alpha|F,\beta)$. For the lowest two orders, for example, we find

$$T_-^{(1)}(E,\alpha|E',\alpha')$$
$$=\iiint K(E,\alpha|E',\alpha'||F,\beta|F,\beta')$$
$$\times T_-(F,\beta|F,\beta')dFd\beta d\beta', \quad (7.17)$$

$$T_-^{(2)}(E,\alpha|E',\alpha')$$
$$=\iiiii K(E,\alpha|E',\alpha'||F,\beta|F,\beta')$$
$$\times T_-^{(1)}(F,\beta|G,\gamma)\gamma_-^*(G-F)$$
$$\times T_-^{*(1)}(F,\beta|G,\gamma)dFd\beta d\beta' d\gamma$$
$$+\iiiii K(E,\alpha|D,\delta||F,\beta|F,\beta')$$
$$\times T_-^{(1)}(F,\beta|F,\beta')\gamma_-(E'-D)$$
$$\times T_-^{(1)}(D,\delta|E',\alpha')dFd\beta d\beta' dDd\delta. \quad (7.18)$$

JOURNAL OF APPLIED PHYSICS VOLUME 27, NUMBER 12 DECEMBER, 1956

Reflectionless Transmission through Dielectrics and Scattering Potentials*

I. KAY AND H. E. MOSES
Institute of Mathematical Sciences, Division of Electromagnetic Research, New York University, New York 3, New York

(Received June 5, 1956)

A complete solution is given for the problem of constructing a plane stratified dielectric medium having the property that at a fixed frequency and polarization a plane wave at any angle of incidence will be transmitted without reflection by the medium. This problem can also be interpreted quantum-mechanically. The dielectric medium of the first interpretation becomes, in the second interpretation, a potential for the one-dimensional Schrödinger equation such that a particle which is sent in from $-\infty$ is transmitted with probability one of getting through to $+\infty$, no matter what the initial kinetic energy of the particle may be.

In addition, the results can be applied to the problem of finding perfectly matched nonuniform transmission lines.

1. INTRODUCTION

A DIFFICULT problem of practical as well as theoretical interest is the construction of a solid dielectric medium which is perfectly transparent to electromagnetic radiation. Generally the index of refraction of the medium is required to vary between set minimum and maximum values, and the medium itself must be confined to a slab of finite thickness bounded on either side by a vacuum.

The problem as stated, however, is an ideal one which seems impossible to solve. The most one can hope to do is to guarantee that a large percentage of the incident energy in a given frequency band over a given range of incidence angles will be transmitted with a correspondingly small percentage of the energy reflected by the medium.

As a start toward investigating this restricted, but still difficult, version of the problem it seems reasonable to look for dielectric media which are transparent in an even more limited sense. Thus, we shall consider an infinite class of dielectric media having the following properties:

(1) Each medium in the class will have a plane stratified variable index of refraction extending from $-\infty$ to $+\infty$;

(2) At all angles of incidence but at a fixed frequency and polarization a plane wave incident from $-\infty$ will be transmitted to $+\infty$ without reflection;

(3) The index of refraction of each medium will vary continuously throughout and will approach unity at both $-\infty$ and $+\infty$.

There is some hope that the results to be given here can be applied to the more complicated practical problem after further investigation. That is, in the infinite class of dielectrics we shall obtain there may be one or more which satisfies all requirements of the original problem well enough to be of use.

We start with the wave equation in a plane stratified medium with an index of refraction $n(x)$, $-\infty<x<\infty$,

$$\Delta w(x,y,z,k)+k^2n^2(x)w(x,y,z,k)=0. \tag{1.1}$$

It is convenient for physical reasons to assume that the index of refraction has the form

$$n^2(x)=1+\mu^2(x). \tag{1.2}$$

For an incident plane wave we can assume that the solution of (1.1) has the form

$$w(x,y,z,k)=u(x,k)\exp[ik(y\cos\beta+z\cos\gamma)], \tag{1.3}$$

if the propagation direction is given by the direction cosines: $\cos\alpha$, $\cos\beta$, $\cos\gamma$. Using (1.2) and (1.3) in (1.1) we arrive at

$$\frac{d^2u(x,E)}{dx^2}+[E-V(x)]u(x,E)=0, \tag{1.4}$$

where

$$E=k^2\cos^2\alpha \quad\text{and}\quad V(x)=-k^2\mu^2(x). \tag{1.5}$$

We note that $V(x)$ must be negative or zero for all x for any physically realizable medium.

Our problem now is to find a function $V(x)$ such that for a solution $u(x,E)$ of (1.4) having the property that

$$u(x,E)\sim\exp(i\sqrt{E}x) \quad\text{as}\quad x\rightarrow-\infty,$$

we shall have

$$u(x,E)\sim t(E)\exp(i\sqrt{E}x) \quad\text{as}\quad x\rightarrow+\infty. \tag{1.6}$$

We shall see in what follows that an infinite class of such functions $V(x)$ can be constructed. In addition, it will be proved that each $V(x)$ is continuous and everywhere negative for all real x and that $V(x)$ approaches zero exponentially as x approaches $+\infty$ or $-\infty$. Every $V(x)$ of the infinite class of functions, as well as the corresponding solution $u(x,E)$ of (1.4), will turn out to be an elementary function of x. It will be shown in Appendix 1 that these functions $V(x)$ are the only possible ones satisfying our conditions.

Note that (1.4) is the one-dimensional Schrödinger equation. Another physical interpretation of the functions $V(x)$ we shall construct is therefore possible: each

* The research reported in this article has been made possible through support and sponsorship extended by the U. S. Air Force, Air Force Cambridge Research Center, under Contract No. AF 19(122)-42.

such $V(x)$ is a potential which has the property that it will allow an incident particle to pass through it with probability one, no matter what the particle's initial kinetic energy may be.[1]

Another possible application of the results might also be mentioned at this point. There exists a transformation[2] which changes an equation of the form (1.4) into the transmission-line equations for a two-wire line with variable capacitance and inductance. In this case the transmission line corresponding to any $V(x)$ we obtain having properties just described will be matched perfectly at all frequencies, so that a voltage or current wave sent in from the left will travel toward the right with no reflection. The necessary transformation will be given in Appendix 2.

2. CONSTRUCTION OF $V(x)$

To obtain a $2N$-parameter family of functions $V(x)$ each of which has the desired properties we first choose N arbitrary positive constants A_n and N arbitrary positive constants κ_n. We shall assume that a corresponding solution $u(x,E)$ of (1.4) has the form

$$u(x,E)=\{1+\sum_{n=1}^{N}[f_n(x)/(\kappa_n+i\sqrt{E})]\exp(\kappa_n x)\}\times\exp(i\sqrt{E}x). \quad (2.1)$$

The functions $f_n(x)$ are defined to be solutions of the N linear algebraic equations

$$A_n\exp(\kappa_n x)\sum_{\nu=1}^{N}[\exp(\kappa_\nu x)/(\kappa_n+\kappa_\nu)]f_\nu(x)+f_n(x)+A_n\exp(\kappa_n x)=0,\quad n=1,\cdots,N. \quad (2.2)$$

It will turn out that the differential equation (1.4) has N discrete eigenvalues, $E=-\kappa_n^2$, and the functions $f_n(x)$ are the corresponding eigenfunctions.

Before we can prove the main result it is necessary to show that the system (2.2) has a unique solution for the set of N functions $f_n(x)$.

Lemma 1: The system (2.2) has a nonsingular coefficient matrix whose determinant is positive for all finite x.

Proof: The coefficient matrix of (2.2) can be written $I+A$, where

$$I=\|\delta_{n\nu}\|,\text{ the unit matrix,}$$
and
$$A=\|A_n\exp(\kappa_n+\kappa_\nu)/(\kappa_n+\kappa_\nu)\|. \quad (2.3)$$

We apply the diagonal transformation

$$D=\|A_n^{-\frac{1}{2}}\delta_{n\nu}\|, \quad (2.4)$$

to (2.2), obtaining the result

$$\sum_{\nu=1}^{N}\{A_n^{\frac{1}{2}}A_\nu^{\frac{1}{2}}\exp[(\kappa_n+\kappa_\nu)x]/(\kappa_n+\kappa_\nu)\}g_\nu(x)+g_n(x)+A_n^{\frac{1}{2}}\exp(\kappa_m x)=0, \quad (2.5)$$

where we define $g_n(x)=A_n^{-\frac{1}{2}}f_n(x)$. For the symmetric matrix of (2.5) we write

$$\hat{A}=\|A_n^{\frac{1}{2}}A_\nu^{\frac{1}{2}}\exp[(\kappa_n+\kappa_\nu)x]/(\kappa_n+\kappa_\nu)\|=DAD^{-1}. \quad (2.6)$$

Since, according to (2.6), the matrices A and $\hat{A}$ are similar, we have

$$\det A=\det\hat{A}, \quad (2.7)$$

and also

$$\det(A+I)=\det(\hat{A}+I).$$

We can now work with the symmetric coefficient matrix $\hat{A}+I$ of (2.5).

To prove that $\det(\hat{A}+I)>0$, it is sufficient to prove that $\det\hat{A}>0$. We shall show, in fact, that $\hat{A}$ is a positive definite matrix; that is, given any set of N numbers y_n,

$$\sum_{n,\nu=1}^{N}y_n\bar{y}_\nu A_n^{\frac{1}{2}}A_\nu^{\frac{1}{2}}\exp[(\kappa_n+\kappa_\nu)x]/(\kappa_n+\kappa_\nu)>0, \quad (2.8)$$

unless all the y_n are zero.

The left side of (2.8) can be written

$$\int_{-\infty}^{x}\sum_{n,\nu=1}^{N}A_n^{\frac{1}{2}}A_\nu^{\frac{1}{2}}y_n\bar{y}_\nu\exp[(\kappa_n+\kappa_\nu)z]dz=\int_{-\infty}^{x}\left|\sum_{n=1}^{N}A_n^{\frac{1}{2}}y_n\exp(\kappa_n z)\right|^2dz. \quad (2.9)$$

Expression (2.9) is always positive unless

$$\sum_{n=1}^{N}A_n^{\frac{1}{2}}y_n\exp(\kappa_n z)=0,$$

for all z between $-\infty$ and x. It follows at once that (2.9) is positive unless every y_n is zero, and thus *Lemma 1* is proved.

The main result can now be proved:

Theorem 1: The function $u(x,E)$ defined in (2.1) and (2.2) satisfies the differential equation

$$d^2u/dx^2+[E-V(x)]u=0, \quad (2.10)$$

where

$$V(x)=-2d^2\{\log\det(\hat{A}+I)\}/dx^2, \quad (2.11)$$

and $\hat{A}$ is the symmetric matrix of (2.6).

Proof: For convenience we define two symbols L and M_n to represent the differential operators

$$L=d^2/dx^2+E-V(x),$$

[1] The situation here resembles one described by Bargmann [Revs. Modern Phys. **21**, 3, 488–493 (1949)] for the radial Schrödinger equation with zero angular momentum. Bargmann has produced an example of a potential for the radial equation such that the corresponding zero angular momentum asymptotic phase shift is zero. In spite of their superficial resemblance, however, the problem considered here and the example of Bargmann are quite different in several important respects.

[2] I. Kay, The inverse scattering problem, Research Rept. EM-74, Inst. Math. Sci., New York University (1955), pp. 26–38.

and

$$M_n = d^2/dx^2 - [\kappa_n^2 + V(x)].$$

We substitute the expression for $u(x,E)$ given by (2.1) into the expression on the left in (2.10). We obtain

$$Lu = \{\sum_{n=1}^{N} [M_n f_n(x)] \exp(\kappa_n x)/(\kappa_n + i\sqrt{E}) + 2d[\sum_{n=1}^{N} f_n(x) \exp(\kappa_n x)]/dx - V(x)\} \times \exp(i\sqrt{E}x). \quad (2.12)$$

Obviously (2.12) is zero if $f_n(x)$ satisfies

$$M_n f_n(x) = 0, \quad (2.13)$$

and also

$$V(x) = 2d\{\sum_{n=1}^{N} f_n(x) \exp(\kappa_n x)\}/dx. \quad (2.14)$$

We apply the operator M_n to the nth equation of (2.2) and substitute the expression for $V(x)$ given by (2.14). The result is

$$M_n f_n(x) + A_n \exp(\kappa_n x) \cdot \sum_{\nu=1}^{N} [M_\nu f_\nu(x)/(\kappa_n + \kappa_\nu)] \exp(\kappa_\nu x) = 0. \quad (2.15)$$

Equations (2.15) are a homogeneous system of linear algebraic equations for the quantity $M_n f_n(x)$ with coefficient matrix $A+I$. By *Lemma 1*, this matrix has an inverse, and therefore $M_n f_n(x) = 0$. Thus, (2.13) is satisfied by the $f_n(x)$ as defined in (2.2) when $V(x)$ has the definition (2.14).

At this stage we know that for $V(x)$ given by (2.14) the $u(x,E)$ defined by (2.1) and (2.2) satisfies (2.10). It remains to show that (2.14) and (2.11) are equivalent statements. To do this we write the solutions $f_n(x)$ of (2.2) in determinant form and put them together in the expression

$$\sum_{n=1}^{N} f_n(x) \exp(\kappa_n x).$$

By inspection it can be seen that

$$\sum_{n=1}^{N} f_n(x) \exp(\kappa_n x) = -d\{\log \det(A+I)\}/dx = -d\{\log \det(\hat{A}+I)\}/dx. \quad (2.16)$$

The theorem follows at once from (2.14).

3. PROPERTIES OF $V(x)$

It follows from *Lemma 1* and (2.11) that $V(x)$ is bounded and continuous for all x.

Now we shall demonstrate two important properties of the function $V(x)$ constructed in *Theorem 1*. To do so we shall need another lemma.

Lemma 2: $\det\{\hat{A}+I\}$ *has the form*

$$\det\{\hat{A}+I\} = 1 + \sum_n \alpha_n \exp(\beta_n x), \quad (3.1)$$

where each α_n *and* β_n *is a positive constant.*

Proof: We recall that

$$\hat{A} = \|A_n^{\frac{1}{2}} A_\nu^{\frac{1}{2}} \exp[(\kappa_n + \kappa_\nu)x]/(\kappa_n + \kappa_\nu)\|, \quad (3.2)$$

and thus we have

$$\det\hat{A} = (\det\|1/(\kappa_n + \kappa_\nu)\|)(\prod_{n=1}^{N} A_n) \times \exp[2(\sum_{n=1}^{N} \kappa_n x)]. \quad (3.3)$$

Also, since $\|1/(\kappa_n + \kappa_\nu)\|$ has the form of $\hat{A}$, by *Lemma 1* it has a positive determinant. Therefore, by (3.3), $\det\hat{A}$ has the form $\alpha \exp(\beta x)$ with α and β positive constants. The same result holds for every principle minor of $\hat{A}$ since the principle minors all have the same form as $\det\hat{A}$.

Now the characteristic polynomial of $\hat{A}$ is

$$\det\{\hat{A}+\lambda I\} = \lambda^N + a_1\lambda^{N-1} + \cdots + a_N \quad (3.4)$$

where a_r is the sum of the principle minors of $\hat{A}$ of order r (e.g. $a_N = \det\hat{A}$ and $a_1 = \text{tr}\hat{A}$). If we set $\lambda = 1$, the Lemma follows.

One would expect, physically, that if $V(x)$ were ever positive, there would have to be a nonzero reflection for values of E below the positive maximum of $V(x)$. This is contrary to the fact. obvious from *Theorem 1*, that $u(x,E)$ contains no reflected part. Actually we can make a slightly stronger statement.

Theorem 2: $V(x)$ *is negative for all finite* x.

Proof: Writing

$$\Delta = \det\{\hat{A}+I\}$$

we have

$$V(x) = -2(\Delta\Delta'' - \Delta'^2)/\Delta^2 \quad (3.5)$$

from (2.11). The primes mean differentiation with respect to x. We must show, then, that

$$\Delta\Delta'' - \Delta'^2 > 0$$

for all finite x. We have by Lemma 2,

$$\Delta\Delta'' - \Delta'^2 = [1 + \sum_n \alpha_n \exp(\beta_n x)][\sum_n \beta_n^2 \alpha_n \exp(\beta_n x)] - [\sum_n \beta_n \alpha_n \exp(\beta_n x)]^2 = \sum_n \beta_n^2 \alpha_n \exp(\beta_n x) + \tfrac{1}{2} \sum_{n,\nu} \alpha_n \alpha_\nu (\beta_n - \beta_\nu)^2 \exp[(\beta_n + \beta_\nu)x]. \quad (3.6)$$

By *Lemma 2* the α_n and β_n are positive; therefore, (3.6) is positive. This completes the proof of *Theorem 2*.

Finally, let us examine the asymptotic behavior of $V(x)$. We state

Theorem 3: $V(x)$ approaches zero exponentially as x approaches either $+\infty$ or $-\infty$.

Proof: From the proof of *Theorem 2* we have for the form of $V(x)$:

$$V(x)=-2\{\sum_n \beta_n{}^2\alpha_n \exp(\beta_n x)$$
$$+\tfrac{1}{2}\sum_{n,\nu} \alpha_n\alpha_\nu(\beta_n-\beta_\nu)^2 \exp[(\beta_n+\beta_\nu)x]\}/$$
$$\{1+2\sum_n \alpha_n \exp(\beta_n x)$$
$$+\sum_{n,\nu}\alpha_n\alpha_\nu \exp[(\beta_n+\beta_\nu)x]\}. \quad (3.7)$$

Now suppose β_μ is the largest of the constants β_n. Then the term in the denominator of (3.7) of largest exponential growth as x approaches $+\infty$ is that for which $n=\nu=\mu$, that is, for which the exponent is $2\beta_\mu$. This term is obviously missing from the numerator of (3.7); thus, the denominator grows faster than the numerator, and $V(x)$ approaches zero as x approaches $+\infty$. As x approaches $-\infty$, $V(x)$ obviously approaches zero.

4. EXAMPLE

Consider the case of one constant A_1 and one constant κ_1. The system (2.2) will consist of a single equation:

$$A_1\exp(\kappa_1 x)+f_1(x)+[A_1\exp(2\kappa_1 x)/2\kappa_1]f_1(x)=0. \quad (4.1)$$

Thus

$$f_1(x)=-A_1\exp(\kappa_1 x)/[1+A_1\exp(2\kappa_1 x)/2\kappa_1], \quad (4.2)$$

$$u(x,E)=\{1-A_1\exp(2\kappa_1 x)/[1+A_1\exp(2\kappa_1 x)/2\kappa_1] \times[\kappa_1+i\sqrt{E}]\}\exp(i\sqrt{E}x), \quad (4.3)$$

and from (2.11), since $\det\{I+\hat{A}\}=1+A_1\exp(2\kappa_1 x)/2\kappa_1$, we have

$$V(x)=-4\kappa_1 A_1\exp(2\kappa_1 x)/[1+A_1\exp(2\kappa_1 x)/2\kappa_1]^2. \quad (4.4)$$

We observe that $V(x)$ has a single finite minimum and no finite maximum. It is interesting to find the position and value of the minimum of $V(x)$ in terms of the two parameters A_1 and κ_1. We have for $dV/dx|_{x=x_0}=0$:

$$-8\kappa_1{}^2A_1\exp(2\kappa_1 x_0)[1+A_1\exp(2\kappa_1 x_0)/2\kappa_1] +8\kappa_1 A_1{}^2\exp(4\kappa_1 x_0)=0,$$

for the position x_0 of the minimum. Solving for x_0 we have

$$x_0=[\log(2\kappa_1/A_1)]/2\kappa_1. \quad (4.5)$$

At its minimum position,

$$V(x_0)=-2\kappa_1{}^2. \quad (4.6)$$

We can obtain a more convenient form for $V(x)$ by using the parameters κ_1 and x_0 explicitly. Thus, substituting

$$A_1=2\kappa_1\exp(-2\kappa_1 x_0), \quad (4.7)$$

in (4.4) we obtain

$$V(x)=-2\kappa_1{}^2\,\mathrm{sech}^2\kappa_1(x-x_0). \quad (4.8)$$

Equation (4.8) is plotted in Fig. 1 for several values of κ_1.

The physical meaning of the parameters is evident. The position of minimum $V(x)$ is x_0, while κ_1 alone determines the minimum value of $V(x)$. Obviously $V(x)$ is symmetrical about its minimum point.

The solution (4.3) will now have the form

$$u(x,E)=\{[i\sqrt{E}-\kappa_1\tanh\kappa_1(x-x_0)]/(\kappa_1+i\sqrt{E})\} \times\exp(i\sqrt{E}x). \quad (4.9)$$

5. CONCLUSION AND REMARKS

Formulas (2.6) and (2.11) enable us to determine a $2N$-parameter family of functions $V(x)$ which in turn determines a $2N$-parameter family of dielectrics through (1.2). Formula (2.1), along with (1.3) and (1.4), demonstrates that the dielectric media are all transparent in the restricted sense described in the Introduction.

Since $V(x)$ is everywhere negative there must be at least one discrete eigenvalue in the spectrum of (1.4). We can actually say more than this: since in the proof of *Theorem 1* we say that $M_n f_n(x)=0$, where M_n was defined to be the operator,

$$d^2/dx^2+E-V(x),$$

with E replaced by $-\kappa_n{}^2$, and since each $f_n(x)$ is obviously a square integrable function in the interval $-\infty<x<+\infty$, it follows that there must be N discrete eigenvalues

$$E=-\kappa_n{}^2, \quad n=1,\cdots,N$$

with corresponding eigenfunctions $f_n(x)$.

The example of Sec. 4 had a single eigenvalue. Now we can make the minimum of $V(x)$ in that example as

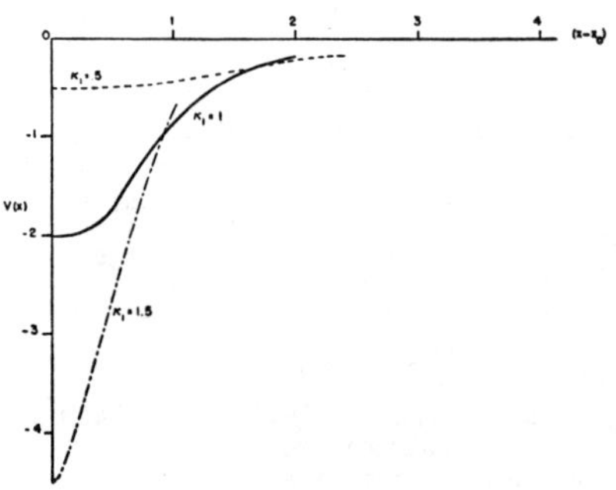

FIG. 1. A family of reflectionless potentials. The curves are symmetric about the $V(x)$ axis.

large as we please in absolute value, according to (4.6), and the larger we make it the narrower will be the interval on which $V(x)$ differs appreciably from zero. This is true because the area between $V(x)$ and the x axis is necessarily limited by the presence of only one eigenvalue. To get a "broader" $V(x)$ we would have to consider a case with more parameters A_n and κ_n.

From (4.9) we see that

$$u(x,E)\sim\exp(i\sqrt{E}x) \quad \text{as } x\rightarrow-\infty, \tag{5.1}$$

while as $x\rightarrow+\infty$

$$u(x,E)\sim[(i\sqrt{E}-\kappa_1)/(i\sqrt{E}+\kappa_1)]\exp(i\sqrt{E}x).$$

The transmission coefficient [the bracketed expression in (5.1)] has the absolute value unity, as it should. Moreover, it has a pole at $E=-\kappa_n^2$ (the eigenvalue), as standard theory tells us it should. It can be seen easily that in general the transmission coefficient is always independent of the constants A_n; i.e., it is a function of the κ_n alone.

For practical considerations one must consider dielectric slabs of finite extent rather than the infinite ones considered here. If we cut off any one of our infinite transparent media so that the corresponding $V(x)$ is required to vanish outside a given finite interval, the reflection coefficient of the resulting medium will no longer be zero. However, the reflection coefficient can be calculated explicitly by matching across the discontinuities in $V(x)$. By a proper choice of parameters in $V(x)$ we can arrive at a finite slab as transparent as we desire. Of course the practical problem is still not solved unless certain other conditions are satisfied at the same time. For example, an index of refraction which is always greater than some minimum value may be required. Thus it will be necessary to examine the transparent media more closely before a direct practical application of the results can be attempted. Fortunately, all of the transparent media described here are physically realizable in the sense that they all have indices of refraction greater than one at any finite point.

ACKNOWLEDGMENTS

The authors wish to express appreciation to Professor J. B. Keller for his discussions concerning this work and for his very helpful suggestions.

APPENDIX 1. PROOF THAT THE $V(x)$ WE HAVE CONSTRUCTED ARE THE ONLY ONES FOR WHICH THERE IS NO REFLECTION

It has been shown[3] that a necessary condition for a suitable function $V(x)$ in (1.4) is

$$V(x)=2dK(x,x)/dx, \tag{A.1}$$

where $K(x,y)$ is a function of x and y which satisfies an integral equation

$$0=R(x+y)+K(x,y)+\int_{-\infty}^{x}R(y+z)K(x,z)dz, \tag{A.2}$$

for all real values of x and for values of y between $-\infty$ and x. The function $R(x+y)$ is defined by

$$R(x+y)=(2\pi)^{-1}\int_{-\infty}^{\infty}r(k)\exp[-ik(x+y)]dk \\ -i\sum_{n=1}^{N}r_n\exp[-ik_n(x+y)], \tag{A.3}$$

where $r(k)$ is the complex reflection coefficient of (1.4), $E=k_n^2=-\kappa_n^2$ are the discrete eigenvalues of (1.4), and the r_n are certain numbers on the positive imaginary axis. Since in our problem $r(k)\equiv0$, (A.2) becomes, on writing $r_n=iA_n$,

$$0=\sum_{n=1}^{N}A_n\exp[\kappa_n(x+y)]+K(x,y) \\ +\sum_{n=1}^{N}A_n\int_{-\infty}^{x}K(x,z)\exp[\kappa_n(y+z)]dz. \tag{A.4}$$

It is obvious from (A.4) that $K(x,y)$ has the form

$$K(x,y)=\sum_{n=1}^{N}f_n(x)\exp(\kappa_n y). \tag{A.5}$$

Substitution of (A.5) into (A.4) produces the system of equations (2.2) for the functions $f_n(x)$. By *Lemma 1* the system has a unique solution; therefore, $K(x,y)$ and hence $V(x)$ are unique for a given $R(x+y)$ of the form assumed. When $r(k)\equiv0$ this form is the only one possible for $R(x+y)$.

APPENDIX 2

Consider the inhomogeneous transmission-line equations with variable capacitance $C(z)$ and variable inductance $L(z)$:

$$dV(z)/dz=i\omega L(z)I(z), \\ dI(z)/dz=i\omega C(z)V(z), \tag{A.6}$$

where $V(z)$ is the voltage, $I(z)$ the current, and ω the circular frequency. We can define a local characteristic impedance

$$Z(z)=[L(z)/C(z)]^{\frac{1}{2}} \tag{A.7}$$

and a local wave velocity

$$v(z)=[L(z)C(z)]^{-\frac{1}{2}}. \tag{A.8}$$

Let

$$\Psi(z)=[Z(z)]^{-\frac{1}{2}}. \tag{A.9}$$

Then with the changes of variables

$$V(z)=[Z(z)]^{\frac{1}{2}}U(z),$$

[3] I. Kay and H. E. Moses, Nuovo cimento Ser. X, 3, 276–304 (1955).

and

$$x=\int_{-\infty}^{z}[1/v(z)]dz, \tag{A.10}$$

we have

$$d^2U/dx^2+[\omega^2-P(x)]U=0, \tag{A.11}$$

where $P(x)$ is related to $\Psi(z(x))$ by

$$d^2\Psi/dx^2-P(x)\Psi=0. \tag{A.12}$$

Note that (A.11) has exactly the same form as (1.4) with $P(x)$ replacing the "potential" $V(x)$ there. Moreover, (A.12) is the same as (A.11) with $\omega=0$ and U replaced by Ψ.

To use this result we observe that if $x\rightarrow-\infty$ when $z\rightarrow-\infty$ and if $x\rightarrow+\infty$ when $z\rightarrow+\infty$, x being a monotonic function of z (these conditions must be satisfied), then when $P(x)$ represents a reflectionless potential the transmission line given by (A.6) is reflectionless. To get the characteristic impedance we choose the solution Ψ of (A.12) such that $\Psi\rightarrow Z_0^{-\frac{1}{2}}$ as $x\rightarrow-\infty$, where Z_0 is any desired limiting value of the characteristic impedance at $-\infty$, and $d\Psi/dx$ approaches zero as $x\rightarrow-\infty$. We have shown how to construct a solution of (A.11) as well as the $P(x)$; thus, we can obtain Ψ by considering the special case of (A.11) for which $\omega=0$.

Note that our procedure determines only the characteristic impedance Z. The wave velocity v can be chosen more or less arbitrarily provided it is such that x and z are monotonically related as assumed.

IL NUOVO CIMENTO VOL. XIII, N. 1 1° Luglio 1959

Potentials with Zero Scattering Phase (*).

H. E. MOSES

Institute for Mathematical Sciences, New York University - N. Y.

SAN FU TUAN (**)

Department of Mathematics, University of California - Berkeley, Cal.

(ricevuto il 20 Marzo 1959)

Summary — Let us consider the class of spherically symmetric potentials which are so smooth that the functions $f(k)$ are analytic in the lower half plane. The Gel'fand-Levitan equation is used to construct all potentials of this class which, for a given angular momentum, have scattering phases which are identically zero for all energy. In these cases the potentials support point eigenvalues which are either zero or are positive and embedded in the continuous spectrum. In the case $l=0$, the potential which supports the zero point eigenvalue was given earlier by Bargmann who used a special technique to find it.

1. – The Gel'fand-Levitan equation for zero angular momentum.

The Gel'fand-Levitan equation was introduced in reference (1) as a device for obtaining scattering potentials from the spectral measure function. In

(*) The work of H.E.M. was sponsored by the Geophysics Research Directorate, Air Force Cambridge Research Center, Air Research and Development Command under Contract no. 926 AF 19(604) 4555.

The work of S.F.T. was supported in part by the Office of Naval Research and by the National Science Foundation (69520-12-20426).

(**) Present address: Enrico Fermi Institute for Nuclear Physics, University of Chicago, Chicago 37, Ill.

(1) I. M. GEL'FAND and B. M. LEVITAN: *Isv. Akad. Nauk SSSR Math. Series*, **15**, 309 (1951).

1769

reference (²) it was shown how the spectral measure function may be obtained from the scattering phase and hence how one may obtain the scattering potential from the scattering phases and the point eigenvalues and normalizations of the eigenfunctions belonging to the point eigenvalues.

Explicitly, the Gel'fand-Levitan equation for $l=0$ is

$$(1.1)\quad \begin{cases} K(r,r') = -D(r,r') - \displaystyle\int_0^r K(r,r'')\,D(r'',r')\,\mathrm{d}r''\,, \\ \text{where} \\ D(r,r') = \dfrac{2}{\pi}\displaystyle\int_0^\infty [|f(k)|^{-2}-1]\sin kr \sin kr'\,\mathrm{d}k + \sum_{i=1}^{n} C_i^{-1} g(r,E_i)\cdot g^*(r',E_i)\,. \end{cases}$$

The values E_i correspond to prescribed point eigenvalues. The functions $g(r,E_i)$ are given by

$$(1.1a)\quad \begin{cases} g(r,E_i) = A_i \exp[i\pi/4]\,\dfrac{\sinh\sqrt{-E_i}\,r}{\sqrt{\pi}(-E_i)^{\frac14}}\,, & (E_i<0), \\ \qquad\quad\; = \dfrac{A_i \sin\sqrt{E_i}\,r}{\sqrt{\pi}\,(E_i)^{\frac14}}\,, & (E_i>0), \\ \qquad\quad\; = \dfrac{A_i}{\sqrt{\pi}}\,r\,, & (E_i=0). \end{cases}$$

The constants C_i are prescribed normalizations for the eigenstates corresponding to the point eigenvalues, while the constants A_i are arbitrary complex numbers which one must also prescribe.

GEL'FAND and LEVITAN show that if one prescribes the function $|f(k)|$, the point eigenvalues E_i and the normalization constants C_i, and if one solves (1.1) for $K(r,r')$, then on writing

$$(1.2)\quad \begin{cases} V(r) = 2\cdot\dfrac{\mathrm{d}}{\mathrm{d}r}K(r,r)\,, \\ \Psi(r,k) = \sqrt{\dfrac{2}{\pi}}\sin kr + \sqrt{\dfrac{2}{\pi}}\displaystyle\int_0^r K(r,r')\sin kr'\,\mathrm{d}r', \\ \Psi(r,E_i) = g(r,E_i) + \displaystyle\int_0^r K(r,r')\,g(r',E_i)\,\mathrm{d}r'\,. \end{cases}$$

(²) R. JOST and W. KOHN: *Det. Kgl. Dan. Vidensk. Selsk., Mat.-fys. Medd.*, **27**, no. 9 (1953).

The real functions $\Psi(r, k)$ are eigenfunctions of the operator $-(\mathrm{d}^2/\mathrm{d}r^2)+V(r)$ belonging to the continuous spectrum corresponding to the eigenvalues k^2, whereas the generally complex functions $\Psi(r, E_i)$ are eigenfunctions corresponding to the point eigenvalue E_i.

Furthermore, the eigenfunctions satisfy the orthogonality conditions

$$
\left\{
\begin{aligned}
&\int_0^\infty \Psi(r, k)\Psi(r, k')\,\mathrm{d}r = |f(k)|^2\,\delta(k-k')\,,\\
&\int_0^\infty \Psi^*(r, E_i)\Psi(r, E_j)\,\mathrm{d}r = C_i\delta_{ij}\,,\\
&\int_0^\infty \Psi(r, E_i)\Psi(r, k)\,\mathrm{d}r = 0\,,
\end{aligned}
\right.
\tag{1.3}
$$

and the completeness relations

$$
\int_0^\infty \Psi(r, k)\,|f(k)|^{-2}\,\Psi(r', k)\,\mathrm{d}k + \sum_{i=1}^{n} C_i^{-1}\Psi(r, E_i)\Psi^*(r', E_i) = \frac{2}{\pi}\int_0^\infty \sin kr\,\sin kr'\,\mathrm{d}k = \delta(r-r')\,.
\tag{1.4}
$$

The function $f(k)$ is related to the phase δ in the following way for $k>0$:

$$
\exp\left[2i\,\delta(k)\right] = \frac{f(k)}{f^*(k)}\,,
\tag{1.5}
$$

or

$$
f(k) = |f(k)|\exp\left[i\delta(k)\right]\,.
\tag{1.5a}
$$

It can also be shown that

$$
\lim_{r\to\infty}\ \Psi(r, k) = \sqrt{\frac{2}{\pi}}\,|f(k)|\sin\left(kr+\delta\right).
\tag{1.6}
$$

Hence, on using (1.6) and (1.5a), one could obtain $f(k)$.

In Reference (2), in order that one may obtain $f(k)$ uniquely from the phase by means of (1.5), the following assumptions are made:

1) $f(-k) = f(k)^*$ for k real.

2) $f(k)$ is analytic in the lower half plane.

3) $\lim\limits_{|k|\to\infty} f(k) = 1$, k in the lower half plane.

4) The point eigenvalues of the operator $-(d^2/dr^2)+V(r)$ are given by the zeros of $f(k)$ on the negative imaginary axis.

A class of potentials which lead to such functions $f(k)$ are discussed in reference ([3]).

It should be pointed out that these conditions are not necessary to obtain the potential $V(r)$ from the spectral measure function or, equivalently, from $|f(k)|$, C_i and E_i. In reference ([4]) KAY and MOSES have shown that it is only necessary for the Gel'fand-Levitan equation to have a solution in order to reproduce the prescribed spectral measure function. It is only for the purpose of obtaining spectral measure functions from the phase δ that one must impose some extra conditions.

For the cases in which the phase δ is identically zero for all k, the conditions above are too severe to admit any potential but the trivial one $V = 0$. Therefore, to get non-trivial results we shall relax the conditions on $f(k)$. Specifically we shall drop condition 4) which relates the zeros of $f(k)$ to the point eigenvalues. In doing so we are following a technique which was used by KAY and MOSES ([5]) to obtain reflectionless potentials in the one-dimensional case ($-\infty < x < \infty$). We shall be led to potentials which, while in a certain sense pathological, are perfectly smooth and die sufficiently rapidly at infinity to permit scattering to be defined. Thus we shall consider the class of potentials which lead to functions $f(k)$ which satisfy the first three requirements and find all the potentials of this class which lead to scattering with zero phase.

We shall now proceed to derive the phaseless potentials.

From (1.5) we see that if $\delta(k)$ is identically zero, for real $k > 0$, $f(k)$ is real on the positive real axis. On using conditions 1), 2), and 3) above and the reflection theorem of analysis, we come to the conclusion that a necessary and sufficient condition on $f(k)$ is that

$$f(k) = 1 \text{ on the whole complex } k \text{ plane.} \tag{1.7}$$

If we accepted condition 4) on $f(k)$, namely that bound states correspond to zeros of $f(k)$ on the negative imaginary axis, we see that, since in the present case there are no zeros, there should be no bound states. Under these

([3]) N. LEVINSON: *Det. Kgl. Dan. Vidensk. Selsk., Mat.-fys. Medd.*, **25**, no. 9 (1949).

([4]) I. KAY and H. E. MOSES: *Nuovo Cimento*, **2**, 917 (1955); **3**, 66 (1955).

([5]) I. KAY and H. E. MOSES: *Journ. App. Phys.*, **27**, 1503 (1956).

conditions we would have $D(r, r') = 0$ and hence also $K(r, r') = 0$ and $V(r) = 0$. We shall therefore assume the possibility of bound states in order to obtain non-trivial potentials. We note, however, that in (1.1) only the absolute value of $f(k)$ appears. If we admit bound states, we shall expect to obtain potentials which reproduce the given spectral measure function and find in particular that $|f(k)| = 1$. In general, however, the phase will *not* be zero identically. Hence, it will be our objective to find those bound states or combinations of bound states which will lead to zero phase.

Let us first consider the case in which a single bound state exists with a corresponding point eigenvalue E_1. In such cases (1.1) can be solved by separation of variables, *i.e.*, by setting $K(r, r') = H(r) \sinh \sqrt{-E_1}\, r'$ and solving for the function $H(r)$.

For $E_1 < 0$, we have

$$D(r, r') = C_1^{-1} |A_1|^2 \frac{\sinh \sqrt{-E_1}\, r \sinh \sqrt{-E_1}\, r'}{\pi \sqrt{-E_i}}, \tag{1.8}$$

$$K(r, r') = \frac{-\sinh \sqrt{-E_1}\, r \sinh \sqrt{-E_1}\, r'}{\left[\mu + \frac{1}{2}\left(\frac{\sinh 2\sqrt{-E_1}\, r}{2\sqrt{-E_1}} - r\right)\right]}, \tag{1.9}$$

where

$$\mu = \frac{\pi C_1 \sqrt{-E_1}}{|A_1|^2}, \tag{1.9a}$$

$$V(r) = 2 \frac{\mathrm{d}}{\mathrm{d}r} K(r, r) = \tag{1.10}$$

$$= \frac{2 \sinh \sqrt{-E_1}\, r \left[2\sqrt{-E_1}\left(\frac{r}{2} - \mu\right) \cosh \sqrt{-E_1}\, r - \sinh \sqrt{-E_1}\, r\right]}{\left[\mu + \frac{1}{2}\left(\frac{\sinh 2\sqrt{-E_1}\, r}{2\sqrt{-E_1}} - r\right)\right]^2},$$

$$\Psi(r, k) = \sqrt{\frac{2}{\pi}} \sin kr - \tag{1.11}$$

$$-\sqrt{\frac{2}{\pi}} \frac{\sinh \sqrt{-E_1}\, r \left(\sqrt{-E_1} \sin kr \cosh \sqrt{-E_1}\, r - k \cos kr \sinh \sqrt{-E_1}\, r\right)}{\left[\mu + \frac{1}{2}\left(\frac{\sinh 2\sqrt{-E_1}\, r}{2\sqrt{-E_1}} - r\right)\right] (k^2 - E_1)},$$

$$\Psi(r, E_1) = \frac{A_1 \exp[i\pi/4] \sinh \sqrt{-E_1}\, r}{\sqrt{\pi}\, (-E)^{\frac{1}{4}}} \frac{\mu}{\left[\mu + \frac{1}{2}\left(\frac{\sinh 2\sqrt{-E_1}\, r}{2\sqrt{-E_1}} - r\right)\right]}. \tag{1.12}$$

On inspecting the results, it is clear that the potential decays exponentially for large r. Furthermore, it can be shown by direct substitution that the eigenfunctions (1.11) and (1.12) really are eigenfunctions of the operator $-(d^2/dr^2)+V(r)$.

However, we see that the scattering phase is not zero, for

$$\lim_{r\to\infty} \Psi(r,k) = \sqrt{\frac{2}{\pi}}\left[\sin kr + \frac{2}{(k^2-E_1)}(E_1 \sin kr + k\sqrt{-E_1}\cos kr)\right] = \sqrt{\frac{2}{\pi}}\sin(kr+\delta)\,, \tag{1.13}$$

where

$$\cos\delta = \frac{(k^2+E_1)}{(k^2-E_1)}\,,\qquad \sin\delta = \frac{2k\sqrt{-E_1}}{(k^2-E_1)}\,. \tag{1.14}$$

In the present case the absolute value of $f(k)$ is 1 on the real k axis as we would expect from the Gel'fand-Levitan equation. However $f(k)$ itself is not 1. In fact on the real axis

$$f(k) = \exp[i\delta] = \cos\delta + i\sin\delta = \frac{(k+i\sqrt{-E_1})^2}{(k^2-E_1^2)} = \frac{k+i\sqrt{-E_1}}{k-i\sqrt{-E_1}}\,. \tag{1.15}$$

From (1.15) we see that the analytic continuation of $f(k)$ is zero when $k=-i\sqrt{-E_1}$, as is required by the usual scattering theory.

As the next case, let us take $E_1=0$. We have, on solving (1.1) by separation of variables,

$$D(r,r') = C_1^{-1}\frac{|A_1|^2}{\pi}rr'\,, \tag{1.16}$$

$$K(r,r') = \frac{-3rr'}{r^3+\lambda}\,, \tag{1.17}$$

where

$$\lambda = \frac{3C_1\pi}{|A_1|^2}\,, \tag{1.17a}$$

$$V = \frac{6r(r^3-2\lambda)}{(r^3+\lambda)^2}\,, \tag{1.18}$$

$$\Psi(r,k) = \sqrt{\frac{2}{\pi}}\left[\sin kr + \frac{3r^2}{r^3+\lambda}\frac{\cos kr}{k} - \frac{3r}{r^3+\lambda}\frac{\sin kr}{k^2}\right]\,, \tag{1.19}$$

and

$$\Psi(r,E_1) = \frac{A_1\lambda}{\sqrt{\pi}}\frac{r}{r^3+\lambda}\,. \tag{1.20}$$

In the present case we see

$$\lim_{r\to\infty} \Psi(r, k) = \sqrt{\frac{2}{\pi}} \sin kr \,. \tag{1.21}$$

Hence the scattering phase is zero for all k, and we have accomplished our objective of obtaining such a potential. In the present case $f(k)$ is identically 1 on the whole complex k plane as is required.

It should be mentioned that the potential (1.18) was derived earlier by BARGMANN ([6]) in his discussion of the inverse problem for the special case where the phase is a rational function of k.

Let us now consider the case for which $E_1 > 0$. Then

$$D(r, r') = \frac{C_1^{-1} |A_1|^2}{\pi\sqrt{E_1}} \sin\sqrt{E_1}\, r \sin\sqrt{E_1}\, r' \,. \tag{1.22}$$

$$K(r, r') = \frac{-\sin\sqrt{E_1}\, r \sin\sqrt{E_1}\, r'}{\lambda + \frac{1}{2}\left(r - \frac{\sin 2\sqrt{E_1}\, r}{2\sqrt{E_1}}\right)} \,, \tag{1.23}$$

where

$$\lambda = \frac{\pi C_1 \sqrt{E_1}}{|A_1|^2} \,, \tag{1.23a}$$

$$V(r) = \frac{2\sin\sqrt{E_1}\, r}{\left[\lambda + \frac{1}{2}\left(r - \frac{\sin 2\sqrt{E_1}\, r}{2\sqrt{E_1}}\right)\right]^2} \left[\sin\sqrt{E_1}\, r - 2\sqrt{E_1}\left(\lambda + \frac{r}{2}\right)\cos\sqrt{E_1}\, r\right] , \tag{1.24}$$

$$\Psi(r, k) = \sqrt{\frac{2}{\pi}} \sin kr - \sqrt{\frac{2}{\pi}}\, \frac{\sin\sqrt{E_1}\, r\, (\sqrt{E_1} \sin kr \cos\sqrt{E_1}\, r - k \sin\sqrt{E_1}\, r \cos kr)}{\left[\lambda + \frac{1}{2}\left(r - \frac{\sin 2\sqrt{E_1}\, r}{2\sqrt{E_1}}\right)\right](k^2 - E_1)} \,, \tag{1.25}$$

$$\Psi(r, E_1) = \frac{A_1 \lambda}{\sqrt{\pi}\,(E_1)^{\frac{1}{4}}}\, \frac{\sin\sqrt{E_1}\, r}{\lambda + \frac{1}{2}\left(r - \frac{\sin 2\sqrt{E_1}\, r}{2\sqrt{E_1}}\right)} \,. \tag{1.26}$$

In the present case it is clear that

$$\lim_{r\to\infty} \Psi(r, k) = \sqrt{\frac{2}{\pi}} \sin kr \,. \tag{1.27}$$

([6]) V. BARGMANN: *Rev. Mod. Phys.*, **21**, 488 (1949).

Hence again the scattering phase is zero. In the present case the potential behaves like $(\sin 2\sqrt{E_1}r)/r$ for large r. Thus it is not a member of the class of potential discussed in references ([2]) and ([3]).

We can now discuss the case where there are several proper eigenstates present. In the manner similar to that above one can show generally that solutions of the Gel'fand-Levitan equation for which $|f(k)|=1$, and for which a finite number of *non-negative* point eigenvalues are assumed to exist lead to potentials with phaseless scattering. Conversely, of all potentials which support a finite number of bound states, these are the only non-trivial ones which lead to zero scattering phase for all k.

2. – Higher angular momenta.

The formulas for the radial equation corresponding to the general angular momentum l are close to those for zero angular momentum.

The Gel'fand-Levitan equation is

(2.1)
$$\begin{cases} K(r,r') = -D(r,r') - \displaystyle\int_0^r K(r,r'')\,D(r'',r')\,\mathrm{d}r'' , \\ \text{where} \\ D(r,r') = \dfrac{2rr'}{\pi}\displaystyle\int_0^\infty [\,|f(k)|^{-2}-1]\,k^2 j_l(kr)\,j_l(kr')\,\mathrm{d}k - \sum_{i=1}^{n} C_i^{-1} g(r,E_i)\,g^*(r',E_i) , \end{cases}$$

where the real numbers, E_i, correspond to the point eigenvalues. The functions $g(r, E_i)$ are given by

(2.1a)
$$\begin{cases} g(r,E_i) = \dfrac{A_i}{\sqrt{\pi}}(-E_i)^{\frac14}\exp[i\pi/4]\,r j_l(i\sqrt{-E_i}\,r) , & (E_i<0), \\ \qquad\quad\;\; = \dfrac{A_i}{\sqrt{\pi}}(E_i)^{\frac14} r j_l(\sqrt{E_i}\,r) , & (E_i>0), \\ \qquad\quad\;\; = \dfrac{A_i}{\sqrt{\pi}}\, r^{l+1} , & (E_i=0). \end{cases}$$

As before, the constants C_i are the prescribed normalizations for the eigenstates corresponding to the point eigenvalues, while the constants A_i are arbitrary complex numbers which one must also prescribe.

Also, as before, one prescribes $|f(k)|$, the point eigenvalues E_i, and the normalization constants C_i and solves (2.1) for $K(r, r')$. One can find the potential $V(r)$, the eigenfunctions corresponding to the continuous spectrum

$\Psi(r, k)$ and the eigenfunctions corresponding to the point spectrum $\Psi(r, E_i)$ from

$$
(2.2)\qquad \begin{cases} V(r) = 2\dfrac{\mathrm{d}}{\mathrm{d}r}K(r, r)\,, \\[2ex] \Psi(r, k) \;= \sqrt{\dfrac{2}{\pi}}\, krj_l(kr) + \sqrt{\dfrac{2}{\pi}}\, k\displaystyle\int_0^r K(r, r')r'j_l(kr')\,\mathrm{d}r'\,, \\[2ex] \Psi(r, E_i) = g(r, E_i) + \displaystyle\int_0^r K(r, r')g(r', E_i)\,\mathrm{d}r'\,. \end{cases}
$$

The eigenfunctions satisfy the orthogonality and completeness relations (1.3) and (1.4) and are eigenfunctions of the operator

$$
-\frac{\mathrm{d}^2}{\mathrm{d}r^2} + \frac{l(l+1)}{r^2} + V(r)\,.
$$

The scattering phase is related to $f(k)$ as in (1.5). The asymptotic form of $\Psi(r, k)$ is given by

$$
(2.3)\qquad \lim_{r\to\infty} \Psi(r, k) = \sqrt{\frac{2}{\pi}}\,|f(k)|\sin\left(kr + \delta - \frac{l\pi}{2}\right).
$$

By the arguments of the previous Section, zero scattering phase leads to a condition $f(k) = 1$, but since only $|f(k)|$ is used in the Gel'fand-Levitan equations, we must examine all solutions of the Gel'fand-Levitan equation for which $|f(k)| = 1$ to see which solutions lead to $f(k) \equiv 1$ or equivalently to zero phase potentials.

As before, we shall obtain phaseless potentials, if we assume a finite number of non-negative point eigenvalues. These will be the only potentials which are phaseless and which support a finite number of point eigenvalues.

The expressions for the potentials and eigenfunctions are calculated in a straightforward way, but they are somewhat cumbersome. Hence we shall write down only the simplest reflectionless potential and the corresponding eigenfunctions. This potential arises when one assumes a single point eigenvalue with the value zero and is analogous to the Bargmann potential of the previous section,

$$
(2.4)\qquad \begin{cases} V(r) = \dfrac{2(2l+3)r^{2l+1}\left[r^{2l+3} - 2\lambda(l+1)\right]}{(r^{2l+3}+\lambda)^2}\,, \\[2ex] \Psi(r, k) \;= \sqrt{\dfrac{2}{\pi}}\, rkj_l(rk) + \sqrt{\dfrac{2}{\pi}}\,\dfrac{1}{k}\,\dfrac{(2l+3)r^{2(l+1)}}{(r^{2l+3}+\lambda)}\left[krj_l'(kr) - l\,j_l(kr)\right], \\[2ex] \Psi(r, E_1) = \dfrac{A_1\lambda r^{l+1}}{\sqrt{\pi}(r^{2l+3}+\lambda)}\,, \end{cases}
$$

where

$$\lambda = \frac{(2l+3)C_1\pi}{|A_1|^2}\,. \tag{2.4a}$$

One might ask the question whether it is possible to choose a potential such that the phases for two or more angular momenta might not be identically zero. It is possible to show that in the class of potentials of the type considered in the present paper, namely for which $f(k)$ is analytic in the lower half plane, this is not possible.

RIASSUNTO (*)

Consideriamo la classe di quei potenziali a simmetria sferica che si raccordino così dolcemente che le funzioni $f(k)$ siano analitiche nel semipiano inferiore. Si usa l'equazione di Gel'fand-Levitan per costruire tutti i potenziali di questa classe che, per un dato momento angolare, hanno fasi di scattering identicamente nulle per ogni energia. In questi casi i potenziali sostengono autovalori puntuali o nulli o positivi inclusi nello spettro continuo. Nel caso $l=0$ il potenziale che sostiene l'autovalore nullo fu dato precedentemente da Bargmann, che per trovarlo ricorse a una tecnica speciale.

(*) *Traduzione a cura della Redazione.*

COMMUNICATIONS ON PURE AND APPLIED MATHEMATICS, VOL. XIII, 371–393 (1960)

The Inverse Scattering Problem When the Reflection Coefficient is a Rational Function*

IRVIN KAY

1. Introduction

In quantum mechanics and in electromagnetic theory the so-called scattering problem is of fundamental importance. Usually the aim is to determine the asymptotic scattered field in terms of an incident field and some prescribed potential or medium which is responsible for the scattering. In this article we shall be concerned with the inverse problem of determining the scatterer from a knowledge of the asymptotic scattered field. The present discussion will be limited to a one-dimensional situation for which an explicit solution can be obtained.

Associated with a differential equation of the form

$$\frac{d^2u(x,k)}{dx^2}+\{k^2-V(x)\}u(x,k)=0 \tag{1.1}$$

for which a complex reflection coefficient $r(k)$ can be defined is the inverse scattering problem: to determine from a knowledge of $r(k)$ for all real k the function $V(x)$ for all real x. It will be shown in the sections which follow how the function $V(x)$ and the solutions $u(x, k)$ can be determined explicitly in closed form in terms of elementary functions if $r(k)$ is a rational function of k.[1]

The condition

$$V(x)=0, \qquad x<0, \tag{1.2}$$

will be assumed throughout this paper, although the results presented here either hold as they stand or can be generalized without much difficulty when less severe conditions on $V(x)$ replace (1.2).

Some of the general well-known properties of the scattering matrix

*The research reported in this paper has been sponsored by the Electronics Research Directorate of the Air Force Cambridge Research Center, Air Research and Development Command, under Contract No. AF 19(604)5238. Reproduction in whole or in part permitted for any purpose of the United States Government.

[1]It was found by V. A. Bargmann [3] that potentials for the radial Schrödinger equation of quantum mechanics, as well as the corresponding wave functions, can be expressed in terms of elementary functions under similar circumstances.

associated with the differential equation (1.1) are described in Section 2 for convenient reference later.

Section 3 is a discussion of the scattering matrix elements when the reflection coefficient is a rational function of k.

Section 4 is devoted to proving a theorem connecting the reflection coefficient with other elements of the scattering matrix associated with (1.1) under the assumption that the reflection coefficient is a rational function of k. At the same time an explicit closed form expression for $V(x)$ in terms of elementary functions is derived. It will be seen also that the solutions $u(x, k)$ of (1.1) can be given explicitly in terms of elementary functions.

In the appendix a derivation will be given of a general method for reconstructing $V(x)$ from the reflection coefficient. Although the method has already been presented in other places [1], [2], the new derivation given here is more physically motivated and somewhat more general than the one previously used in that the new derivation does not require the existence of the entire scattering matrix—just the existence of the reflection coefficient. The method itself is an outgrowth of that of I. M. Gelfand and B. M. Levitan [4] for the solution of a similar class of problems.

2. The Scattering Matrix[2]

If $V(x)$ in the differential equation (1.1) is piecewise continuous and vanishes sufficiently rapidly at $x = \pm\infty$, any solution $u(x, k)$ has the asymptotic behavior

$$\begin{aligned} &\lim_{x\to+\infty} \{A_+ e^{ikx} + B_+ e^{-ikx} - u(x, k)\} = 0, \\ &\lim_{x\to-\infty} \{A_- e^{ikx} + B_- e^{-ikx} - u(x, k)\} = 0, \end{aligned} \tag{2.1}$$

where the coefficients $A_\pm$, $B_\pm$ are independent of x but may depend on k.

The differential equation (1.1) can be interpreted physically as an equation for the steady state solution of a time dependent wave equation, where the time dependence is assumed to be periodic with angular frequency ω. In this interpretation k is the wave number given by

$$k = \frac{\omega}{c},$$

where c is the constant velocity of any transient or pulse wave front in the

[2]Specific reference to the literature in this section will not be made because most of the material comes from a commonly held body of knowledge difficult to attribute to any individual.

medium of propagation. The function $V(x)$ is a term which causes dispersion but does not affect the transient wave front velocity.

According to (2.1), then, the asymptotic forms of $u(x, k)$ at large positive and negative values of x can be regarded as simple waves moving to the left and right in free space where the function $V(x)$ vanishes. On the basis of the convention that the suppressed periodic time factor in the waves is $e^{-i\omega t}$ a *left moving wave* can be defined as one asymptotically proportional to e^{-ikx} and a *right moving wave* as one asymptotically proportional to e^{ikx}. In addition, an *incoming wave* can be defined as a *right moving wave* at $-\infty$ or a *left moving wave* at $+\infty$, while an *outgoing wave* can be defined as a *right moving wave* at $+\infty$ or a *left moving wave* at $-\infty$.

The set of all *incoming waves* can be regarded as a two dimensional vector space of column vectors such that the first component of each vector is the amplitude of the *right moving* part and the second component is the amplitude of the *left moving* part. Similarly, the *outgoing waves* can be represented by a two dimensional column vector space for which the first component of each vector is the amplitude of the *right moving* part and the second component is the amplitude of the *left moving* part.

Corresponding to every differential equation of the form (1.1) whose solutions have the asymptotic behavior (2.1) there exists a matrix, called the *scattering matrix*, which transforms each vector in the *incoming wave vector space* into a vector of the *outgoing wave vector space*. If the *incoming wave vector* contains the amplitudes of the incoming part of $u(x, k)$ in the asymptotic limits (2.1), the corresponding *outgoing wave vector* obtained by means of the scattering matrix will contain the amplitudes of the outgoing part. The scattering matrix has the form

$$S = \begin{pmatrix} t(k), & \rho(k) \\ r(k), & \tau(k) \end{pmatrix}; \tag{2.2}$$

the elements of S are called

the transmission coefficient on the right: $t(k)$,
the transmission coefficient on the left: $\tau(k)$,
the reflection coefficient on the right: $\rho(k)$,
the reflection coefficient on the left: $r(k)$.

If (2.1) holds, then

$$\begin{pmatrix} A_+ \\ B_- \end{pmatrix} = S \begin{pmatrix} A_- \\ B_+ \end{pmatrix},$$

or equivalently,

$$\begin{aligned} A_+ &= A_- t(k) + B_+ \rho(k), \\ B_- &= A_- r(k) + B_+ \tau(k). \end{aligned} \tag{2.3}$$

Several facts about the scattering matrix are well known. If $V(x)$ in (1.1) satisfies the conditions

$$\int_{-\infty}^{\infty} |V(x)| dx < \infty$$

and

$$\int_{-\infty}^{\infty} |x||V(x)| dx < \infty,$$

the elements of S are analytic functions of the wave number k in the complex plane, and these functions are regular on the real axis. Furthermore, if there are no proper eigenfunctions, under the condition (1.2) on $V(x)$ the scattering coefficients $r(k)$ and $t(k)$ are regular in the upper half-plane and on the real axis, and in the same region they obey the asymptotic conditions for large k:

$$r(k) = O\left(\frac{1}{k}\right), \quad t(k) = 1+O\left(\frac{1}{k}\right). \tag{2.4}$$

The following relations hold in general for real k when $V(x)$ is real for real x:

$$r(-k) = \overline{r(k)}, \quad t(-k) = \overline{t(k)}, \quad \rho(-k) = \overline{\rho(k)}, \quad \tau(-k) = \overline{\tau(k)}, \tag{2.5}$$

where the bar over a number indicates the complex conjugate. It is also known that S is a unitary matrix when k is real and $V(x)$ is real for real x:

$$S^*(k) = S^{-1}(k), \qquad k \text{ real.} \tag{2.6}$$

In addition, the reciprocity law holds:

$$t(k) = \tau(k) \tag{2.7}$$

for all k.

Relations (2.5), (2.6) and (2.7) imply that for real k

$$\begin{aligned} |r(k)|^2+|t(k)|^2 &= r(k)r(-k)+t(k)t(-k) = 1 \quad \text{(energy conservation)}, \\ \rho(k)\overline{t(k)}+\overline{r(k)}t(k) &= \rho(k)t(-k)+r(-k)t(k) = 0 \quad \text{(phase law)}. \end{aligned} \tag{2.8}$$

3. The Case in Which $r(k)$ is a Rational Function

If the reflection coefficient $r(k)$ is known for all real values of k, or even for just the positive real values because of (2.5), the function $V(x)$ and any solution $u(x, k)$ of (1.1) can be determined as indicated in the appendix. In principle, it is then possible to find the other elements of the scattering matrix in terms of $r(k)$. As a practical matter, however, the relationship between the elements of the scattering matrix thus exhibited is quite complicated and indirect. On the other hand, if the given $r(k)$ is a rational function of k, an explicit relation between $r(k)$ and $t(k)$ or between $r(k)$ and $\rho(k)$ can be found as well as a closed form expression for $V(x)$. These relations between the scattering matrix elements will now be discussed.

Consider the general form which $r(k)$ can take if it is a rational function of k:

$$r(k) = \frac{r_0 \prod_{i=1}^{m} (k-\mu_i)}{\prod_{i=1}^{n} (k-\lambda_i)}, \tag{3.1}$$

where r_0, the μ_i and the λ_i are complex constants, and the integers m and n satisfy the inequality

$$m \leqq n+2. \tag{3.2}$$

The condition (3.2) is necessary if a piecewise continuous $V(x)$ is desired. Thus, for example, if $r(k)$ is not $o(1/k)$ for large k, $V(x)$ will be an improper function such as a delta function. This can be seen by examining the method of the appendix. The complex constants r_0, λ_i, μ_i must be such that the reality condition on $r(k)$ given by (2.5) is satisfied. Also, to comply with condition (1.2), assuming that no proper eigenfunctions exist [2], $r(k)$ must be regular in the upper half of the complex k-plane.[3] In other words, the constants λ_i are all in the lower half-plane:

$$\mathscr{I}m\, \lambda_i < 0, \qquad i = 1, 2, \cdots, n. \tag{3.3}$$

Finally, it is convenient to impose the condition

$$|r(k)| < 1, \qquad \text{for all real } k. \tag{3.4}$$

Condition (3.4) is slightly stronger than the necessary one

$$|r(k)| \leqq 1, \qquad \text{for all real } k, \tag{3.5}$$

implied by (2.8), but it is not essentially more restrictive since the $r(k)$ satisfying (3.5) can be multiplied by a parameter ε such that

$$0 < \varepsilon < 1$$

and the correct result obtained as a limiting case when ε approaches one. In the same spirit, it will be assumed that the numbers κ_i, which will be introduced later as zeros of $1-r(k)r(-k)$ in the upper half-plane, are all distinct, the case of multiple zeros being treated as a limiting case in which two or more zeros coalesce.

If some *a priori* assumptions are made about $t(k)$, by a standard argument the connection between $t(k)$ and $r(k)$ can be given even for reflection

[3]See the Appendix, relation (A.6).

coefficients $r(k)$ which are not rational functions of k. Suppose, in fact, it is assumed that $t(k)$ has no zeros or singularities in the upper half of the complex k-plane. For real k, by (2.8) and (2.5),

$$|t(k)| = \sqrt{1-|r(k)|^2} = \sqrt{1-r(k)r(-k)}. \tag{3.6}$$

The function

$$\log t(k) = \log |t(k)|+i\theta(k), \qquad k \text{ real,} \tag{3.7}$$

is analytic on the real axis and in the upper half-plane according to the assumptions made about $t(k)$. The phase $\theta(k)$ of $t(k)$ is an odd function of k and the modulus $|t(k)|$ is an even function of k when k is real, according to (2.5). Also by (2.4)

$$\lim_{k\to\infty} \log t(k) = 0, \tag{3.8}$$

if the limit is approached on the real axis or in the upper half-plane. By Cauchy's integral theorem, integrating over an infinite semi-circle in the upper half-plane and over the real axis, and by (3.8)

$$\log t(k) = \frac{1}{2\pi i}\int_{-\infty}^{\infty} \frac{\log t(\sigma)}{\sigma--k}\, d\sigma, \tag{3.9}$$

k in the upper half plane. Moreover, since $\bar{k}$ is outside the closed contour

$$0 = \frac{1}{2\pi i}\int_{-\infty}^{\infty} \frac{\log t(\sigma)}{\sigma-\bar{k}}\, d\sigma, \tag{3.10}$$

k in the upper half-plane. On taking the complex conjugate of both sides, (3.10) becomes

$$0 = -\frac{1}{2\pi i}\int_{-\infty}^{\infty} \frac{\overline{\log t(\sigma)}}{\sigma-k}\, d\sigma, \tag{3.11}$$

k in the upper half-plane. Subtracting (3.11) from (3.9) gives the result:

$$\log t(k) = \frac{1}{2\pi i}\int_{-\infty}^{\infty} \frac{\log\{1-r(\sigma)r(-\sigma)\}}{\sigma-k}\, d\sigma, \tag{3.12}$$

k in the upper half-plane, where (3.7) and (3.6) have been used. Equation (3.12) provides the general relation between $t(k)$ and $r(k)$ when $t(k)$ is regular and has no zeros in the upper half-plane.

If these conditions are violated so that $t(k)$ has zeros and poles in the upper half-plane but the zeros and poles of $t(k)$ are known, the argument just given can be modified to produce the desired relationship between $r(k)$ and $t(k)$. If the zeros and poles of $t(k)$ in the upper half-plane are not known, the argument of this section fails completely.

If (3.12) is applied to the $r(k)$ given by (3.1), an explicit formula for $t(k)$ can be obtained. The evaluation of the integral in (3.12) can be carried out by considering first

$$\frac{d}{dk}\log t(k) = \frac{1}{2\pi i}\int_{-\infty}^{\infty}\frac{\log\{1-r(\sigma)r(-\sigma)\}}{(\sigma-k)^2}\,d\sigma, \tag{3.13}$$

k in the upper half-plane. After an integration by parts (3.13) becomes

$$\frac{d}{dk}\log t(k) = -\frac{1}{2\pi i}\int_{-\infty}^{\infty}\frac{\dfrac{d}{d\sigma}\{r(\sigma)r(-\sigma)\}}{(\sigma-k)\{1-r(\sigma)r(-\sigma)\}}\,d\sigma, \tag{3.14}$$

k in the upper half-plane. The integrand of the integral on the right of (3.14) is now a rational function of σ and, therefore, can be evaluated by the method of residues. The result is a rational function of k equal to the derivative of $\log t(k)$. The integration of that function is easily carried out and the constant of integration is determined by condition (3.8). The result is

$$t(k) = \frac{\prod_{i=1}^{n}(k+\kappa_i)}{\prod_{i=1}^{n}(k-\lambda_i)}, \tag{3.15}$$

where the n complex constants κ_i are the n roots in the upper half-plane of the equation

$$1-r(k)r(-k) = 0. \tag{3.16}$$

That there are exactly n roots of (3.16) in the upper half-plane is clear from the invariance of (3.16) when k is replaced by $-k$ and from condition (3.4). The numbers κ_i are assumed to be distinct, i.e., it is assumed that there are no multiple roots of the equation (3.16). If this is not so, $r(k)$ can be modified slightly so that the multiple roots of (3.16) split up into simple roots, and one can consider the original case as a limiting form in which two or more roots of (3.16) coalesce. It is obvious that all conditions on $t(k)$ are satisfied by (3.15) in any case, even if one ignores its derivation.

The phase law of (2.8) provides the means for obtaining an expression for the function $\rho(k)$ once $t(k)$ is assumed to be given by (3.15):

$$(3.17) \qquad \rho(k) = \frac{(-1)^{m-n+1} r_0 \prod_{i=1}^{m} (k+\mu_i) \prod_{j=1}^{n} (k+\kappa_j)}{\prod_{i=1}^{n} (k-\lambda_i) \prod_{j=1}^{n} (k-\kappa_j)}.$$

Notice that $\rho(k)$, unlike $r(k)$, has poles in the upper half-plane at κ_j. Since no proper eigenfunctions can exist it is clear from the Appendix that the function $V(x)$ does not vanish identically for all values of x larger than some finite number; i.e., $V(x)$ must be different from zero in the positive x direction at arbitrarily large values of x.

Unfortunately the results (3.15) and (3.17) depend upon the *a priori* assumptions made on the $t(k)$ that it has no zeros or singularities in the upper half-plane. An independent derivation of (3.15) and (3.17), without these assumptions, which will at the same time provide a closed form expression for $V(x)$, will be given in Section 4. This derivation will start with the formal solution obtained by the method of the Appendix. However, the solution as well as the relations (3.15) and (3.17) will be verified directly, so that there will be no need to make the argument of the Appendix rigorous. If an attempt is made to begin with a solution based on the reflection coefficient $r(k)$ given by (3.1), it will be found that the formal result is rather complicated. On the other hand, if the solution is based on $\rho(k)$ given by (3.17), the result is considerably simpler and can be expressed in closed form. The reason why the form of the solution based on $\rho(k)$ is so much simpler than the form based on $r(k)$ is that $r(k)$ leads to a function $V(x)$ which vanishes for all negative values of x while $\rho(k)$ leads to a function $V(-x)$ which is regular and does not vanish on any finite interval in the interval $-\infty < x < 0$. These facts become evident upon direct examination of the formal solutions obtained by the method of the Appendix.

When $\rho(k)$ is used it is convenient to replace x by $-x$ in the differential equation (1.1) so that $\rho(k)$ becomes the reflection coefficient on the left, $u(x, k)$ is replaced by $u(-x, k)$ and $V(x)$ is replaced by $V(-x)$. The formal solution which would be obtained by the method of the Appendix if it were based on $r(k)$ as the reflection coefficient on the left is such that $V(x)$ vanishes for all negative x. Thus, it can be expected that $V(-x)$ will vanish for all positive x, and, in the region of positive x, $u(-x, k)$ will therefore be a linear combination of waves e^{ikx} and e^{-ikx} of free space. The relations (3.15) and (3.17) can then be verified by matching the limits of an appropriate solution $u(-x, k)$ and its derivative as x approaches zero from the left with the limits of the corresponding linear combination of free space waves and its derivative as x approaches zero from the right.

Temporarily it is convenient to define

$$\hat{\varrho}(k) = \frac{(-1)^{m-n+1} r_0 \prod_{i=1}^{m} (k+\mu_i) \prod_{j=1}^{n} (k+\kappa_j)}{\prod_{i=1}^{n} (k-\lambda_i) \prod_{j=1}^{n} (k-\kappa_j)} \tag{3.18}$$

and

$$\hat{t}(k) = \frac{\prod_{i=1}^{n} (k+\kappa_i)}{\prod_{j=1}^{n} (k-\lambda_i)}. \tag{3.19}$$

LEMMA. *For all real k*

$$\begin{gathered} \hat{\varrho}(-k) = \overline{\hat{\varrho}(k)}, \quad \hat{t}(-k) = \overline{\hat{t}(k)}, \\ |r(k)| = |\hat{\varrho}(k)|, \quad |r(k)|^2 + |\hat{t}(k)|^2 = 1, \\ \overline{r(k)}\hat{t}(k) + \hat{\varrho}(k)\overline{\hat{t}(k)} = 0. \end{gathered} \tag{3.20}$$

Proof: Consider

$$P_1(k) = \hat{t}(k)\hat{t}(-k) \prod_{i=1}^{n} (k-\lambda_i) \prod_{j=1}^{n} (-k-\lambda_j) \tag{3.21}$$

and

$$P_2(k) = \{1-r(k)r(-k)\} \prod_{i=1}^{n} (k-\lambda_i) \prod_{j=1}^{n} (-k-\lambda_j). \tag{3.22}$$

From (3.1) and (3.19) it can be seen that $P_1(k)$ and $P_2(k)$ are each polynomials in k of degree $2n$. The roots of $P_1(k)$ are the $2n$ numbers $\pm\kappa_i$, $i = 1, \cdots, n$. By the definition of the κ_i in connection with (3.16) and the fact that replacing k by $-k$ in (3.16) leaves the equation unchanged, it is obvious that the numbers $\pm\kappa_i$, $i = 1, \cdots, n$, are also the roots of $P_2(k)$. After dividing (3.21) and (3.22) by $\prod_{i=1}^{n} (k-\lambda_i) \prod_{j=1}^{n} (-k-\lambda_j)$ and allowing k to approach ∞ it is found that each of the expressions approaches one. The conclusion is

$$\hat{t}(k)\hat{t}(-k) = 1-r(k)r(-k). \tag{3.23}$$

It follows from (3.23), (3.4) and (2.5) that $\hat{t}(k)\hat{t}(-k)$ is real and positive for real k.

By (3.1) and (2.5), for real k,

$$r(-k) = \frac{\bar{r}_0 \prod_{i=1}^{m} (k-\bar{\mu}_i)}{\prod_{i=1}^{n} (k-\bar{\lambda}_i)}. \tag{3.24}$$

Analytic continuation implies the validity of (3.24) for all k. From (3.1),

(3.23) and (3.24) it can be seen that the $2n$ poles of $\hat{t}(k)\hat{t}(-k)$ are λ_i, $i = 1, \cdots, n$, and $\bar{\lambda}_i$, $i = 1, \cdots, n$. Each λ_i is distinct from all the $\bar{\lambda}_i$ since the λ_i are given in the lower half-plane. It can be seen then that the $2n$ roots of (3.16) consist of the $2n$ numbers, $\pm\kappa_i$, $i = 1, \cdots, n$, as well as the $2n$ numbers $\pm\bar{\kappa}_i$, $i = 1, \cdots, n$. Of course (3.16) has at most $2n$ distinct roots altogether; by the assumption of this article that the roots are all simple, in fact, it has exactly $2n$ distinct roots. By (3.23), therefore, $\hat{t}(k)\hat{t}(-k)$ has as zeros the $\pm\kappa_i$, $i = 1, \cdots, n$, as well as the $\pm\bar{\kappa}_i$, $i = 1, \cdots, n$. Thus, the rational function

$$\hat{t}(k)\overline{\hat{t}(k)} = \frac{\prod_{i=1}^{n} (k+\kappa_i) \prod_{j=1}^{n} (k+\bar{\kappa}_j)}{\prod_{i=1}^{n} (k-\lambda_i) \prod_{j=1}^{n} (k-\bar{\lambda}_j)}, \qquad k \text{ real,}$$

has the same poles and zeros as $\hat{t}(k)\hat{t}(-k)$, and both expressions approach one as k approaches ∞. It follows that

$$\hat{t}(-k) = \overline{\hat{t}(k)}, \qquad k \text{ real.} \tag{3.25}$$

From (3.18) and (3.19),

$$\hat{\rho}(k) = \frac{-r(-k)\hat{t}(k)}{\hat{t}(-k)}. \tag{3.26}$$

The relations (3.20) of the lemma follow at once from (3.25), (3.26) and (3.23).

The arguments just presented concerning the zeros of $\hat{t}(k)\hat{t}(-k)$ demonstrate at the same time that the n zeros of $\hat{t}(-k)$ are the κ_i and the $-\bar{\kappa}_i$, $i = 1, \cdots, n$. It follows from (3.26) that the n poles of $\hat{\rho}(k)$ in the upper half-plane are the κ_i and the $-\bar{\kappa}_i$, $i = 1, \cdots, n$, since the λ_i are all given in the lower half-plane. By the Mittag-Leffler theorem the function $\hat{\rho}(k)$ can be written in the form

$$\hat{\rho}(k) = \sum_{\nu=1}^{n} \frac{\rho_\nu}{k-\lambda_\nu} + \sum_{\nu=1}^{n} \frac{\rho_{n+\nu}}{k-\kappa_\nu}. \tag{3.27}$$

By considering the analytic continuation of $\overline{\hat{\rho}(k)}$, k real, and comparing with $\hat{\rho}(-k)$, using (3.20), it may be observed that (3.27) can be expressed as the sum of pairs of terms of the form

$$\frac{\rho_i}{k-\alpha_i} + \frac{\rho_j}{k+\alpha_i} = -\frac{\bar{\rho}_i}{k+\alpha_i} - \frac{\bar{\rho}_j}{k-\alpha_i}, \qquad k \text{ real.}$$

It follows at once then that for each pair in this sum

$$\rho_j = -\bar{\rho}_i. \tag{3.28}$$

Hence, to every residue of $\hat{\rho}(k)$ there is another which is its reflection in the imaginary axis.

4. $V(x)$ when $r(k)$ is a Rational Function

The heuristic results of the Appendix are useful for the purpose of motivating the analysis of this section. They play no role in the proofs of the theorems presented here, however.

Consider the problem of determining $V(-x)$ in (1.1) with $-x$ replacing x when the function $\hat{\rho}(k)$, given by (3.18), is the reflection coefficient from the left.[4] According to the Appendix the problem can be solved by solving the integral equation[5]

$$\text{(A.20a)} \qquad 0 = R(x+t)+K(x, t) + \int_{-\infty}^{x} K(x, z)R(z+t)dz$$

for $K(x, t)$ as a function of t in the interval $-\infty < t < x$ for each value of x. The function $R(x+t)$ is given by

$$\text{(4.1)} \qquad R(z+t) = \frac{1}{2\pi}\int_{-\infty}^{\infty} \hat{\rho}(k)e^{-ik(z+t)}\,dk = i\sum_{\nu=1}^{n} \rho_\nu e^{-i\kappa_\nu(z+t)}, \quad z+t<0,$$

where the ρ_ν are the residues of $\hat{\rho}(k)$ at its poles in the upper half-plane. From the lemma of Section 3 it can be concluded that $R(x+t)$ is real for real values of x and t. The function $V(-x)$ can then be determined by

$$V(-x) = 2\,\frac{dK(x, x)}{dx}.$$

A wave solution $u(-x, k)$ can be determined in terms of a free space wave $u_0(-x, k)$ by

$$u(-x, k) = u_0(-x, k) + \int_{-\infty}^{x} K(x, z)u_0(-z, k)dz.$$

To solve the integral equation (A.20a) it is convenient to assume that when x is negative $K(x, t)$ has the form

$$\text{(4.2)} \qquad K(x, t) = \sum_{\nu=1}^{n} f_\nu(x)e^{-i\kappa_\nu t}.$$

With this assumption the integral equation (A.20a) takes the form

---

[4]It will be assumed throughout that the poles of $\hat{\rho}(k)$ in the upper half-plane, the κ_i, are distinct. If there are multiple poles, the reflection coefficient is to be replaced by one for which the κ_i are distinct, the poles which were formerly multiple being replaced by simple poles very near each other. The case in question then becomes a limiting case in which certain simple poles coalesce to form multiple poles.

[5]The equation (A.20a) corresponds to (A.20) of the appendix. The function R in (A.20) vanishes for negative values of its argument, whereas R in (A.20a) does not; hence the difference in the lower limits of (A.20) and (A.20a).

$$(4.3) \qquad -\rho_\nu \sum_{\mu=1}^{n} \frac{e^{-i(\kappa_\nu+\kappa_\mu)x}}{\kappa_\nu+\kappa_\mu} f_\mu(x)+f_\nu(x)+i\rho_\nu e^{-i\kappa_\nu x} = 0, \quad \nu = 1, \cdots, n,$$

which holds when $x < 0$.

The uniqueness of the solution of the system of equations (4.3) can be established in exactly the same way in which the uniqueness of the solution of the integral equation (A.20a) can be proved [2].

LEMMA 1. *The system* (4.3) *has a unique solution for the* $f_\nu(x)$, $\nu = 1, \cdots, n$.

Proof: Multiply the homogeneous system of algebraic equations corresponding to (4.3) by $e^{-i\kappa_\nu t}$ and sum over ν:

$$(4.4) \qquad \sum_{\nu,\mu=1}^{n} \frac{\rho_\nu e^{-i(\kappa_\nu+\kappa_\mu)x}}{\kappa_\nu+\kappa_\mu} e^{-i\kappa_\nu t} f_\mu^0(x) + \sum_{\nu=1}^{n} e^{-i\kappa_\nu t} f_\nu^0(x) = 0, \quad t < x < 0.$$

Equation (4.4) can be written as

$$(4.5) \qquad \begin{aligned} &\int_{-\infty}^{x} dz \left\{ \sum_{\nu=1}^{n} i\rho_\nu e^{-i\kappa_\nu(t+z)} \sum_{\mu=1}^{n} e^{-i\kappa_\mu z} f_\mu^0(x) \right\} + \sum_{\nu=1}^{n} e^{-i\kappa_\nu t} f_\nu^0(x) \\ &= \int_{-\infty}^{x} dz \left\{ \frac{1}{2\pi} \int_{-\infty}^{\infty} \hat{\rho}(k) e^{-ik(z+t)} \, dk \sum_{\mu=1}^{n} e^{-i\kappa_\mu z} f_\mu^0(x) \right\} + \sum_{\nu=1}^{n} e^{-i\kappa_\nu t} f_\nu^0(x) = 0, \end{aligned}$$

for $x+t < 0$, $t < x$. On setting

$$(4.6) \qquad U^0(x, t) = \sum_{\nu=1}^{n} e^{-i\kappa_\nu t} f_\nu^0(x),$$

equation (4.5) becomes

$$(4.7) \qquad \int_{-\infty}^{x} R(t+z) U^0(x, z) dz + U^0(x, t) = 0, \quad x+t < 0, \; t < x.$$

Equation (4.7) can be written as

$$(4.8) \qquad \lim_{N\to\infty} \int_{-\infty}^{x} \left\{ \frac{1}{2\pi} \int_{-N}^{N} \hat{\rho}(k) e^{-ik(z+t)} \, dk + \frac{1}{2\pi} \int_{-N}^{N} e^{-ik(z-t)} \, dk \right\} U^0(x, z) dz = 0,$$

for $x+t < 0$, $t < x$, since $U^0(x, t)$ is obviously absolutely integrable in the interval $-\infty$ to x.

Using the lemma of Section 3 it can be verified in a straightforward way that

$$(4.9) \qquad \begin{aligned} &\int_{-N}^{N} \{\hat{\rho}(k) e^{-ik(z+t)} + e^{-ik(z-t)}\} \, dk \\ &= \int_{0}^{N} \{1-|\hat{\rho}(k)|^2\} e^{ik(t-z)} \, dk + \int_{-N}^{0} \{e^{ikt} + \hat{\rho}(k) e^{-ikt}\} \overline{\{e^{ikz} + \hat{\rho}(k) e^{-ikz}\}} \, dk. \end{aligned}$$

After substituting (4.9) into (4.8), multiplying by $\overline{U^0(x, t)}$ and integrating over t from $-\infty$ to x, there results

$$\lim_{N\to\infty}\int_0^N \{1-|\hat{\rho}(k)|^2\}\left\{\overline{\int_{-\infty}^{x} U^0(x,t)e^{-ikt}dt}\right\}\left\{\int_{-\infty}^{x} U^0(x,z)e^{-ikz}dz\right\}dk$$

$$(4.10)\qquad +\lim_{N\to\infty}\int_{-N}^{0}\left\{\int_{-\infty}^{x}[e^{ikt}+\hat{\rho}(k)e^{-ikt}]\overline{U^0(x,t)}dt\right\}$$

$$\cdot\left\{\overline{\int_{-\infty}^{x}[e^{ikz}+\hat{\rho}(k)e^{-ikz}]\overline{U^0(x,z)}dz}\right\}dk = 0, \qquad x<0.$$

According to (3.20), conservation of energy for $\hat{\rho}(k)$ and $\hat{t}(k)$ holds when k is real; i.e., $|\hat{\rho}| \leqq 1$ when k is real. It follows that each term on the left of (4.10) is positive for any function $\int_{-\infty}^{x} U^0(x,t)e^{-ikt}dt$ unless it is identically zero when $x<0$, and so equation (4.10) implies that $U^0(x,t)=0$ if $x+t<0$. From (4.6), then

$$(4.11)\qquad f_\nu^0(x)=0, \qquad x<0.$$

Thus, the only solution of the homogeneous system (4.4) is the trivial solution $f_\nu^0\equiv 0$, and, therefore, the system (4.3) has a unique solution.

To establish a solution of the inverse problem two more lemmas are required.

LEMMA 2. *The functions $f_\nu(x)$, $\nu=1,\cdots,n$, satisfying the system* (4.3) *also satisfy the differential equations*

$$(4.12)\qquad \frac{d^2f_\nu}{dx^2}+\{\kappa_\nu^2-W(x)\}f_\nu=0, \qquad \nu=1,\cdots,n,$$

where

$$(4.13)\qquad W(x)=2\sum_{\mu=1}^{n}\left\{\frac{df_\mu}{dx}-i\kappa_\mu f_\mu\right\}e^{-i\kappa_\mu x}.$$

Proof: Apply the differential operator

$$L=\frac{d^2}{dx^2}+\kappa_\nu^2-W(x)$$

to the system (4.3). With the aid of (4.13) the resulting system of equations can be put into the form of the homogeneous system of equations corresponding to (4.3) with the functions Lf_ν, $\nu=1,\cdots,n$, replacing the f_ν, $\nu=1,\cdots,n$. By Lemma 1 of this section (4.12) follows.

LEMMA 3. *If the $f_\nu(x)$, $\nu=1,\cdots,n$, satisfy the system of equations* (4.3), *then the functions $v(x,\pm k)$ given by*

$$(4.14)\qquad v(x,\pm k)=e^{\mp ikx}+ie^{\mp ikx}\sum_{\nu=1}^{n}\frac{f_\nu(x)}{\kappa_\nu\pm k}e^{-i\kappa_\nu x}$$

are solutions of the differential equation

$$(4.15)\qquad \frac{d^2 v}{dx^2}+\{k^2-W(x)\}v=0.$$

Proof: Substitute from (4.14) into the left side of (4.15) and use (4.12) and (4.13) of Lemma 2. Relation (4.15) follows.

It should be observed that

$$(4.16)\qquad \lim_{x\to-\infty}\{v(x,\pm k)-e^{\mp ikx}\}=0.$$

For a reflection coefficient $r(k)$ given by (3.1), satisfying the general conditions of Section 2 as well as the special conditions (3.2), (3.3) and (3.4), relations for the other elements $t(k)$ and $\rho(k)$ of the scattering matrix can be given. In fact,

THEOREM 1. *The functions $\hat{\rho}(k)$ given by* (3.18) *and $\hat{t}(k)$ given by* (3.19) *are the scattering matrix elements $\rho(k)$ and $t(k)$ associated with $r(k)$ given by* (3.1).

Proof: Suppose $-x$ is substituted for x in the differential equation (1.1), so that $V(-x)$ replaces $V(x)$ and $u(-x,k)$ replaces $u(x,k)$. Set

$$(4.17)\qquad W(x)=V(-x),\quad v(x,k)=u(-x,k).$$

The reflection coefficient on the left, $r(k)$, corresponding to $V(x)$ becomes the reflection coefficient on the right corresponding to $W(x)$. The reflection coefficient on the left corresponding to $W(x)$ is $\rho(k)$, and the transmission coefficient on either side is $t(k)$.

Let $v(x,k)$ be the solution which on the right is given by

$$(4.18)\qquad v(x,k)=e^{-ikx}+r(k)e^{ikx},\qquad x>0,$$

and on the left by

$$(4.19)\qquad \lim_{x\to-\infty}\{v(x,k)-t(k)e^{-ikx}\}=0.$$

Let $\hat{v}(x,k)$ be the solution which is determined on the left by

$$(4.20)\qquad \lim_{x\to-\infty}\{\hat{v}(x,k)-\hat{t}(k)e^{-ikx}\}=0.$$

To prove the theorem it is only necessary to prove that

$$(4.21)\qquad \hat{v}(x,k)=e^{-ikx}+r(k)e^{ikx},\qquad x>0,$$

or equivalently that

$$(4.22)\qquad \hat{v}(0,k)=1+r(k)\quad\text{and}\quad \left.\frac{d\hat{v}}{dx}\right|_{x=0}=ik\{r(k)-1\}.$$

From (4.20), Lemma 3 and (4.16), the following expression for $\hat{v}(x, k)$ is obtained:

$$\hat{v}(x, k) = \hat{t}(k)e^{-ikx} + i\hat{t}(k)e^{-ikx} \sum_{\nu=1}^{n} \frac{f_\nu(x)e^{-i\kappa_\nu x}}{\kappa_\nu + k}, \qquad x < 0. \tag{4.23}$$

Relations (4.22) can now be written as

$$\hat{t}(k) + i\hat{t}(k) \sum_{\nu=1}^{n} \frac{f_\nu(0)}{\kappa_\nu + k} = 1 + r(k) \tag{4.24}$$

and

$$-ik\hat{t}(k) + \hat{t}(k) \sum_{\nu=1}^{n} f_\nu(0) + i\hat{t}(k) \sum_{\nu=1}^{n} \frac{f'_\nu(0)}{\kappa_\nu + k} = ik\{r(k) - 1\}. \tag{4.25}$$

In order to demonstrate (4.24) it is convenient to define the complex numbers g_ν by the equation

$$\sum_{\nu=1}^{n} \frac{g_\nu}{k + \kappa_\nu} = -i \left\{ \frac{r(k)}{\hat{t}(k)} + \frac{1}{\hat{t}(k)} - 1 \right\}. \tag{4.26}$$

The expression on the left of (4.26) is a legitimate representation of the rational function of k on the right by the Mittag-Leffler theorem, and the coefficients g_ν are given by

$$g_\nu = -i \lim_{k \to -\kappa_\nu} (k + \kappa_\nu) \left\{ \frac{r(k)}{\hat{t}(k)} + \frac{1}{\hat{t}(k)} - 1 \right\} = -i \left\{ \frac{r(-\kappa_\nu)}{\hat{t}_{-\nu}} + \frac{1}{\hat{t}_{-\nu}} \right\}, \qquad \nu = 1, \cdots, n, \tag{4.27}$$

where the residue of $1/\hat{t}(k)$ at the pole $-\kappa_\nu$ has been designated by $1/\hat{t}_{-\nu}$. From (4.3),

$$f_\nu(0) = \rho_\nu \left\{ \sum_{\mu=1}^{n} \frac{f_\mu(0)}{\kappa_\nu + \kappa_\mu} - i \right\}, \qquad \nu = 1, \cdots, n. \tag{4.28}$$

The numbers g_ν of (4.27) will satisfy the system of equations (4.28) when substituted for the numbers $f_\nu(0)$. To see this, consider, using (4.26),

$$\rho_\nu \left\{ \sum_{\mu-1}^{n} \frac{g_\mu}{\kappa_\nu \mid \kappa_\mu} - i \right\} = -i\rho_\nu \left\{ \frac{r(\kappa_\nu)}{\hat{t}(\kappa_\nu)} + \frac{1}{\hat{t}(\kappa_\nu)} \right\}, \qquad \nu = 1, \cdots, n. \tag{4.29}$$

The right side of (4.29) can be written as

$$\begin{aligned} &\lim_{k \to \kappa_\nu} -i(k - \kappa_\nu)\hat{\rho}(k) \left\{ \frac{r(k)}{\hat{t}(k)} + \frac{1}{\hat{t}(k)} \right\} \\ &= \lim_{k \to \kappa_\nu} i(k - \kappa_\nu) \left\{ \frac{1}{\hat{t}(-k)} - \hat{t}(k) + \frac{r(-k)}{\hat{t}(-k)} \right\} \\ &= -i \left\{ \frac{r(-\kappa_\nu)}{\hat{t}_{-\nu}} + \frac{1}{\hat{t}_{-\nu}} \right\}, \qquad \nu = 1, \cdots, n, \end{aligned} \tag{4.30}$$

making use of the lemma relations (3.20). From (4.27), (4.29) and (4.30),

$$g_\nu = \rho_\nu \left\{ \sum_{\mu=1}^{n} \frac{g_\mu}{\kappa_\nu + \kappa_\mu} - i \right\}, \qquad \nu = 1, \cdots, n; \tag{4.31}$$

i.e., the g_ν satisfy the system of equations (4.28).

Since by Lemma 1 of this section, the system (4.28) must have a unique solution, it follows that

$$f_\nu(0) = g_\nu, \qquad \nu = 1, \cdots, n. \tag{4.32}$$

After substituting from (4.32) into (4.26) the relation (4.24) follows.

To demonstrate the remaining relation (4.25) it is necessary to consider the derivative of (4.3) at $x = 0$:

$$\kappa_\nu \rho_\nu + f_\nu'(0) + i\rho_\nu \sum_{\mu=1}^{n} f_\mu(0) - \rho_\nu \sum_{\mu=1}^{n} \frac{f_\mu'(0)}{\kappa_\nu + \kappa_\mu} = 0, \qquad \nu = 1, \cdots, n. \tag{4.33}$$

Let the numbers h_ν be defined by the equation

$$i\hat{t}(k) \sum_{\nu=1}^{n} \frac{h_\nu}{k + \kappa_\nu} = ik\{r(k) - 1\} + ik\hat{t}(k) - \hat{t}(k) \sum_{\nu=1}^{n} f_\nu(0). \tag{4.34}$$

From (4.26), (4.32) and (3.2) it follows that

$$\sum_{\nu=1}^{n} f_\nu(0) = \lim_{k \to \infty} ik \left\{ \frac{\hat{t}(k) - 1}{\hat{t}(k)} \right\}. \tag{4.35}$$

Relation (4.35) is consistent with (4.34) as can be seen by allowing k to approach ∞ in (4.34). Since the poles on the left side of (4.34) are the same as those on the right, (4.34) is entirely consistent. The numbers h_ν are then given by

$$\begin{aligned} h_\nu &= \lim_{k \to -\kappa_\nu} (k + \kappa_\nu) \left\{ \frac{kr(k) + k\hat{t}(k) - k}{\hat{t}(k)} + i \sum_{\mu=1}^{n} f_\mu(0) \right\} \\ &= \frac{-\kappa_\nu\{r(-\kappa_\nu) - 1\}}{\hat{t}_{-\nu}}, \qquad \nu = 1, \cdots, n. \end{aligned} \tag{4.36}$$

Also, from (4.34),

$$\sum_{\mu=1}^{n} \frac{h_\mu}{\kappa_\nu + \kappa_\mu} = \frac{\kappa_\nu\{r(\kappa_\nu) + \hat{t}(\kappa_\nu) - 1\}}{\hat{t}(\kappa_\nu)} + i \sum_{\mu=1}^{n} f_\mu(0), \qquad \nu = 1, \cdots, n. \tag{4.37}$$

After substituting from (4.37) and (4.36) into the expression

$$\rho_\nu \sum_{\mu=1}^{n} \frac{h_\mu}{\kappa_\nu + \kappa_\mu} - h_\nu \tag{4.38}$$

there results

$$
\begin{aligned}
&\frac{\rho_\nu \kappa_\nu\{r(\kappa_\nu)+\hat{t}(\kappa_\nu)-1\}}{\hat{t}(\kappa_\nu)} + i\rho_\nu \sum_{\mu=1}^{n} f_\mu(0) + \frac{\kappa_\nu\{r(-\kappa_\nu)-1\}}{t_{-\nu}} \\
&= \lim_{k\to\kappa_\nu} (k-\kappa_\nu)\hat{\rho}(k)k\left\{\frac{r(k)+\hat{t}(k)-1}{\hat{t}(k)}\right\} \\
&\qquad + i\rho_\nu \sum_{\mu=1}^{n} f_\mu(0) - \lim_{k\to\kappa_\nu}(k-\kappa_\nu)k\left\{\frac{r(-k)}{\hat{t}(-k)} - \frac{1}{\hat{t}(-k)}\right\} \\
&= \lim_{k\to\kappa_\nu}(k-\kappa_\nu)k\left\{-\frac{r(k)r(-k)}{\hat{t}(-k)} + \hat{\rho}(k) + \frac{r(-k)}{\hat{t}(-k)}\right\} \\
&\qquad + i\rho_\nu \sum_{\mu=1}^{n} f_\mu(0) - \lim_{k\to\kappa_\nu}(k-\kappa_\nu)k\left\{\frac{r(-k)}{\hat{t}(-k)} - \frac{1}{\hat{t}(-k)}\right\} \\
&= \kappa_\nu \rho_\nu + i\rho_\nu \sum_{\mu=1}^{n} f_\mu(0), \qquad \nu = 1, \cdots, n,
\end{aligned} \tag{4.39}
$$

using the relations (3.20) of the lemma in Section 3. The identity of the expressions (4.38) and (4.39) means that the numbers h_ν when substituted for the $f'_\nu(0)$ satisfy the system (4.33). Since (4.33) has a unique solution,

$$f'_\nu(0) = h_\nu, \qquad \nu = 1, \cdots, n. \tag{4.40}$$

From (4.40) and (4.34) it follows that the relation (4.25) is satisfied. The remarks made concerning (4.24) and (4.25) show that

$$t(k) = \hat{t}(k). \tag{4.41}$$

The fact that the scattering matrix is unitary implies then that

$$\rho(k) = \hat{\rho}(k). \tag{4.42}$$

It is now possible to obtain a closed form expression for the function $V(x)$ corresponding to $r(k)$. This result is contained in

THEOREM 2. *Given a reflection coefficient $r(k)$ of the form (3.1) subject to the general conditions on the scattering matrix elements and the special condition (3.2). The corresponding function $V(x)$ of (1.1) is then given by*

$$V(x) = -2\frac{d^2}{dx^2}\log \Delta(x), \qquad x > 0, \tag{4.43}$$

where $\Delta(x)$ is the determinant of the matrix whose elements are

$$\delta_{\nu\mu} - \frac{\rho_\nu e^{i(\kappa_\nu+\kappa_\mu)x}}{\kappa_\nu+\kappa_\mu}, \qquad \nu, \mu = 1, \cdots, n. \tag{4.44}$$

Here the quantity $\delta_{\mu\nu}$ is the Kronecker delta

$$\delta_{\mu\nu} = \begin{cases} 0 & \text{for } \mu \neq \nu, \\ 1 & \text{for } \mu = \nu, \end{cases} \tag{4.45}$$

and the κ_ν are roots in the upper half-plane of the equation

$$1 - r(k)r(-k) = 0. \tag{4.46}$$

Proof: If the solutions $f_\nu(x)$ of the system of equations (4.3) are represented by Cramer's rule as ratios of determinants, it can easily be seen that

$$\sum_{\mu=1}^{n} e^{-i\kappa_\mu x} f_\mu(x) = - \frac{d \log \Delta(-x)}{dx}, \qquad x < 0. \tag{4.47}$$

It follows at once that in (4.13)

$$W(x) = -2 \frac{d^2 \log \Delta(-x)}{dx^2}, \qquad x < 0, \tag{4.48}$$

and from the relation (4.17) between $V(x)$ and $W(x)$ the equation (4.43) follows.

Finally, it should be noted that according to Lemma 3, two linearly independent solutions[6] of the differential equation (1.1) are given by

$$u(x, \pm k) = e^{\pm ikx} + ie^{\pm ikx} \sum_{\nu=1}^{n} \frac{f_\nu(-x)e^{i\kappa_\nu x}}{\kappa_\nu \pm k}. \tag{4.49}$$

5. Conclusion

For $r(k)$ a reflection coefficient at $x = 0$ which is a rational function of k regular in the upper half-plane the transmission coefficient $t(k)$ exists, is regular on the real axis and in the upper half-plane and has no zeros on the real axis or in the upper half-plane. The expressions for the scattering matrix elements $t(k)$ and $\rho(k)$ in terms of $r(k)$ are given in Section 3 in (3.19) and (3.18) when $r(k)$ is given by (3.1). The expression for the corresponding $V(x)$ is given in Section 4 in (4.43). Two linearly independent solutions of the original differential equation (1.1) are given in (4.49).

Appendix

The Inverse Problem

In this section a derivation of the method for solving the inverse problem described in Section 1 will be given. The results are not actually needed for the demonstration of Section 4, the major interest of this paper, but

[6]Consider the Wronskian for large k.

are useful for motivating those demonstrations. Since, in addition, the solution of the inverse problem is generally valid if one allows the use of improper functions, even when various functions which occur are not well-behaved, the derivation presented here will be largely heuristic. There would be no real difficulty in making the argument rigorous by imposing restrictions implied by the requirements of the standard theory of linear integral equations.[7]

In the differential equation (1.1) the conditions

$$V(x) = 0,\ x < 0 \text{ and } V(x) \text{ real in the interval } -\infty < x < \infty \tag{A.1}$$

will be assumed.

Because of (A.1) the asymptotic limit of $u(x, k)$ given by (2.1) for $x \to -\infty$ is identical with $u(x, k)$ itself for negative values of x. If in (2.3)

$$A_- = 1, \qquad B_+ = 0,$$

then

$$\begin{aligned} u(x, k) &= e^{ikx} + r(k)e^{-ikx}, \qquad x < 0, \\ \lim_{x\to\infty} \{u(x, k) - &t(k)e^{ikx}\} = 0. \end{aligned} \tag{A.2}$$

The asymptotic conditions (A.2) mean that a wave of unit amplitude is incident from the left, producing a reflected wave of amplitude $r(k)$ and a transmitted wave of amplitude $t(k)$. Assuming that $r(k)$ is given for all positive real values of k, the central problem of this section is to find the function $V(x)$ in the differential equation (1.1).

It will be assumed that no proper eigenfunctions of (1.1) exist. Because of this fact and (A.1) it can be shown that $r(k)$ is regular in the upper half-plane [2].

The Fourier transform $U(x, t)$ of $u(x, k)$ is a transient wave solution of the time dependent differential equation

$$\frac{\partial^2 U(x, t)}{\partial x^2} - \frac{\partial^2 U(x, t)}{\partial t^2} - V(x)U(x, t) = 0. \tag{A.3}$$

If $u(x, k)$ satisfies the asymptotic conditions (A.2), the Fourier transform of $u(x, k)$ does not exist as a proper function; i.e., it contains a delta function in the region $x < 0$. However, if the incident steady state wave of unit amplitude is replaced by one having an absolutely integrable amplitude over $-\infty < k < +\infty$, the corresponding transient solution of (A.3) is a proper function. The transient solution of (A.3) corresponding to an incident steady-state continuous wave of unit amplitude may therefore be regarded as a

[7] E.g., see [4].

limiting case of well-behaved transients—useful because of its formal simplicity.

The reflected transient due to the incident wave

$$U_{\text{inc}} = \delta(x-t), \qquad x < 0, \quad t < 0,$$

is

(A.4) $$R(x+t) = \frac{1}{2\pi}\int_{-\infty}^{\infty} r(k)e^{-ik(x+t)}\,dk.$$

In order to obtain (A.4) when $r(k)$ is known for positive real k it is necessary to extend $r(k)$ to the negative real axis by using (2.5). At the same time it can be seen that $R(x+t)$ is real for all real x and t. The total solution $U(x, t)$ of (A.3) in the region $x < 0$ is then

(A.5) $$U(x, t) = \delta(x-t)+R(x+t), \qquad x < 0,$$

and $U(x, t)$ is real for all x and t.

The function $R(z)$ has the property

(A.6) $$R(z) = 0, \qquad z < 0,$$

because of (2.4) and the regularity of $r(k)$ in the upper half-plane. The physical meaning of (A.6) in connection with (A.5) is easily understood; for, the reflected transient should not appear at any point x until the appropriate amount of time has elapsed after the incident transient has struck the edge of the reflecting medium at $x = 0$.

Experience in the perturbation theory of linear hermitian operators indicates that a solution $u(x, k)$ of the equation (1.1) should be related to some solution $u_0(x, k)$ of the "unperturbed" equation

(A.7) $$\frac{d^2 u_0}{dx^2} + k^2 u_0 = 0$$

by means of a linear transformation which is independent of the spectral parameter k. On this basis one might venture the guess that there exists a function $K(x, z)$ such that

(A.8) $$u(x, k) = u_0(x, k) + \int_{-\infty}^{+\infty} K(x, z)u_0(z, k)dz.$$

The Fourier transform of (A.8) is

(A.9) $$U(x, t) = U_0(x, t) + \int_{-\infty}^{\infty} K(x, z)U_0(z, t)dz.$$

Now, if $U(x, t)$ is a right moving transient in the sense that

(A.10) $$U(x, t) = 0, \qquad x > t,$$

then it would seem natural to suppose that $U(x, t)$ will depend only on

$U_0(x, t)$ at those values of the space variable x which the non-zero part of the traveling wave would have had time to reach at the time t. In other words, one might guess that

$$K(x, z) = 0, \qquad z > x. \tag{A.11}$$

Assuming (A.10) and (A.11), (A.9) becomes

$$U(x, t) = U_0(x, t) + \int_{-\infty}^{x} K(x, z)U_0(z, t)dz. \tag{A.12}$$

If $U_0(x, t)$ is the particular transient corresponding to (A.2),

$$U_0(x, t) = \delta(x-t)+R(x+t), \tag{A.13}$$

then (A.12) becomes

$$U(x, t) = \delta(x-t)+R(x+t)+K(x, t)\eta(x-t) + \int_{-t}^{x} K(x, z)\eta(x-z)R(z+t)dz, \tag{A.14}$$

where

$$\eta(z) = \begin{cases} 0 & \text{for } z < 0, \\ 1 & \text{for } z > 0. \end{cases}$$

Assuming that

$$K(x, t) = 0, \qquad t \leqq -x, \tag{A.15}$$

it follows from (A.14) that

$$U(x, t) = U_0(x, t), \qquad x < 0.$$

It will now be shown directly that there exists a function $K(x, z)$ satisfying (A.11) and (A.15) such that (A.12) holds [2]. Applying (A.15) to (A.12) one obtains

$$U(x, t) = \begin{cases} U_0(x, t) + \int_{-x}^{x} K(x, z)U_0(z, t)dz & \text{for } x \geqq 0, \\ U_0(x, t) & \text{for } x \leqq 0. \end{cases} \tag{A.16}$$

After substituting from (A.16) into (A.3), it is easy to verify that (A.16) provides a solution $U(x, t)$ of (A.3) if $K(x, z)$ satisfies

$$\frac{\partial^2 K}{\partial x^2} - \frac{\partial^2 K}{\partial z^2} - V(x)K = 0, \qquad z < x, \tag{A.17}$$

subject to the conditions

$$K(x, -x) = 0, \quad 2\,\frac{dK(x, x)}{dx} = V(x). \tag{A.18}$$

Since it is known from the theory of the linear second order hyperbolic partial differential equation that the Cauchy problem defined by (A.17) and (A.18)[8] has a solution, the existence of $K(x, z)$ satisfying the required conditions follows and thus the expression (A.16) for $U(x, t)$ [2].

Now equation (A.16) can be used to reconstruct $V(x)$. First it is necessary to obtain $K(x, t)$.

Consider (A.16) with $U_0(x, t)$ given by (A.13). Assume that the corresponding $U(x, t)$ is a right moving wave; i.e.,

$$U(x, t) = 0, \qquad x > t. \tag{A.19}$$

Then (A.16) implies

$$0 = R(x+t)+K(x, t) + \int_{-x}^{x} K(x, z)R(z+t)dz, \qquad x > t. \tag{A.20}$$

Since, according to (A.6), $R(z+t)$ is zero for $z < -t$, the lower limit in the integral of (A.20) can be replaced by $-t$, and it follows from (A.20) that

$$K(x, t) = 0, \qquad t < -x. \tag{A.21}$$

Equation (A.20) is a Fredholm integral equation for $K(x, t)$ considered as a function of t (with x regarded as a parameter) with the kernel $R(z+t)$ over the range $-x < t < x$. When $R(z+t)$ is given through (A.4), the integral equation (A.20) can be solved for $K(x, t)$ at every value of x. The function $V(x)$ can then be obtained from the second relation in (A.18).

The existence of a solution $K(x, t)$ of (A.20) can be established by means of a standard argument in the Fredholm theory: one simply proves the uniqueness of the solution.[9] At the same time the uniqueness of the solution $K(x, t)$ of (A.20) can be used to establish the fact that $K(x, t)$ satisfies (A.17) when $V(x)$ is defined by (A.18). Thus, consider now the integral equation (A.20). The first relation of (A.18) follows at once from (A.20) and (A.6). Moreover, the fact that the solution of (A.20), $K(x, t)$, satisfies (A.17) follows without much difficulty on applying the differential operator

$$\frac{\partial^2}{\partial x^2} - \frac{\partial^2}{\partial t^2} - V(x)$$

to (A.20). The result is

---

[8]Conditions (A.18) and continuity of $K(x, z)$ in the region $z \leq x$ imply

$$K(x, x) = \begin{cases} \frac{1}{2} \int_0^x V(x)dx & \text{for } x \geq 0, \\ 0 & \text{for } x \leq 0. \end{cases}$$

[9]See Section 4 and [2].

$$\text{(A.22)}\quad 0=\left\{\frac{\partial^2 K}{\partial x^2}-\frac{\partial^2 K}{\partial t^2}-V(x)K\right\}+\int_{-x}^{x}\left\{\frac{\partial^2 K}{\partial x^2}-\frac{\partial^2 K}{\partial t^2}-V(x)K\right\}R(z+t)dz,\quad x>t.$$

Equation (A.22) is the homogeneous integral equation corresponding to (A.20) for the function

$$\frac{\partial^2 K}{\partial x^2}-\frac{\partial^2 K}{\partial t^2}-V(x)K.$$

If the solution of (A.20) is unique, the solution of (A.22) must vanish:

$$\frac{\partial^2 K}{\partial x^2}-\frac{\partial^2 K}{\partial t^2}-V(x)K=0,$$

as required by (A.17).

Bibliography

[1] Kay, I., and Moses, H., *The determination of the scattering potential from the spectral measure function*, Nuovo Cimento, Vol. 2, 1955, pp. 917–961; also Div. Electromag. Res., Inst. Math. Sci., New York Univ., Res. Rep. No. CX-18, 1955.

[2] Kay, I., *The inverse scattering problem*, Div. Electromag. Res., Inst. Math. Sci., New York Univ., Res. Rep. No. EM-74, 1955.

[3] Bargman, V., *On the connection between phase shifts and scattering potential*, Rev. Modern Physics, Vol. 21, 1949, pp. 488–493.

[4] Gelfand, I. M., and Levitan, B. M., *On the determination of a differential equation by its spectral function*, Amer. Math. Soc. Transl., Ser. 2, Vol. 1, 1955, pp. 253–304.

Received August, 1959.

COMMUNICATIONS ON PURE AND APPLIED MATHEMATICS, VOL. XIV, 435–445 (1961)

K. O. Friedrichs anniversary issue

A Simple Verification of the Gelfand-Levitan Equation for the Three-Dimensional Scattering Problem*

I. KAY and H. E. MOSES

Institute of Mathematical Sciences, New York University — *Polytechnic Institute of Brooklyn*

1. The Gelfand-Levitan Equation for the Three-Dimensional Problem

In [1] we derived the Gelfand-Levitan equation for the three-dimensional problem from the general approach of [2]. We shall summarize some results.

Let $\psi_0(\boldsymbol{x}|\boldsymbol{p})$ and $\psi_0(\boldsymbol{x}|p, \theta, \phi)$ be defined by

$$(1.1)\qquad \begin{aligned} \psi_0(\boldsymbol{x}|\boldsymbol{p}) &= \frac{1}{(2\pi)^{3/2}} e^{i\boldsymbol{p}\cdot\boldsymbol{x}}, \\ \psi_0(\boldsymbol{x}|p, \theta, \phi) &= p\sqrt{\sin\theta}\,\psi_0(\boldsymbol{x}|\boldsymbol{p}), \end{aligned}$$

where

$$\begin{aligned} p &= |\boldsymbol{p}|, \\ \theta &= \text{angle between } \boldsymbol{p} \text{ and } z\text{-axis} \\ \phi &= \text{angle between projection of } \boldsymbol{p} \text{ in } x, y\text{-plane and } x\text{-axis.} \end{aligned}$$

It is clear that both $\psi_0(\boldsymbol{x}|\boldsymbol{p})$ and $\psi_0(\boldsymbol{x}|p, \theta, \phi)$ are eigenfunctions of $H_0 = -\nabla^2$ with the eigenvalue $E = p^2$.

Let us define $\psi_0(\boldsymbol{x}|-p, \theta, \phi)$ by

$$(1.2)\qquad \psi_0(\boldsymbol{x}|-p, \theta, \phi) = -\psi_0^*(\boldsymbol{x}|p, \theta, \phi),$$

where the asterisk means complex conjugate. The functions $\psi_0(\boldsymbol{x}|-p, \theta, \phi)$ are likewise eigenfunctions of H_0 with eigenvalue p^2.

Let $b_p(\theta, \phi|\theta', \phi')$ be a function of $p, \theta, \phi, \theta', \phi'$ such that

*The work of H. E. Moses was supported by the Office of Naval Research, Contract No. Nonr-839(30), Project No. NR 013-106. Reproduction in whole or in part permitted for any purpose of the United States Government.

$$b_{-p}(\theta, \phi|\theta', \phi) = [b_p(\theta, \phi|\theta', \phi')]^*,$$
(1.3)
$$b_p(\theta, \phi|\theta', \phi') = b_p(\pi-\theta', \Phi'|\pi-\theta, \Phi),$$

where

$$\Phi = \begin{cases} \phi+\pi & \text{if } 0 \leqq \phi < \pi, \\ \phi-\pi & \text{if } \pi \leqq \phi < 2\pi, \end{cases}$$

Φ' is defined similarly in terms of ϕ'.

Then the function $\Omega(\boldsymbol{x}|\boldsymbol{x}')$ given by

$$\Omega(\boldsymbol{x}|\boldsymbol{x}') = \int_{-\infty}^{+\infty} dp \int_0^{\pi/2} d\theta \int_{\pi/2}^{\pi} d\theta' \int_0^{2\pi} d\phi \int_0^{2\pi} d\phi' \cdot \psi_0(\boldsymbol{x}|p, \theta, \phi) b_p(\theta, \phi|\theta', \phi') \psi_0^*(\boldsymbol{x}'|p, \theta', \phi') \tag{1.4}$$

can be shown to be real.

The Gelfand-Levitan theorem for three-dimensional scattering is the following. Let us write

$$\psi_0(r, \lambda, \sigma|p, \theta, \phi) \equiv \psi_0(\boldsymbol{x}|p, \theta, \phi),$$
$$\Omega(r, \lambda, \sigma|r', \lambda', \sigma') \equiv \Omega(\boldsymbol{x}|\boldsymbol{x}'),$$

where $r = |\boldsymbol{x}|$, and λ, σ are polar angles of $\boldsymbol{x}$ analogous to θ, ϕ, respectively, and a similar statement holds for primed quantities.

Let the equation for $K(r, \lambda, \sigma|r', \lambda', \sigma')$

$$K(r, \lambda, \sigma|r', \lambda', \sigma') = -\Omega(r, \lambda, \sigma|r', \lambda', \sigma') - \int_r^{\infty} dr'' \int_0^{\pi} \sin \lambda'' d\lambda'' \int_0^{2\pi} d\sigma'' K(r, \lambda, \sigma|r'', \lambda'', \sigma'') \Omega(r'', \lambda'', \sigma''|r', \lambda', \sigma') \tag{1.5}$$

have a unique solution for $r \leqq r'$. For $r > r'$ the definition of $K(r, \lambda, \sigma|r', \lambda', \sigma')$ is immaterial and can be arbitrary. Then $K(r, \lambda, \sigma|r', \lambda', \sigma')$ is real for $r \leqq r'$ and $\psi(\boldsymbol{x}|p, \theta, \phi) \equiv \psi(r, \lambda, \sigma|p, \theta, \phi)$, where

$$\psi(r, \lambda, \sigma|p, \theta, \phi) = \psi_0(r, \lambda, \sigma|p, \theta, \phi) + \int_r^{\infty} dr' \int_0^{\pi} \sin \lambda' \int_0^{2\pi} d\sigma' K(r, \lambda, \sigma|r', \lambda', \sigma') \times \psi_0(r', \lambda', \sigma'|p, \theta, \phi) \tag{1.6}$$

is an eigenfunction of the operator $H = H_0+V$, V being an integral operator of the form

$$\int V(\boldsymbol{x}|\boldsymbol{x}') f(\boldsymbol{x}') d\boldsymbol{x}' = r^2 \int_0^{\pi} \sin \lambda' \, d\lambda' \int_0^{2\pi} d\sigma' V(r|\lambda, \sigma; \lambda', \sigma') f(r, \lambda', \sigma') \tag{1.7}$$

in the $\boldsymbol{x}$-representation. The function $f(\boldsymbol{x}) \equiv f(r, \lambda, \sigma)$ is a state in the $\boldsymbol{x}$-representation and

$$V(r|\lambda, \sigma; \lambda', \sigma') = -2 \frac{1}{r^2} \frac{\partial}{\partial r} r^2 K(r, \lambda, \sigma|r, \lambda', \sigma'). \tag{1.8}$$

Furthermore, the eigenfunctions $\psi(\boldsymbol{x}|p, \theta, \phi) = -\psi^*(\boldsymbol{x}|-p, \theta, \phi)$ satisfy the completeness relation

$$\begin{aligned} &\int_0^\infty dp \int_0^\pi d\theta \int_0^{2\pi} d\phi \psi(\boldsymbol{x}|p, \theta, \phi)\psi^*(\boldsymbol{x}'|p, \theta, \phi) \\ &\quad + \int_{-\infty}^{+\infty} dp \int_0^{\pi/2} d\theta \int_{\pi/2}^{\pi} d\theta' \int_0^{2\pi} d\phi \int_0^{2\pi} d\phi' \\ &\quad \times \psi(\boldsymbol{x}|p, \theta, \phi) b_p(\theta, \phi|\theta', \phi')\psi^*(\boldsymbol{x}'|p, \theta', \phi') = \delta(\boldsymbol{x}-\boldsymbol{x}'). \end{aligned} \tag{1.9}$$

It will be our objective in the following two sections of the paper to prove the above theorem. If equation (1.5) has a unique solution for K, then $\psi(\boldsymbol{x}|p, \theta, \phi)$ are eigenfunctions of H corresponding to the eigenvalues $E = p^2$ and satisfy the completeness relation (1.9).

2. Equations for the Eigenfunctions; Expression for the Potential

From the Gelfand-Levitan equation (1.5) and (1.6) we have

$$\begin{aligned} (H_0-p^2)\psi(\boldsymbol{x}|p, \theta, \phi) &= H_0 \int_r^\infty r'^2 dr' \int_0^\pi \sin\lambda' d\lambda' \int_0^{2\pi} d\sigma' K(r, \lambda, \sigma|r', \lambda', \sigma') \\ &\quad \times \psi_0(r', \lambda', \sigma'|p, \theta, \phi) - \int_0^\infty r'^2 dr' \int_0^\pi \sin\lambda' d\lambda' \int_0^{2\pi} d\sigma' \\ &\quad \times K(r, \lambda, \sigma|r', \lambda', \sigma') H_0' \psi_0(r', \lambda', \sigma'|p, \theta, \phi). \end{aligned} \tag{2.1}$$

In (2.1), H_0 and H_0' operate on the primed and unprimed variables respectively. On using the polar coordinate form for the Laplacian

$$H_0 = -\left[\frac{1}{r^2}\frac{\partial}{\partial r}\left(r^2\frac{\partial}{\partial r}\right) + \frac{1}{r^2}\frac{1}{\sin\lambda}\frac{\partial}{\partial\lambda}\left(\sin\lambda\frac{\partial}{\partial\lambda}\right) + \frac{1}{r^2}\frac{1}{\sin^2\lambda}\frac{\partial^2}{\partial\sigma^2}\right],$$

we have

$$\begin{aligned} &H_0 \int_r^\infty r'^2 dr' \int_0^\pi \sin\lambda' d\lambda' \int_0^{2\pi} d\sigma' K(r, \lambda, \sigma|r', \lambda', \sigma')\psi_0(r', \lambda', \sigma'|p, \theta, \phi) \\ &= \int_r^\infty r'^2 dr' \int_0^\pi \sin\lambda' d\lambda' \int_0^{2\pi} d\sigma' H_0 K(r, \lambda, \sigma|r', \lambda', \sigma') \\ &+ 4r \int_0^\pi \sin\lambda' d\lambda' \int_0^{2\pi} d\sigma' K(r, \lambda, \sigma|r, \lambda', \sigma')\psi_0(r, \lambda', \sigma'|p, \theta, \phi) \\ &+ r^2 \Bigl\{\int_0^\pi \sin\lambda' d\lambda' \int_0^{2\pi} d\sigma' \left[\frac{\partial}{\partial r} K(r, \lambda, \sigma|r, \lambda', \sigma') + K_r(r, \lambda, \sigma|r, \lambda', \sigma')\right] \\ &\times \psi_0(r, \lambda', \sigma'|p, \theta, \phi) + \int_0^\pi \sin\lambda' d\lambda' \int_0^{2\pi} d\sigma' K(r, \lambda, \sigma|r, \lambda', \sigma') \\ &\times \frac{\partial}{\partial r}\psi_0(r, \lambda', \sigma'|p, \theta, \phi)\Bigr\}. \end{aligned} \tag{2.2}$$

In (2.2), $K_r(r, \lambda, \sigma|r, \lambda', \sigma')$ is defined by

$$K_r(r, \lambda, \sigma|r, \lambda', \sigma') = \frac{\partial}{\partial r} K(r, \lambda, \sigma|r', \lambda, \sigma)\Big|_{r'=r}. \tag{2.3}$$

Later we shall use $K_{r'}(r, \lambda, \sigma|r, \lambda', \sigma')$ which is defined by

$$K_{r'}(r, \lambda, \sigma|r, \lambda', \sigma') = \frac{\partial}{\partial r'} K(r, \lambda, \sigma|r', \lambda', \sigma')\Big|_{r'=r}. \tag{2.4}$$

On using Green's theorem over the obvious volume and surfaces in the primed variables and assuming

$$\lim_{r'\to\infty} K(r, \lambda, \sigma|r', \lambda', \sigma') = \lim_{r'\to\infty} \frac{\partial}{\partial r'} K(r, \lambda, \sigma|r', \lambda', \sigma') = 0,$$

we have

$$\begin{aligned}
&\int_r^\infty r'^2 dr' \int_0^\pi \sin\lambda' d\lambda' \int_0^{2\pi} d\sigma' K(r, \lambda, \sigma|r', \lambda', \sigma') H_0' \psi(r', \lambda', \sigma'|p, \theta, \phi) \\
&= \int_r^\infty r'^2 dr' \int_0^\pi \sin\lambda' dr' \int_0^{2\pi} d\sigma' \Big[H_0' K(r, \lambda, \sigma|r', \lambda', \sigma')\Big] \psi_0(r', \lambda', \sigma'|p, \theta, \phi) \\
&\quad + r^2 \Big\{\int_0^\pi \sin\lambda' d\lambda' \int_0^{2\pi} d\sigma' [K(r, \lambda, \sigma|r, \lambda', \sigma')] \frac{\partial}{\partial r} \psi_0(r, \lambda', \sigma'|p, \theta, \phi) \\
&\quad - K_{r'}(r, \lambda, \sigma|r, \lambda', \sigma') \psi_0(r, \lambda', \sigma'|p, \theta, \phi)\Big\}.
\end{aligned} \tag{2.5}$$

Substituting (2.2) and (2.5) into (2.1) we have

$$\begin{aligned}
(H_0 - p^2)\psi(\mathbf{x}|p, \theta, \phi) &= -r^2 \int_0^\pi \sin\lambda' d\lambda' \int_0^{2\pi} d\sigma' V(r|\lambda, \sigma; \lambda', \sigma') \psi_0(r, \lambda', \sigma'|p, \theta, \phi) \\
&+ \int_r^\infty r'^2 dr' \int_0^\pi \sin\lambda' d\lambda' \int_0^{2\pi} d\sigma' [(H_0 - H_0') K(r, \lambda, \sigma|r', \lambda', \sigma')] \psi_0(r', \lambda', \sigma'|p, \theta, \phi).
\end{aligned} \tag{2.6}$$

In obtaining (2.6) with $V(r|\lambda, \sigma; \lambda', \sigma')$ given by (1.8) we have used the chain rule of differentiation:

$$\frac{\partial}{\partial r} K(r, \lambda, \sigma|r, \lambda', \sigma') = K_r(r, \lambda, \sigma|r, \lambda', \sigma') + K_{r'}(r, \lambda, \sigma|r, \lambda', \sigma'). \tag{2.7}$$

We see from (2.6) that we must determine

$$(H_0 - H_0') K(r, \lambda, \sigma|r', \lambda', \sigma').$$

We note first that

$$H_0 \Omega(\mathbf{x}|\mathbf{x}') = H_0' \Omega(\mathbf{x}|\mathbf{x}'). \tag{2.8}$$

Hence on using the integral equation for K (1.5), we see that

$$(H_0-H_0')K(r,\lambda,\sigma|r',\lambda',\sigma') = -H_0\int_r^\infty r''^2dr''\int_0^\pi \sin\lambda''d\lambda''\int_0^{2\pi}d\sigma''$$
$$\times K(r,\lambda,\sigma|r'',\lambda'',\sigma'')\Omega(r'',\lambda'',\sigma''|r',\lambda',\sigma')+\int_r^\infty r''^2dr''\int_0^\pi \sin\lambda''d\lambda'' \tag{2.9}$$
$$\times\int_0^{2\pi}d\sigma''K(r,\lambda,\sigma|r'',\lambda'',\sigma'')H_0''\Omega(r'',\lambda'',\sigma''|r',\lambda',\sigma').$$

Now by exactly the same procedure as we used to obtain the right-hand side of (2.1), we can obtain the right-hand side of (2.9). We find the equation analogous to (2.6)

$$(H_0-H_0')K(r,\lambda,\sigma|r',\lambda',\sigma') = r^2\int_0^\pi \sin\lambda''d\lambda''\int_0^{2\pi}d\sigma''V(r|\lambda,\sigma;\lambda'',\sigma'')$$
$$\times\Omega(r,\lambda'',\sigma''|r',\lambda',\sigma')-\int_r^\infty r''^2dr''\int_0^\pi\sin\lambda''d\lambda''\int_0^{2\pi}d\sigma'' \tag{2.10}$$
$$\times[(H_0-H_0'')K(r,\lambda,\sigma|r'',\lambda'',\sigma'')]\Omega(r'',\lambda'',\sigma''|r',\lambda',\sigma').$$

Equation (2.10) can be considered an integral equation for

$$(H_0-H_0')K(r,\lambda,\sigma|r',\lambda',\sigma').$$

It has the same kernel and almost the same inhomogeneous term as the equation (1.5) for K. In fact using (1.5), it is easy to show that a solution of (2.10) is

$$(H_0-H_0')K(r,\lambda,\sigma|r',\lambda',\sigma') = -r^2\int_0^\pi\sin\lambda''d\lambda''\int_0^{2\pi}d\lambda''V(r|\lambda,\sigma;\lambda'',\sigma'')$$
$$\times K(r,\lambda'',\sigma''|r',\lambda',\sigma'). \tag{2.11}$$

From the assumption that the solution of the Gelfand-Levitan equation is unique we see that the solution of (2.10) is unique. On substituting (2.11) into (2.6) and using (1.6) we have

$$(H_0-p^2)\psi(x|p,\theta,\phi)$$
$$= -r^2\int_0^\pi\sin\lambda'd\lambda'\int_0^{2\pi}d\sigma'V(r|\lambda,\sigma;\lambda',\sigma')\psi(r,\lambda',\sigma'|p,\theta,\phi) \tag{2.12}$$

which is what we wished to show.

3. Proof of the Completeness Relation

We shall now prove the completeness relation (1.9). Toward this end it will be useful to obtain two expressions for the function $K(r,\lambda,\sigma|r',\lambda',\sigma')$ for $r'>r$. From the integral equation for K (1.5) and the expression (1.4) for $\Omega(x|x')$, we have

$$K(r,\lambda,\sigma|r',\lambda',\sigma') = -\Omega(r,\lambda,\sigma|r',\lambda',\sigma')$$
$$) -\int_r^\infty r''^2dr''\int_0^\pi\sin\lambda''d\lambda''\int_0^{2\pi}d\sigma''\int_{-\infty}^{+\infty}dp\int_0^{\pi/2}d\theta\int_{\pi/2}^\pi d\theta'\int_0^{2\pi}d\phi\int_0^{2\pi}d\phi'$$
$$\times K(r,\lambda,\sigma|r'',\lambda'',\sigma'')\psi_0(r'',\lambda'',\sigma''|p,\theta,\phi)b_p(\theta,\phi|\theta',\phi')\psi_0^*(r',\lambda',\sigma'|p,\theta',\phi').$$

In (3.1) let us integrate over r'', λ'', σ'' first. We note that this integral is just the second term of the expression for $\psi(r, \lambda, \sigma|p, \theta, \phi)$ given by (1.6). We substitute for this integral $\psi(r, \lambda, \sigma|p, \theta, \phi)-\psi_0(r, \lambda, \sigma|p, \theta, \phi)$ and our expression (3.1) becomes

$$K(r, \lambda, \sigma|r', \lambda', \sigma') = -\Omega(r, \lambda, \sigma|r', \lambda', \sigma')-\int_{-\infty}^{+\infty} dp \int_0^{\pi/2} d\theta \int_{\pi/2}^{\pi} d\theta' \int_0^{2\pi} d\phi \int_0^{2\pi} d\phi' \times \psi(r, \lambda, \sigma|p, \theta, \phi) b_p(\theta, \phi|\theta', \phi')\psi_0^*(r', \lambda', \sigma'|p, \theta', \phi') + \int_{-\infty}^{+\infty} dp \int_0^{\pi/2} d\theta \int_{\pi/2}^{\pi} d\theta' \int_0^{2\pi} d\phi \int_0^{2\pi} d\phi' \psi_0(r, \lambda, \sigma|p, \theta, \phi) \times b_p(\theta, \phi|\theta', \phi')\psi_0^*(r', \lambda', \sigma'|p, \theta', \phi').$$

Now the third term on the right is just the negative of the first term (see (1.4)) and therefore it cancels the first term. Hence

$$K(r, \lambda, \sigma|r', \lambda', \sigma') = -\int_{-\infty}^{+\infty} dp \int_0^{\pi/2} d\theta \int_{\pi/2}^{\pi} d\theta' \int_0^{2\pi} d\phi \int_0^{2\pi} d\phi' \times \psi(r, \lambda, \sigma|p, \theta, \phi) b_p(\theta, \phi|\theta', \phi')\psi_0^*(r', \lambda', \sigma'|p, \theta', \phi') \tag{3.2}$$

for $r' > r$.

Now in the right-hand member of (3.2) we shall replace $\psi_0^*(r', \lambda', \sigma'|p, \theta', \phi')$ as follows:

$$\psi_0^*(r', \lambda', \sigma'|p, \theta', \phi') = \psi^*(r', \lambda', \sigma'|p, \theta', \phi')-\int_r^{\infty} r''^2 dr'' \int_0^{\pi} \sin \lambda'' dr'' \int_0^{2\pi} d\sigma'' \times K(r', \lambda', \sigma'|r'', \lambda'', \sigma'')\psi_0^*(r'', \lambda'', \sigma''|p, \theta', \phi').$$

This equation follows from (1.6) and the fact that K is real. Then

$$K(r, \lambda, \sigma|r', \lambda', \sigma') = -\int_{-\infty}^{+\infty} dp \int_0^{\pi/2} d\theta \int_{\pi/2}^{\pi} d\theta' \int_0^{2\pi} d\phi \int_0^{2\pi} d\phi' \psi(r, \lambda, \sigma|p, \theta, \phi) \times b_p(\theta, \phi|\theta', \phi')\psi^*(r', \lambda', \sigma'|p, \theta', \phi') + \int_r^{\infty} r''^2 dr'' \int_0^{\pi} \sin \lambda'' dr'' \int_0^{2\pi} d\sigma'' \int_{-\infty}^{+\infty} dp \int_0^{\pi/2} d\theta \int_{\pi/2}^{\pi} d\theta' \int_0^{2\pi} d\phi \int_0^{2\pi} d\phi' \times K(r', \lambda', \sigma'|r'', \lambda'', \sigma'')\psi(r, \lambda, \sigma|p, \theta, \phi) b_p(\theta, \phi|\theta', \phi') \times \psi_0^*(r'', \lambda'', \sigma''|p, \theta', \phi').$$

Or if we use (3.2) in the second term on the right, we have

$$K(r, \lambda, \sigma|r', \lambda', \sigma') = -\int_{-\infty}^{+\infty} dp \int_0^{\pi/2} d\theta \int_{\pi/2}^{\pi} d\theta' \int_0^{2\pi} d\phi \int_0^{2\pi} d\phi' \psi(r, \lambda, \sigma|p, \theta, \phi) \times b_p(\theta, \phi|\theta', \phi')\psi^*(r', \lambda', \sigma'|p, \theta', \phi')-\int_r^{\infty} r''^2 dr'' \int_0^{\pi} \sin \lambda'' d\lambda'' \int_0^{2\pi} d\sigma'' \times K(r, \lambda, \sigma|r'', \lambda'', \sigma'')K(r', \lambda', \sigma'|r'', \lambda'', \sigma''), \qquad r' > r. \tag{3.3}$$

Hence

$$\begin{aligned}(3.4)\quad &\int_{-\infty}^{+\infty} dp \int_0^{\pi/2} d\theta \int_{\pi/2}^{\pi} d\theta' \int_0^{2\pi} d\phi \int_0^{2\pi} d\phi' \psi(r, \lambda, \sigma|p, \theta, \phi) b_p(\theta, \phi|\theta', \phi') \psi^*(r', \lambda', \sigma'|p, \theta', \phi') \\ &= -K(r, \lambda, \sigma|r', \lambda', \sigma') - \int_r^{\infty} r''^2 dr'' \int_0^{\pi} \sin \lambda'' d\lambda'' \\ &\times \int_0^{2\pi} d\sigma'' K(r, \lambda, \sigma|r'', \lambda'', \sigma'') K(r', \lambda', \sigma'|r'', \lambda'', \sigma''), \qquad r' > r.\end{aligned}$$

We want now to find

$$\int_{-\infty}^{+\infty} dp \int_0^{\pi/2} d\theta \int_{\pi/2}^{\pi} d\theta' \int_0^{2\pi} d\phi \int_0^{2\pi} d\phi' \psi(r, \lambda, \sigma|p, \theta, \phi) b_p(\theta, \phi|\theta', \phi') \psi^*(r', \lambda', \sigma'|p, \theta', \phi')$$

when $r' < r$. Toward this end we shall have to use the relation

$$(3.5)\qquad \psi_0(x|p, \pi-\theta, \Phi) = \psi_0^*(x|p, \theta, \phi),$$

where

$$\Phi = \begin{cases} \phi - \pi & \text{if} \quad \phi > \pi, \\ \phi + \pi & \text{if} \quad \phi < \pi. \end{cases}$$

Equations (3.5) and (1.6) lead to

$$(3.6)\qquad \psi(x|p, \pi-\theta, \Phi) = \psi^*(x|p, \theta, \phi).$$

The use of (3.6) and (1.3) leads to

$$\begin{aligned}(3.7)\quad &\int_{-\infty}^{+\infty} dp \int_0^{\pi/2} d\theta \int_{\pi/2}^{\pi} d\theta' \int_0^{2\pi} d\phi \int_0^{2\pi} d\phi' \psi(x|p, \theta, \phi) b_p(\theta, \phi|\theta', \phi') \psi^*(x'|p, \theta', \phi') \\ &= \int_{-\infty}^{+\infty} dp \int_0^{\pi/2} d\theta \int_{\pi/2}^{\pi} d\theta' \int_0^{2\pi} d\phi \int_0^{2\pi} d\phi' \, \psi(x'|p, \theta, \phi) b_p(\theta, \phi|\theta', \phi') \psi^*(x|p, \theta', \phi').\end{aligned}$$

Hence on using (3.7) and (3.4) we have for $r' < r$

$$\begin{aligned}&\int_{-\infty}^{+\infty} dp \int_0^{\pi/2} d\theta \int_{\pi/2}^{\pi} d\theta' \int_0^{2\pi} d\phi \int_0^{2\pi} d\phi' \psi(r, \lambda, \sigma|p, \theta, \phi) b_p(\theta, \phi|\theta', \phi') \psi^*(r', \lambda', \sigma'|p, \theta', \phi') \\ &= -K(r', \lambda', \sigma'|r, \lambda, \sigma) - \int_{r'}^{\infty} r''^2 dr'' \int_0^{\pi} \sin \lambda'' d\lambda'' \int_0^{2\pi} d\sigma'' K(r, \lambda, \sigma|r'', \lambda'', \sigma'') \\ &\times K(r', \lambda', \sigma'|r'', \lambda'', \sigma''),\end{aligned}$$

or generally

$$\begin{aligned}&\int_{-\infty}^{+\infty} dp \int_0^{\pi/2} d\theta \int_{\pi/2}^{\pi} d\theta' \int_0^{2\pi} d\phi \int_0^{2\pi} d\phi' \psi(r, \lambda, \sigma|p, \theta, \phi) b_p(\theta, \phi|\theta', \phi') \psi^*(r', \lambda', \sigma'|p, \theta', \phi') \\ (3.8)\quad &= -\eta(r'-r) K(r, \lambda, \sigma|r', \lambda', \sigma') - \eta(r-r') K(r', \lambda', \sigma'|r, \lambda, \sigma) \\ &-\eta(r'-r) \int_r^{\infty} r''^2 dr'' \int_0^{\pi} \sin \lambda'' d\lambda'' \int_0^{2\pi} d\sigma'' K(r, \lambda, \sigma|r'', \lambda'' \, \sigma'') K(r', \lambda', \sigma'|r'', \lambda'', \sigma'') \\ &-\eta(r-r') \int_{r'}^{\infty} r''^2 dr'' \int_0^{\pi} \sin \lambda'' d\lambda'' \int_0^{2\pi} d\sigma'' K(r, \lambda, \sigma|r'', \lambda'', \sigma'') K(r', \lambda', \sigma'|r'', \lambda'', \sigma'').\end{aligned}$$

In (3.8) the function $\eta(x)$ is the step function

(3.9) $$\eta(x) = \begin{cases} 1 & \text{if} \quad x > 0, \\ 0 & \text{if} \quad x < 0. \end{cases}$$

Let us now evaluate $\int_0^\infty dp \int_0^\pi d\theta \int_0^{2\pi} d\phi\, \psi(\boldsymbol{x}|p, \theta, \phi)\psi^*(\boldsymbol{x}'|p, \theta, \phi)$. We have from (1.6), on rearranging some orders of integration

$$\begin{aligned}
&\int_0^\infty dp \int_0^\pi d\theta \int_0^{2\pi} d\phi\, \psi(r, \lambda, \sigma|p, \theta, \phi)\psi^*(r', \lambda', \sigma'|p, \theta, \phi) \\
&\quad = \int_0^\infty dp \int_0^\pi d\theta \int_0^{2\pi} d\phi\, \psi_0(r, \lambda, \sigma|p, \theta, \phi)\psi_0^*(r', \lambda', \sigma'|p, \theta, \phi) \\
&\qquad + \int_r^\infty r''^2 dr'' \int_0^\pi \sin\lambda'' d\lambda'' \int_0^{2\pi} d\sigma'' K(r, \lambda, \sigma|r'', \lambda'', \sigma'') \\
&\qquad \times \int_0^\infty dp \int_0^\pi d\theta \int_0^{2\pi} d\phi\, \psi_0(r'', \lambda'', \sigma''|p, \theta, \phi)\psi_0^*(r', \lambda', \sigma'|p, \theta, \phi) \\
&\qquad + \int_{r'}^\infty r''^2 dr'' \int_0^\pi \sin\lambda'' d\lambda'' \int_0^{2\pi} d\sigma'' K(r', \lambda', \sigma'|r'', \lambda'', \sigma'') \\
&\qquad \times \int_0^\infty dp \int_0^\pi d\theta \int_0^{2\pi} d\phi\, \psi_0(r, \lambda, \sigma|p, \theta, \phi)\psi_0^*(r'', \lambda'', \sigma''|p, \theta, \phi) \\
&\qquad + \int_r^\infty r''^2 dr'' \int_0^\pi \sin\lambda'' d\lambda'' \int_0^{2\pi} d\sigma'' \int_{r'}^\infty r'''^2 dr''' \int_0^\pi \sin\lambda''' d\lambda''' \int_0^{2\pi} d\sigma''' \\
&\qquad \times K(r, \lambda, \sigma|r'', \lambda'', \sigma'')K(r', \lambda', \sigma'|r''', \lambda''', \sigma''') \int_0^\infty dp \int_0^\pi d\theta \int_0^{2\pi} d\phi \\
&\qquad \times \psi_0(r'', \lambda'', \sigma''|p, \theta, \phi)\psi_0^*(r''', \lambda''', \sigma'''|p, \theta, \phi).
\end{aligned} \tag{3.10}$$

But

$$\begin{aligned}
&\int_0^\infty dp \int_0^\pi d\theta \int_0^{2\pi} d\phi\, \psi_0(r, \lambda, \sigma|p, \theta, \phi)\psi_0^*(r', \lambda', \sigma'|p, \theta, \phi) \\
&\qquad = \frac{\delta(r-r')\delta(\lambda-\lambda')\delta(\sigma-\sigma')}{r^2 \sin\lambda} = \delta(\boldsymbol{x}-\boldsymbol{x}').
\end{aligned} \tag{3.11}$$

Hence in (3.11) we have evaluated the first term on the right of (3.10). Furthermore, on using (3.11), the second term on the right is

$$\begin{aligned}
&\int_r^\infty r''^2 dr'' \int_0^\pi \sin\lambda'' d\lambda'' \int_0^{2\pi} d\sigma'' K(r, \lambda, \sigma|r'', \lambda'', \sigma'') \frac{\delta(r''-r')\delta(\lambda''-\lambda')\delta(\sigma''-\sigma')}{r''^2 \sin\lambda''} \\
&\qquad = \int_r^\infty dr'' K(r, \lambda, \sigma|r'', \lambda', \sigma')\delta(r''-r') \\
&\qquad = \eta(r'-r)K(r, \lambda, \sigma|r', \lambda', \sigma').
\end{aligned}$$

Similarly, the third term on the right is

$$\int_{r'}^{\infty} r''^2 dr'' \int_0^{\pi} \sin \lambda'' d\lambda'' \int_0^{2\pi} d\sigma'' K(r', \lambda', \sigma' | r'', \lambda'', \sigma'') \frac{\delta(r-r'')\delta(\lambda-\lambda'')\delta(\sigma-\sigma'')}{r''^2 \sin\lambda''}$$

$$= \eta(r-r')K(r', \lambda', \sigma' | r, \lambda, \sigma).$$

The last term is

$$\int_{r}^{\infty} r''^2 dr'' \int_0^{\pi} \sin \lambda'' d\lambda'' \int_0^{2\pi} d\sigma'' \int_{r'}^{\infty} r'''^2 dr''' \int_0^{\pi} \sin \lambda''' d\lambda''' \int_0^{2\pi} d\lambda'''$$

$$\times K(r, \lambda, \sigma | r'', \lambda'', \sigma'') K(r', \lambda', \sigma' | r''', \lambda''', \sigma''') \frac{\delta(r'''-r'')\delta(\lambda'''-\lambda'')\delta(\sigma'''-\sigma'')}{r'''^2 \sin \lambda'''}$$

$$= \int_{r}^{\infty} r''^2 dr'' \int_0^{\pi} \sin \lambda'' d\lambda'' \int_0^{2\pi} d\sigma'' K(r, \lambda, \sigma | r'', \lambda'', \sigma'') K(r', \lambda', \sigma' | r''', \lambda''', \sigma''') \eta(r''-r'$$

$$= \eta(r-r') \int_{r}^{\infty} r''^2 dr'' \int_0^{\pi} \sin \lambda'' d\lambda'' \int_0^{2\pi} d\sigma'' K(r, \lambda, \sigma | r'', \lambda'', \sigma'') K(r', \lambda', \sigma' | r'', \lambda'', \sigma'')$$

$$+ \eta(r'-r) \int_{r'}^{\infty} r''^2 dr'' \int_0^{\pi} \sin \lambda'' d\lambda'' \int_0^{2\pi} d\sigma'' K(r, \lambda, \sigma | r'', \lambda'', \sigma'') K(r', \lambda', \sigma' | r'', \lambda'', \sigma''$$

On replacing the various terms on the right of (3.10) and on adding the resulting equation to (3.8) we obtain the completeness relation (1.9).

4. Significance of the Kernel $b_p(\theta, \phi | \theta', \phi')$

In [1] it was shown that the kernel $b_p(\theta, \phi | \theta', \phi')$ could be considered as the amplitude of a spherical wave corresponding to the asymptotic form of the eigenfunction $\psi(x | p, \theta, \phi)$ defined by

$$\psi_-(x | p, \theta, \phi) = \psi_0(x | p, \theta, \phi) - \frac{1}{4\pi} \int\int \frac{e^{ip|x-x'|}}{|x-x'|} V(x' | x'') \psi_-(x'' | p, \theta, \phi) dx' dx'', \tag{4.1}$$

where $V(x|x')$ is given by (1.7) and (1.8). Specifically

$$\lim_{r\to\infty} \psi_-(x | p, \theta, \phi) - \psi_0(x | p, \theta, \phi) - \frac{i}{\sqrt{2\pi}} \frac{e^{ipr}}{r} \frac{b_p(\lambda, \sigma | \theta, \phi)}{\sqrt{\sin \lambda}}$$

(λ, σ polar angles of x). In physical applications $b_p(\theta, \phi | \theta', \phi')$, which is called the scattering amplitude, is used to construct cross-sections for scattering. Hence the Gelfand-Levitan equation which we have set up gives us the scattering potential from the scattering amplitude.

As was shown in [1], the requirements (1.3) on $b_p(\theta, \phi | \theta', \phi')$ are necessary conditions for this kernel to be a scattering amplitude. Actually only part of the scattering amplitude is needed, i.e., $b_p(\theta, \phi | \theta', \phi')$ is needed only

2 S

for $0 \leqq \theta \leqq \frac{1}{2}\pi$, $\frac{1}{2}\pi \leqq \theta' \leqq \pi$. However, in [3], by using a technique different from the Gelfand-Levitan equation, it was shown that if the scattering potential was completely local, the potential could be obtained from even less, namely from $b_p(\theta, \phi|\pi-\theta, \Phi)$, $0 \leqq \theta \leqq \frac{1}{2}\pi$, and Φ is given by (3.5).

In the next section we shall conjecture a requirement on b_p in order that it gives rise to a purely local potential when used in the Gelfand-Levitan equation and is a necessary condition for a purely local potential. This requirement will arise from the comparison of the results of the present Gelfand-Levitan technique with the method of calculation of [3].

5. Requirement on the Scattering Amplitude for a Purely Local Potential

As mentioned above a very general technique for solving inverse problems was given in [3]. It was applied specifically to the one- and three-dimensional scattering problems. In the one-dimensional problem it was found that the potential, which was assumed to be local, could be reconstructed from the reflection coefficient. The corresponding Gelfand-Levitan equation also depended on the reflection coefficient. On solving for the potential in terms of an expansion depending on the reflection coefficient, it was shown that the first two terms of the expansion which gave the potential were identical to the first two terms of a similar expansion arising from the Gelfand-Levitan equation for one dimension. It thus appeared that the method of [3] and the Gelfand-Levitan equation for one dimension gave the same potential in a term-by-term expansion.

We shall assume that this statement is true in the three-dimensional scattering problem. That is, we shall assume that if the potential is purely local, the Gelfand-Levitan equation of the present paper and that of [3] will give the same results term by term for the scattering potential. In particular, the lowest order expressions of the potentials will be equal. This assumption will lead to a necessary condition on the amplitude of the scattering wave, which, if satisfied, will also assure us that the potential obtained from the Gelfand-Levitan equation will be diagonal in the angular variables as well as the radius variable. The result is

$$\begin{aligned} \frac{\partial}{\partial r}\Big[r^2\int_{-\infty}^{+\infty}dp\int_0^{\pi/2}d\theta\int_{\pi/2}^{\pi}d\theta'\int_0^{2\pi}d\phi\int_0^{2\pi}d\phi'\,\psi_0(r,\lambda,\sigma|p,\theta,\phi) \\ \times b_p(\theta,\phi|\theta',\phi')\psi_0^*(r,\lambda',\sigma'|p,\theta',\phi')\Big] \end{aligned} \tag{5.1}$$

$$= -16\pi^2 i \frac{\delta(\lambda-\lambda')\delta(\sigma-\sigma')}{\sin\lambda} \int_{-\infty}^{+\infty} \operatorname{sgn} p \, dp \int_0^{\pi/2} d\theta \int_0^{2\pi} d\phi$$

$$\times \psi_0(r, \lambda, \sigma | p, \theta, \phi) b_p(\theta, \phi | \pi - \theta, \Phi) \psi_0^*(r, \lambda, \sigma | p, \pi-\theta, \Phi).$$

In (5.1) sgn p is 1 if $p > 0$ and is -1 if $p < 0$.

Bibliography

[1] Kay, I., and Moses, H. E., *The Gelfand-Levitan equation for the three dimensional scattering problem*, Tech. Rep. No. 1, Office of Naval Research Contract Nonr-839(30), Project No. NR 013-106.

[2] Kay, I., and Moses, H. E., *The determination of the scattering potential from the spectral measure function*, Part I, Nuovo Cimento, X, Vol. 2, 1955, pp. 917–961. Part II, Nuovo Cimento, X, Vol. 3, 1956, pp. 66–84. Part III, Nuovo Cimento, X, Vol. 3, 1956, pp. 276–304.

[3] Moses, H. E., *Calculation of scattering potentials from scattering coefficients*, Phys. Rev., Vol. 102, 1956, pp. 559–567.

Received February, 1961.

MASSACHUSETTS INSTITUTE OF TECHNOLOGY
LINCOLN LABORATORY

PROPERTIES OF DIELECTRICS FROM REFLECTION COEFFICIENTS IN ONE DIMENSION

H. E. Moses

C. M. deRidder

Group 37

TECHNICAL REPORT NO. 322

11 JULY 1963

ABSTRACT

This report presents mathematical techniques for calculating dielectrics which give rise to prescribed reflection coefficients in certain problems of one-dimensional electromagnetic propagation. The techniques are, in principle at least, exact. They are based on the use of an equation of the Gel'fand-Levitan type for the one-dimensional Schrödinger equation. Although no practical applications are given, it is hoped that this report will encourage the use of newer techniques in synthesis problems.

LEXINGTON MASSACHUSETTS

TABLE OF CONTENTS

PROPERTIES OF DIELECTRICS FROM REFLECTION COEFFICIENTS IN ONE DIMENSION

I. INTRODUCTION

This report is based on a series of lectures by the senior author and has been only slightly revised for this written presentation. Hence, the report is more informally organized and is probably much longer than would be the case if a thorough revision had been made. Perhaps, however, the great detail will enable a novice in the inverse problem to see through the mathematics more clearly.

The report is concerned with the mathematical problem of determining the character of a dielectric scatterer from the reflection coefficient in one-dimensional electromagnetic scattering. Closely connected with this problem is the "synthesis problem" in which one wishes to construct scatterers with prescribed reflection or transmission characteristics. It is not generally known that powerful techniques are available. The objective of this report is to present these techniques in a comprehensive fashion for the benefit of those who wish such information.

The technique which we use is to map the electromagnetic equations into the one-dimensional Schrödinger equation. We then use the Gel'fand-Levitan algorithm to compute the potential and the wave function from the scattering coefficient. The results for the electromagnetic problem are then obtained from the mapping.

The usual procedure for finding dielectrics with prescribed scattering properties is to consider a family of dielectrics for which the electromagnetic equations can be solved. The parameters of this family are then adjusted until the scattering coefficients of the exactly soluble problem are as close as possible to the prescribed scattering coefficients. This procedure is always approximate. The degree of success, moreover, depends upon one's cleverness in choosing the family of dielectrics.

By contrast, the procedure for finding the dielectrics in the present report is <u>exact</u>, at least in principle. If computing machines are available, one can approximate the dielectrics as closely as one wishes even when one cannot solve the Gel'fand-Levitan equation exactly.

This report contains no very practical solutions because the technique which we propose is very new (it was developed within the past ten years), and most of the work in the past has been concerned with more mathematical aspects of the problem. It is hoped that this report will stimulate research for electromagnetic applications. Indeed, its sole purpose is to stimulate, if possible, research which leads to practical applications of the theory.

II. SCATTERING AND THE ONE-DIMENSIONAL SCHRÖDINGER EQUATION

The time-dependent one-dimensional Schrödinger equation is

$$-\frac{1}{i}\frac{\partial}{\partial t}\psi(x,t) = \left[-\frac{\partial^2}{\partial x^2} + V(x)\right]\psi(x,t) \quad . \tag{1}$$

We require that the solution $\psi(x,t)$ be in Hilbert space and be quadratically integrable:

$$\int_{-\infty}^{\infty} |\psi(x,t)|^2\, dx < \infty \quad . \tag{2}$$

First, let us assume that the potential function V(x) is bounded everywhere and $V(x) \to 0$ as $|x| \to \infty$. Later, we shall make $V(x) = 0$ for $|x| > a$. We proceed to solve the Schrödinger equation in the usual manner by separation of variables.

Let the general solution be

$$\psi(x,t) = \int_{-\infty}^{\infty} \chi(x|p)\, g(p)\, e^{-ip^2 t}\, dp + \sum_i \chi(x|E_i)\, g_i\, e^{-iE_i t} \tag{3}$$

and choose $\chi(x|p)$ and $\chi(x|E_i)$ so that they satisfy the following equations, respectively.

$$\left[-\frac{d^2}{dx^2} + V(x)\right]\chi(x|p) = p^2\,\chi(x|p) \qquad \text{for } p^2 > 0 \quad , \tag{4}$$

$$\left[-\frac{d^2}{dx^2} + V(x)\right]\chi(x|E_i) = E_i\,\chi(x|E_i) \qquad \text{for } E_i < 0 \quad . \tag{5}$$

In general, the solutions of these second-order differential equations are not unique unless boundary conditions are imposed.

The discrete values E_i are called "point eigenvalues," and the theory of Hilbert space requires that they be chosen so that

$$\int_{-\infty}^{\infty} \chi^*(x|E_i)\,\chi(x|E_j) = A_i\delta_{ij} \quad , \tag{6}$$

where $A_i > 0$.

With this "boundary condition," the $\chi(x|E_i)$ become unique except for a normalization constant. When the point eigenvalues E_i are chosen so that (6) is satisfied, the corresponding $\chi(x|E_i)$ are called "proper eigenfunctions." Next we need boundary conditions which will determine the $\chi(x|p)$ uniquely. Since we are interested in solving scattering problems, we wish to impose boundary conditions such that the resulting set of $\chi(x|p)$ will give a "simple" description of scattering problems.

In what follows, let us assume that V(x) is bounded everywhere and that $V(x) = 0$ for $|x| > a$. Let us take the following boundary conditions:

For $p > 0$,

$$\left.\begin{aligned} \chi(x|p) &= \frac{e^{ipx}}{\sqrt{2\pi}} + \frac{b(p)}{\sqrt{2\pi}} e^{-ipx} && \text{for } x < -a \\ \chi(x|p) &= \frac{t(p)}{\sqrt{2\pi}} e^{ipx} && \text{for } x > a \end{aligned}\right\} \quad . \tag{7}$$

For $p < 0$,

$$\left.\begin{aligned} \chi(x|p) &= \frac{e^{ipx}}{\sqrt{2\pi}} + \frac{r(p)}{\sqrt{2\pi}} e^{-ipx} && \text{for } x > a \\ \chi(x|p) &= \frac{s(p)}{\sqrt{2\pi}} e^{ipx} && \text{for } x < -a \end{aligned}\right\} \quad . \tag{8}$$

Let us solve the specific initial value problem for which

$$g_i = 0 \quad \text{for all } i \quad ,$$

$$g(p) = 0 \quad \text{for all } p < 0 \quad .$$

Then from (7) and (3) we obtain a particular solution:

$$\left.\begin{aligned} \psi_p(x,t) &= \int_0^\infty \frac{e^{ipx}}{\sqrt{2\pi}} g(p)\, e^{-ip^2 t}\, dp + \int_0^\infty \frac{e^{-ipx}}{\sqrt{2\pi}} b(p)\, g(p)\, e^{-ip^2 t}\, dp && \text{for } x < -a \\ \psi_p(x,t) &= \int_0^\infty \frac{e^{ipx}}{\sqrt{2\pi}} t(p)\, g(p)\, e^{-ip^2 t}\, dp && \text{for } x > a \end{aligned}\right\} \quad . \tag{9}$$

For any fixed p, the first integral in the expression for $\psi_p(x,t)$ for $x < -a$ and the integral in the expression for $\psi_p(x,t)$ for $x > a$ reduce to expressions representing waves moving in the positive x-direction. The second integral in the expression for $\psi_p(x,t)$ for $x < -a$ reduces to an expression representing a wave moving in the negative x-direction. We may think of the integrals as superpositions of such waves. With this interpretation in mind, the b(p) are called the reflection coefficients for $p > 0$, and the t(p) are the transmission coefficients for $p > 0$. Hence, given the boundary conditions (7) we now have defined $\chi(x|p)$ uniquely for $p > 0$.

In a similar manner, we can solve the initial value problem for which

$$g_i = 0 \quad \text{for all } i \quad ,$$

$$g(p) = 0 \quad \text{for } p > 0 \quad ,$$

and we define $\chi(x|p)$ uniquely for $p < 0$ with the help of the boundary conditions (8). The r(p) and s(p) are called the reflection and transmission coefficients, respectively, for $p < 0$.

Let us call $\chi(x|p)$ defined for $p > 0$, $\chi_+(x|p)$ and $\chi(x|p)$ defined for $p < 0$, $\chi_-(x|p)$. By analytic continuation we may define $\chi_+(x|p)$ for $p < 0$ and $\chi_-(x|p)$ for $p > 0$. It can then be shown that

$$\chi_+(x|-p) = \chi_+^*(x|p) \quad ,$$

$$\chi_-(x|-p) = \chi_-^*(x|p) \quad ,$$

but

$$\chi_+(x|-p) \neq \chi_-(x|p) \quad .$$

Also, by analytic continuation it can be shown that

$$\left.\begin{aligned} b(-p) &= b^*(p) \\ t(-p) &= t^*(p) \end{aligned}\right\} \quad \text{for } p > 0 \quad ,$$

$$\left.\begin{aligned} r(-p) &= r^*(p) \\ s(-p) &= s^*(p) \end{aligned}\right\} \quad \text{for } p < 0 \quad ,$$

and

$$s(p) = t(p) \qquad \text{for all } p \quad .$$

Two other relations satisfied by the reflection and transmission coefficients are

$$|b(p)|^2 + |t(p)|^2 = 1 \qquad \text{for } p > 0 \quad ,$$

$$|r(p)|^2 + |s(p)|^2 = 1 \qquad \text{for } p < 0 \quad .$$

These seem to imply conservation of energy. Also,

$$\lim_{|p|\to\infty} b(p) = 0 \quad ,$$

$$\lim_{|p|\to\infty} t(p) = 1 \quad ,$$

in the upper half plane. This is always true. When V(x) dies down very rapidly, it can be shown in addition that b(p) has poles at the points $p = i\sqrt{-E_i}$; that is, b(p) has poles corresponding to every point eigenvalue E_i. This, however, is true only when V(x) dies down sufficiently rapidly. Examples to illustrate this will be given later.

Once the $\chi(x|E_i)$ and $\chi(x|p)$ have been defined with the help of the boundary conditions (6), (7) and (8), we find that they satisfy the completeness and orthonormality conditions. If chosen differently, the $\chi(x|E_i)$ and $\chi(x|p)$ would not satisfy these conditions.

The orthonormality conditions are

$$\left.\begin{aligned} &\int_{-\infty}^{\infty} \chi^*(x|p)\,\chi(x|p')\,dx = \delta(p-p') \\ &\int_{-\infty}^{\infty} \chi^*(x|E_i)\,\chi(x|E_j)\,dx = A_i\delta_{ij} \\ &\int_{-\infty}^{\infty} \chi^*(x|p)\,\chi(x|E_i)\,dx = 0 \end{aligned}\right\} \qquad (10)$$

The completeness relationship is

$$\int_{-\infty}^{\infty} \chi^*(x|p)\,\chi(x'|p)\,dp + \sum_i \frac{\chi^*(x|E_i)\,\chi(x'|E_i)}{A_i} = \delta(x-x') \qquad (11)$$

The satisfaction of these conditions leads to a complete analogue with Fourier transform theory.

Suppose a function f(x) is in L^2. Then as a result of the completeness theorem we may expand f(x) as

$$f(x) = \int_{-\infty}^{\infty} \chi(x|p)\, g(p)\, dp + \sum \chi(x|E_i)\, g_i \quad , \tag{12}$$

where

$$g(p) = \int_{-\infty}^{\infty} \chi^*(x|p)\, f(x)\, dx \quad ,$$

$$A_i g_i = \int_{-\infty}^{\infty} \chi^*(x|E_i)\, f(x)\, dx \quad .$$

Hence we may write

$$f(x) \Longleftrightarrow g(p);\ g_i \quad .$$

Also suppose that

$$f^1(x) \Longleftrightarrow g^1(p);\ g_i^1 \quad ,$$

$$f^2(x) \Longleftrightarrow g^2(p);\ g_i^2 \quad .$$

Then

$$\int_{-\infty}^{\infty} f^{1*}(x)\, f^2(x)\, dx = \int_{-\infty}^{\infty} g^{1*}(p)\, g^2(p)\, dp + \sum_i g_i^{1*} g_i^2 A_i \quad . \tag{13}$$

But this is simply Parseval's theorem.

In the solution of scattering problems we are interested in the reflection and transmission coefficients. In particular, for $p > 0$, $|b(p)|^2$ and for $p < 0$, $|r(p)|^2$ give us the relative probability that a particle of momentum p will be reflected. Similarly, for $p > 0$, $|t(p)|^2$ and for $p < 0$, $|s(p)|^2$ give us the relative probability that a particle of momentum p will be transmitted. The relations

$$|b(p)|^2 + |t(p)|^2 = 1 \quad ,$$

$$|r(p)|^2 + |s(p)|^2 = 1 \quad ,$$

state that one event or the other will occur. To determine the reflection and transmission coefficients, Eq. (4) must be solved. This can be done in the usual manner by matching boundary conditions where $V(x)$ vanishes.

For an alternate integral equation technique, we write

$$\chi(x|p) = \frac{e^{ipx}}{\sqrt{2\pi}} - \frac{i}{2|p|} \int_{-\infty}^{\infty} e^{i|p|\,|x-x'|}\, V(x')\, \chi(x'|p)\, dx' \quad . \tag{14}$$

This expression for $\chi(x|p)$ satisfies the wave equation and all boundary conditions.

For example, take $p > 0$. Then (14) gives

$$\left.\begin{aligned}
\chi(x|p) &= \frac{e^{ipx}}{\sqrt{2\pi}} - \frac{ie^{-ipx}}{2p}\int_{-\infty}^{\infty} e^{ipx'}\, V(x')\, \chi(x'|p)\, dx' && \text{for } x < -a \\
\chi(x|p) &= \frac{e^{ipx}}{\sqrt{2\pi}}\left\{1 - \sqrt{\frac{\pi}{2}}\,\frac{i}{p}\int_{-\infty}^{\infty} e^{-ipx'}\, V(x')\, \chi(x'|p)\, dx'\right\} && \text{for } x > a
\end{aligned}\right\} . \tag{15}$$

The expressions (15) may be written in the form (7) if we choose

$$\left.\begin{aligned}
b(p) &= -\sqrt{\frac{\pi}{2}}\,\frac{i}{p}\int_{-\infty}^{\infty} e^{ipx'}\, V(x')\, \chi(x'|p)\, dx' \\
t(p) &= 1 - \sqrt{\frac{\pi}{2}}\,\frac{i}{p}\int_{-\infty}^{\infty} e^{-ipx'}\, V(x')\, \chi(x'|p)\, dx'
\end{aligned}\right\} \quad \text{for } p > 0 \quad . \tag{16}$$

By analytic continuation, $b(p)$ and $t(p)$ can be defined for $p < 0$.

Similarly, take $p < 0$. Then (14) gives

$$\left.\begin{aligned}
\chi(x|p) &= \frac{e^{ipx}}{\sqrt{2\pi}} + \frac{i}{2p}\, e^{-ipx}\int_{-\infty}^{\infty} e^{ipx'}\, V(x')\, \chi(x'|p)\, dx' && \text{for } x > a \\
\chi(x|p) &= \frac{e^{ipx}}{\sqrt{2\pi}}\left[1 + \sqrt{\frac{\pi}{2}}\,\frac{i}{p}\int_{-\infty}^{\infty} e^{-ipx'}\, V(x')\, \chi(x'|p)\, dx' \quad \text{for } x < -a\right]
\end{aligned}\right\} . \tag{17}$$

The expression (17) may be written in the form (8) if we choose

$$\left.\begin{aligned}
r(p) &= \sqrt{\frac{\pi}{2}}\,\frac{i}{p}\int_{-\infty}^{\infty} e^{ipx'}\, V(x')\, \chi(x'|p)\, dx' \\
s(p) &= 1 + \sqrt{\frac{\pi}{2}}\,\frac{i}{p}\int_{-\infty}^{\infty} e^{-ipx'}\, V(x')\, \chi(x'|p)\, dx'
\end{aligned}\right\} \quad \text{for } p < 0 \quad . \tag{18}$$

By analytic continuation, $r(p)$ and $s(p)$ can be defined for $p > 0$.

As an example of the solution of the direct problem of the one-dimensional Schrödinger equation, let us consider the delta function potential $V(x) = 2B\delta(x)$, and let us take $B > 0$ so that there are no point eigenvalues. Hence $g_i = 0$ for all i. From (16),

$$\left.\begin{aligned}
b(p) &= -\sqrt{\frac{\pi}{2}}\,\frac{i}{p}\, 2B\chi(0|p) \\
t(p) &= 1 - \sqrt{\frac{\pi}{2}}\,\frac{i}{p}\, 2B\chi(0|p)
\end{aligned}\right\} \quad \text{for } p > 0 \quad .$$

From (18),

$$\left.\begin{aligned}
r(p) &= \sqrt{\frac{\pi}{2}}\,\frac{i}{p}\, 2B\chi(0|p) \\
s(p) &= 1 + \sqrt{\frac{\pi}{2}}\,\frac{i}{p}\, 2B\chi(0|p)
\end{aligned}\right\} \quad \text{for } p < 0 \quad .$$

From (14),

$$\chi(0|p) = \frac{p}{\sqrt{2\pi}\,(p + iB)} \qquad \text{for } p > 0 \quad ,$$

$$\chi(0|p) = \frac{p}{\sqrt{2\pi}\,(p - iB)} \qquad \text{for } p < 0 \quad .$$

Hence,

$$\left.\begin{aligned} b(p) &= \frac{-iB}{p + iB} \\ t(p) &= \frac{p}{p + iB} \end{aligned}\right\} \qquad \text{for } p > 0 \quad .$$

$$\left.\begin{aligned} r(p) &= \frac{iB}{p - iB} \\ s(p) &= \frac{p}{p - iB} \end{aligned}\right\} \qquad \text{for } p < 0 \quad .$$

By analytic continuation,

$$\left.\begin{aligned} b(-p) &= \frac{-iB}{-p + iB} = \frac{iB}{p - iB} = b^*(p) \\ t(-p) &= \frac{-p}{-p + iB} = \frac{p}{p - iB} = t^*(p) \end{aligned}\right\} \qquad \text{for } p > 0 \quad ,$$

$$\left.\begin{aligned} r(-p) &= \frac{iB}{-p - iB} = \frac{-iB}{p + iB} = r^*(p) \\ s(-p) &= \frac{-p}{-p - iB} = \frac{p}{p + iB} = s^*(p) \end{aligned}\right\} \qquad \text{for } p < 0 \quad .$$

Hence $s(p) = t(p)$ for all p and $r(p) = b(p)$ for all p, the latter being so only because of the symmetry of the delta function potential. Also,

$$|b(p)|^2 + |t(p)|^2 = \frac{B^2}{p^2 + B^2} + \frac{p^2}{p^2 + B^2} = 1 \qquad \text{for } p > 0 \quad ,$$

$$|r(p)|^2 + |s(p)|^2 = \frac{B^2}{p^2 + B^2} + \frac{p^2}{p^2 + B^2} = 1 \qquad \text{for } p < 0 \quad .$$

III. EXAMPLES OF ELECTROMAGNETIC PROBLEMS WHICH CAN BE MAPPED INTO A ONE-DIMENSIONAL SCHRÖDINGER EQUATION

We shall next give three examples of one-dimensional electromagnetic theory. It will be shown that with an appropriate mapping each example can be reduced to a one-dimensional Schrödinger equation.

A. Reflection and Transmission of Light of a Fixed Frequency at Varying Angles of Incidence by a Dielectric Slab

Consider a medium in which the dielectric constant $\epsilon = \epsilon(x)$ is a function of one variable x only and is independent of time. In order to make the discussion concrete, we assume that $\epsilon(x) = 1$ for $x < a$ and $x > b$. Many of the results will still hold if $\epsilon(x)$ dies down sufficiently rapidly outside the range $a < x < b$.

Consider a ray of light impinging on this material, the angle of incidence being α. The plane of incidence is the xy-plane; the z-axis projects out of the paper (Fig. 1). We are interested primarily in the amount of energy reflected and transmitted by the material. It will be shown that, for a ray polarized so that the electric field is parallel to the z-axis, the electromagnetic equation

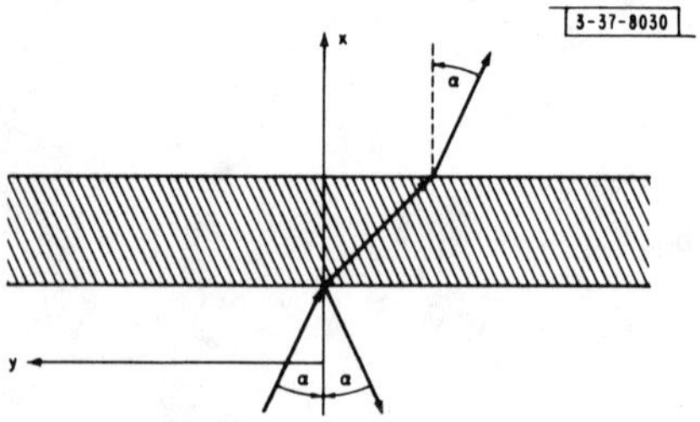

Fig. 1. Ray picture showing reflection and transmission of light by a dielectric slab.

can be mapped into the Schrödinger equation and the b(p) and t(p) of this associated Schrödinger equation will turn out to be essentially the same as the electromagnetic reflection and transmission coefficients.

Using Gaussian units, Maxwell's equations in a source-free region are

$$\underline{\nabla} \cdot \underline{D} = 0 \quad ,$$

$$\underline{\nabla} \cdot \underline{B} = 0 \quad ,$$

$$\underline{\nabla} \times \underline{E} = -\frac{1}{c}\frac{\partial \underline{B}}{\partial t} \quad ,$$

$$\underline{\nabla} \times \underline{H} = \frac{1}{c}\frac{\partial \underline{D}}{\partial t} \quad , \tag{19}$$

where c is the free-space velocity of light. We also have the constitutive equations

$$\underline{B} = \mu \underline{H} \quad ,$$

$$\underline{D} = \epsilon(x)\, \underline{E} \quad . \tag{20}$$

We assume that $\mu = \mu_o$, where μ_o is the permeability of free space and, to eliminate time dependence, write

$$\underline{E}(\underline{x}, t) = \underline{E}(\underline{x})\, e^{-i\omega t} \quad ,$$

$$\underline{H}(\underline{x}, t) = \underline{H}(\underline{x})\, e^{-i\omega t} \quad . \tag{21}$$

The bar under x indicates dependence on both x and y, while x without a bar, such as in $\epsilon(x)$, indicates dependence on the one variable x alone. Substituting in (19) from (20) and (21), we obtain

$$\underline{\nabla} \cdot \epsilon(x)\, \underline{E}(\underline{x}) = 0 \quad ,$$

$$\underline{\nabla} \cdot \underline{H}(\underline{x}) = 0 \quad ,$$

$$\underline{\nabla} \times \underline{E}(\underline{x}) = \frac{i\mu_o \omega}{c}\, \underline{H}(\underline{x}) \quad ,$$

$$\underline{\nabla} \times \underline{H}(\underline{x}) = \frac{-i\epsilon(x)\,\omega}{c}\, \underline{E}(\underline{x}) \quad .$$

Proceeding in the usual manner,

$$\underline{\nabla} \times \underline{\nabla} \times \underline{E}(\underline{x}) = \underline{\nabla}\underline{\nabla} \cdot \underline{E}(\underline{x}) - \underline{\nabla}^2 \underline{E}(\underline{x})$$

$$= \frac{i\mu_o \omega}{c}\, \underline{\nabla} \times \underline{H}(\underline{x}) = \frac{\mu_o \epsilon(x)\,\omega^2}{c^2}\, \underline{E}(\underline{x}) \quad .$$

But

$$\underline{\nabla} \cdot \epsilon(x)\, \underline{E}(\underline{x}) = \underline{E}(\underline{x}) \cdot \underline{\nabla}\epsilon(x) + \epsilon(x)\, \underline{\nabla} \cdot \underline{E}(x) \quad .$$

Hence,

$$\underline{\nabla} \cdot \underline{E}(\underline{x}) = \frac{-\underline{E}(\underline{x}) \cdot \underline{\nabla}\epsilon(x)}{\epsilon(x)} \quad ,$$

and

$$\underline{\nabla}^2 \underline{E}(\underline{x}) + \frac{\mu_o \epsilon(x)\, \omega^2}{c^2}\, \underline{E}(\underline{x}) = -\underline{\nabla}\left[\frac{\underline{E}(\underline{x}) \cdot \underline{\nabla}\epsilon(x)}{\epsilon(x)}\right] \quad .$$

Let us write

$$\underline{E}(\underline{x}) = \underline{E}_n(\underline{x}) + \underline{E}_p(\underline{x}) \quad ,$$

where

$$\underline{E}_n(x) = [0, 0, E_z(\underline{x})]$$

is the component of the electric field normal to the plane of incidence, and

$$\underline{E}_p(\underline{x}) = [E_x(\underline{x}),\ E_y(\underline{x}),\ 0]$$

is the component of the electric field parallel to the plane of incidence. Then $E_z(\underline{x})$ satisfies the equation

$$\underline{\nabla}^2 E_z(\underline{x}) + n^2(x)\, k^2 E_z(\underline{x}) = 0 \tag{22}$$

where

$$k = \frac{\omega}{c} \quad ,$$

$$n(x) = \sqrt{\mu_o \epsilon(x)} \quad .$$

The boundary conditions we wish to impose on $E_z(\underline{x})$ are

$$\left.\begin{aligned} E_z(\underline{x}) &= \frac{e^{i\underline{k}\cdot\underline{x}}}{\sqrt{2\pi}} + \frac{B\, e^{i\underline{k}'\cdot\underline{x}}}{\sqrt{2\pi}} && \text{for } x < a \\ E_z(\underline{x}) &= \frac{T\, e^{i\underline{k}\cdot\underline{x}}}{\sqrt{2\pi}} && \text{for } x > b \end{aligned}\right\} \quad , \tag{23}$$

where $\underline{k}$ and $\underline{k}'$ are vectors given in terms of their components by

$$\left.\begin{aligned} \underline{k} &= (k \cos\alpha,\ k \sin\alpha, 0) \\ \underline{k}' &= (-k \cos\alpha,\ k \sin\alpha, 0) \end{aligned}\right\} \quad . \tag{24}$$

Then the first term in the expression for $E_z(\underline{x})$ for $x > a$ represents a wave moving in the direction $(\cos\alpha,\ \sin\alpha, 0)$. We may consider this to be an incident wave. The second term represents a wave moving in the direction $(-\cos\alpha,\ \sin\alpha, 0)$, i.e., in the direction of specular reflection. It represents a reflected wave. B is the reflection coefficient, and $|B|^2$ gives the relative proportion of energy reflected at the frequency ω. Similarly, the expression for $E_z(\underline{x})$ for $x > b$

represents a wave moving in the direction $(\cos\alpha, \sin\alpha, 0)$. It represents a wave transmitted through the material. T is the transmission coefficient, and $|T|^2$ gives the relative proportion of energy transmitted at the frequency ω.

We shall now show that with the boundary conditions (23) on $E_z(\underline{x})$ we can reduce equation (22) to the Schrödinger equation by an easy mapping and identify B and T with b(p) and t(p), respectively.

Let us consider a fixed frequency $\omega > 0$. To separate spatial variables we write

$$E_z(\underline{x}) = u(x)\, e^{ik \sin\alpha\, y} \quad . \tag{25}$$

We also define

$$\left.\begin{aligned} p &= k \cos\alpha \\ V(x) &= -k^2(n^2 - 1) = -k^2[\epsilon(x)\,\mu_o - 1] \end{aligned}\right\} \quad . \tag{26}$$

Since ω is fixed, p is a function of the angle α. Usually, we will be dealing with $\epsilon(x) \geqslant 1$, and hence $V(x) \leqslant 0$.

Substituting in Eq. (22), we obtain

$$\left[-\frac{d^2}{dx^2} + V(x)\right] u(x) = p^2 u(x) \quad , \tag{27}$$

where $p^2 > 0$, but this is just the time-independent Schrödinger equation.

Suppose the boundary conditions we impose on u(x) are

$$u(x) = \frac{e^{ipx}}{\sqrt{2\pi}} + \frac{b(p)\, e^{-ipx}}{\sqrt{2\pi}} \qquad \text{for } x < a \quad ,$$

$$u(x) = \frac{t(p)\, e^{ipx}}{\sqrt{2\pi}} \qquad \text{for } x > b \quad . \tag{28}$$

Then from (25) and (26) we obtain

$$E_z(\underline{x}) = \frac{e^{i(k\cos\alpha x + k\sin\alpha y)}}{\sqrt{2\pi}} + \frac{b(p)\, e^{i(-k\cos\alpha x + k\sin\alpha y)}}{\sqrt{2\pi}}$$

$$= \frac{e^{i\underline{k}\cdot\underline{x}}}{\sqrt{2\pi}} + \frac{b(p)\, e^{\underline{k}'\cdot\underline{x}}}{\sqrt{2\pi}} \qquad \text{for } x < a \quad ,$$

$$E_z(\underline{x}) = \frac{t(p)\, e^{i(k\cos\alpha x + k\sin\alpha y)}}{\sqrt{2\pi}}$$

$$= \frac{t(p)\, e^{i\underline{k}\cdot\underline{x}}}{\sqrt{2\pi}} \qquad \text{for } x > b \quad .$$

If we identify B with b(p) and T with t(p), these are just the boundary conditions we wish to impose on $E_z(\underline{x})$. Hence, we can solve this particular electromagnetic problem by using the mapping (26) to find the associated Schrödinger equation and identifying its reflection and transmission coefficients with the reflection and transmission coefficients of the electromagnetic problem.

We might note again that in quantum mechanics p represents momentum and $-\infty < p < \infty$. However, p in Eq. (27) as defined by the mapping (26) is a function of the optical angle of incidence α, and this imposes restrictions on the range of p. The technique is valid for complex α and might therefore be used to handle more complicated wave fronts.

B. A Transmission Line Problem

The second example we shall give is a transmission line problem. It will differ from the above example which dealt with a fixed frequency and varying angles of incidence in that now we shall consider a fixed angle for varying frequencies.

Consider a transmission line with distributed impedances L(z) and capacitances C(z). The transmission line equations are

$$\left.\begin{aligned}\frac{\partial V(z,t)}{\partial z} &= -L(z)\,\frac{\partial I(z,t)}{\partial t}\\ \frac{\partial I(z,t)}{\partial z} &= -C(z)\,\frac{\partial V(z,t)}{\partial t}\end{aligned}\right\} , \tag{29}$$

where V(z,t) and I(z,t) are the potential and the current, respectively. For the special case

$$L = L_o \quad ,$$

$$C = C_o \quad ,$$

we obtain at once

$$\frac{\partial^2 V(z,t)}{\partial z^2} = L_o C_o \frac{\partial^2 V(z,t)}{\partial t^2} \quad .$$

This has a solution of the form

$$V(z,t) = F(z - vt) + g(z + vt) \quad ,$$

where

$$v = \frac{1}{\sqrt{L_o C_o}} \quad .$$

Apparently, we have again a wave propagation problem in which reflection and transmission coefficients are of importance. We propose to show that this problem also can be reduced to the Schrödinger equation.

Let us write

$$V(z,t) = V(z)\, e^{-i\omega t} \quad ,$$

$$I(z,t) = I(z)\, e^{-i\omega t} \quad .$$

As above, we take L to be a constant, i.e., $L = L_o$. However, for $a < z < b$ we take C to be a function of z. For $z < a$ and $z > b$ we take C to be a constant $C = C_o$. Then

$$\frac{V(z)}{\partial z} = i\omega L_o I(z) \quad ,$$

$$\frac{\partial I(z)}{\partial z} = i\omega C(z)\, V(z) \quad ,$$

and hence,

$$\frac{\partial^2 V(z)}{\partial z^2} + \omega^2 L_o C(z)\, V(z) = 0 \quad . \tag{30}$$

We define $k^2 = \omega^2 L_o C_o$ and impose the following boundary conditions on the solution

$$V(z) = \frac{e^{ikz}}{\sqrt{2\pi}} + \frac{B(k)\, e^{-ikz}}{\sqrt{2\pi}} \qquad \text{for } z < a \quad ,$$

$$V(z) = \frac{T(k)\, e^{ikz}}{\sqrt{2\pi}} \qquad \text{for } z > b \quad . \tag{31}$$

We shall show that if we can solve a certain Schrödinger equation then we can solve (30) subject to the boundary conditions (31). For convenience, let us rewrite (30) in the form

$$\frac{\partial^2 V(z)}{\partial z^2} + k^2 \epsilon(z)\, V(z) = 0 \quad , \tag{32}$$

where

$$\epsilon(z) = \frac{C(z)}{C_o} \quad .$$

Next let

$$\eta(z) = \epsilon^{1/4}(z) \tag{33}$$

and introduce a new independent variable x related to z by

$$\frac{dz}{dx} = \eta^{-2}(z)$$

or

$$\eta^2(z) = \frac{dx}{dz} \quad . \tag{34}$$

We also introduce a change of dependent variable,

$$u = \eta V \quad . \tag{35}$$

Then

$$\frac{dV}{dz} = \frac{1}{\eta}\frac{du}{dz} - \frac{u}{\eta^2}\frac{d\eta}{dz} \quad .$$

$$\frac{d^2V}{dz^2} = \frac{1}{\eta}\frac{d^2u}{dz^2} - \frac{2}{\eta^2}\frac{d\eta}{dz}\frac{du}{dz} - \frac{u}{\eta^2}\frac{d^2\eta}{dz^2} + \frac{2u}{\eta^3}\left(\frac{d\eta}{dz}\right)^2 \quad .$$

Also,

$$\frac{du}{dz} = \eta^2 \frac{du}{dx} \quad ,$$

$$\frac{d^2u}{dz^2} = \eta^4 \frac{d^2u}{dx^2} + 2\eta^3 \frac{du}{dx}\frac{d\eta}{dx} \quad ,$$

$$\frac{d\eta}{dz} = \eta^2 \frac{d\eta}{dx} \quad ,$$

$$\frac{d^2\eta}{dz^2} = \eta^4 \frac{d^2\eta}{dx^2} + 2\eta^3 \left(\frac{d\eta}{dx}\right)^2 \quad .$$

Substituting in (32), we obtain

$$\frac{1}{\eta}\left[\eta^4 \frac{d^2u}{dx^2} + 2\eta^3 \frac{du}{dx}\frac{d\eta}{dx}\right] - \frac{2}{\eta^2}\left(\eta^2 \frac{d\eta}{dx}\right)\left(\eta^2 \frac{du}{dx}\right)$$

$$- \frac{u}{\eta^2}\left[\eta^4 \frac{d^2\eta}{dx^2} + 2\eta^3 \left(\frac{d\eta}{dx}\right)^2\right] + \frac{2u}{\eta^3}\left(\eta^2 \frac{d\eta}{dx}\right)^2 + k^2\eta^4 \frac{u}{\eta} = 0 \quad .$$

From this we have

$$\frac{d^2u}{dx^2} - \frac{u}{\eta}\frac{d^2\eta}{dx^2} + k^2 u = 0 \quad .$$

We write this as

$$\left[-\frac{d^2}{dx^2} + q(x)\right] u(x) = k^2 u \quad , \tag{36}$$

where

$$q(x) = \frac{1}{\eta}\frac{d^2\eta}{dx^2} \quad . \tag{37}$$

Equation (36) is the Schrödinger equation corresponding to Eq. (30) or (32). Note that $q(x)$ is independent of frequency. Incidentally, the mapping is possible only if $\epsilon(z)$ is continuous.

We shall now turn our attention to an investigation of the mapping and the shape that $q(x)$ will have for a realistically chosen $\epsilon(z)$.

A first observation is that x is a monotonic function of z. Also, since the first-order differential equation relating x to z allows for one constant of integration, we may put $x = A$ when $z = a$; then $z = b$ will correspond to some value of x – say, $x = B$. The mapping is not unique since A can be chosen arbitrarily, but a change in A merely implies a shift in x.

Consider a particular $\epsilon(z)$ such that $\epsilon(z)$ is some function of z for $a < z < b$, and $\epsilon(z) = 1$ for $z < a$ and $z > b$. Then we want to determine first how x depends on z for $z < a$ and $z > b$. Since x is a monotonic function of z,

$z < a$ corresponds to $x < A$,

$z > b$ corresponds to $x > B$.

From (34), we have at once that

$$x - A = z - a \quad \text{for } z < a, \text{ i.e., } x < A \quad ,$$

$$x - B = z - b \quad \text{for } z > b, \text{ i.e., } x > B$$

Suppose Fig. 2(a) represents a typical $\epsilon(z)$. It is essential that $\epsilon(z)$ be continuous for the mapping determined by (33), (34) and (35) to be possible. However, it is not necessary that the derivatives of $\epsilon(z)$ be continuous. Given an $\epsilon(z)$ as shown in Fig. 2(a), $\eta(x)$ would have a shape as shown in

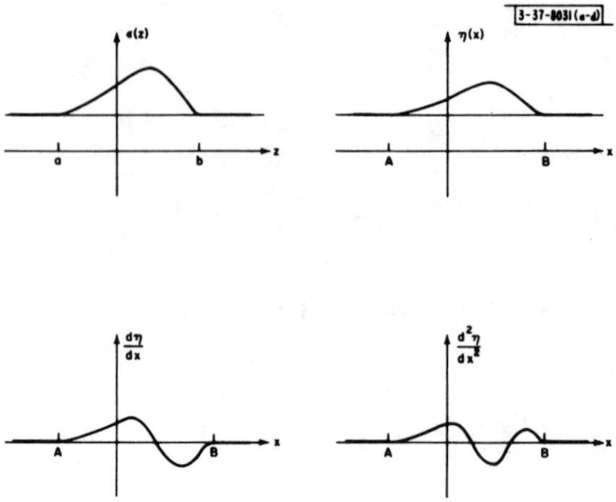

Fig. 2. ε(z) as a function of z and, corresponding to this ε(z), the general shapes taken by η(x), dη(x)/dx and $d^2\eta(x)/dx^2$ as functions of x.

Fig. 2(b). Furthermore, $d\eta/dx$ and $d^2\eta/dx^2$ would have shapes as shown in Figs. 2(c) and (d), respectively. Note that Figs. 2(b), (c) and (d) are not the exact representations of $\eta(x)$, $d\eta/dx$, and $d^2\eta/dx^2$ for $\epsilon(z)$ as shown in Fig. 2(a), but merely are intended to show the general shape of these functions for the given $\epsilon(z)$.

$q(x) = (1/\eta(x))\,(d^2\eta/dx^2)$ will have a shape very similar to that of $d^2\eta/dx^2$, the important thing being that it will have two maxima and one minimum. Hence, if we wish to solve our transmission line problem through its associated Schrödinger equation for a realistic $\epsilon(z)$ as shown in Fig. 2(a), we must solve the Schrödinger equation for a q(x) such as shown in Fig. 3. We might note that, if the bumps in Fig. 3 were replaced by delta function like discontinuities such a q(x) would still correspond to a realistic $\epsilon(z)$. Since $q(x) = 0$ for $x < A$ and $x > B$, we again have a scattering problem and shall look for solutions of the Schrödinger equation which describe scattering. We shall show that if we can solve this scattering problem, we will also have solved the transmission line problem and will be able to relate the coefficients B(k) and T(k) to the transmission and reflection coefficients of the Schrödinger equation. Let us take as solutions of the Schrödinger equation

$$\left.\begin{aligned} u(x) &= \frac{e^{ikx}}{\sqrt{2\pi}} + \frac{b(k)\,e^{-ikx}}{\sqrt{2\pi}} \qquad \text{for } x < A \\ u(x) &= \frac{t(k)\,e^{ikx}}{\sqrt{2\pi}} \qquad \text{for } x > B \end{aligned}\right\} . \tag{38}$$

But when $x < A$ we have $z < a$; $\eta = 1$ and therefore $V(z) = u(x)$. Also, $x - A = z - a$. Hence

$$V(z) = \frac{e^{ikz}\, e^{-ik(a-A)}}{\sqrt{2\pi}} + \frac{b(k)\, e^{-ikz}}{\sqrt{2\pi}}\, e^{ik(a-A)}$$

$$= e^{-ik(a-A)} \left[\frac{e^{ikz}}{\sqrt{2\pi}} + \frac{b(k)\, e^{-ikz}}{\sqrt{2\pi}}\, e^{2ik(a-A)} \right]$$

$$= e^{-ik(a-A)} \left[\frac{e^{ikz}}{\sqrt{2\pi}} + \frac{B(k)\, e^{-ikz}}{\sqrt{2\pi}} \right] ,$$

where

$$B(k) = b(k)\, e^{2ik(a-A)} \quad .$$

Similarly, when $x > B$ we have $z > b$; $\eta = 1$ and therefore $V(z) = u(x)$. Also, now $x - B = z - b$. Hence,

$$V(z) = \frac{t(k)\, e^{ikz}}{\sqrt{2\pi}}\, e^{-ik(b-B)}$$

$$= \frac{t(k)\, e^{ikz}}{\sqrt{2\pi}}\, e^{-ik(a-A)}\, e^{-ik(b-B)+ik(a-A)}$$

$$= \frac{T(k)\, e^{ikz}}{\sqrt{2\pi}}\, e^{-ik(a-A)} ,$$

where

$$T(k) = t(k)\, e^{ik(a-b+B-A)}$$

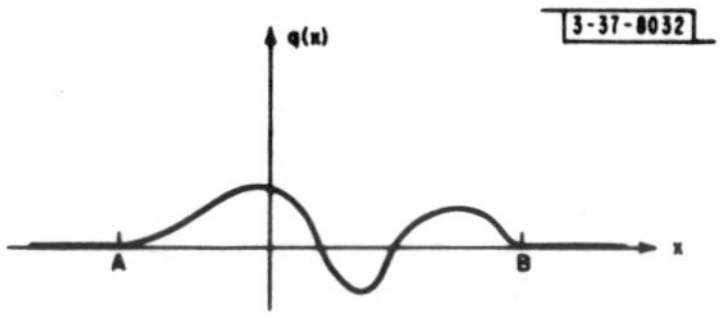

Fig. 3. q(x) as a function of x.

The factor $e^{-ik(a-A)}$ merely changes the incident wave. Hence, we can obtain a solution of our transmission line equation (32) subject to the boundary conditions (31) with the help of the mapping given by (33), (34) and (35). We solve the Schrödinger equation (36) subject to the boundary conditions (38) and obtain the B(k) and T(k) of the transmission line problem from the reflection coefficients b(k) and the transmission coefficients t(k) of the solution of the Schrödinger equation by using the relationships

$$\left. \begin{aligned} B(k) &= b(k)\, e^{2ik(a-A)} \\ T(k) &= t(k)\, e^{ik(a-b+B-A)} \end{aligned} \right\} \quad . \tag{39}$$

In practice, reducing the transmission line equation to the Schrödinger equation involves the following steps. Given $\epsilon(z)$, we use (33) and (34) to find x as a function of z. The choice of the

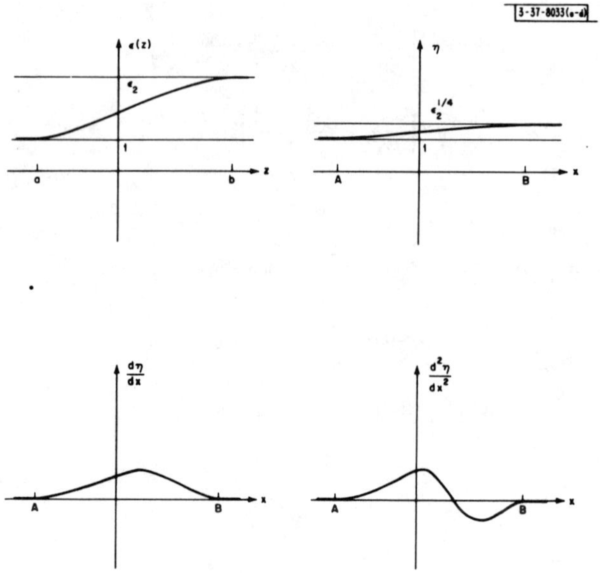

Fig. 4. $\epsilon(z)$ as a function of z, and the corresponding general shapes of $\eta(x)$, $d\eta(x)/dx$ and $d^2\eta(x)/dx^2$ as functions of x.

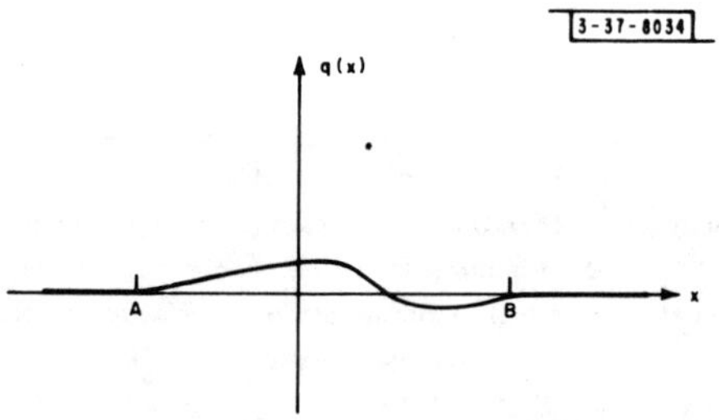

Fig. 5. q(x) as a function of x.

constant of integration A is arbitrary but is usually determined by symmetry considerations. We now have ϵ and η as functions of x and can use (37) to compute q(x).

Next, we shall generalize the above problem slightly. Suppose we again take the transmission line equation (32) and let $\epsilon(z) = 1$ for $z < a$; $\epsilon(z)$ = some function of z for $a < z < b$; but $\epsilon(z) = \epsilon_2$ = constant $\neq 1$ for $z > b$.

We shall require that V(z) satisfy the following boundary conditions:

$$\left.\begin{aligned} V(z) &= \frac{e^{ikz}}{\sqrt{2\pi}} + \frac{B(k)\, e^{ikz}}{\sqrt{2\pi}} \qquad \text{for } z < a \\ V(z) &= \frac{T(k)\, e^{ik\sqrt{\epsilon_2}\, z}}{\sqrt{2\pi}} \qquad \text{for } z > b \end{aligned}\right\} \quad . \tag{40}$$

The mapping is independent of the boundary conditions, and we shall use the same mapping as above. Again, we choose $z = a$ to correspond to $x = A$, and since x is a monotonic function of z we have $z < a$ corresponds to $x < A$. Also, since for $z < a$ we choose $\epsilon(z) = 1$, we have as before $x - A = z - a$ for $z < a$ and $x < A$. $z = b$ will correspond to some value of x – say, $x = B$ – and $z > b$ will correspond to $x > B$. However, for $z > b$ we now have $\epsilon(z) = \epsilon_2$ and hence $\eta = \epsilon_2^{1/4}$. Therefore,

$$\frac{dx}{dz} = \sqrt{\epsilon_2} \quad ,$$

and

$$x - B = \sqrt{\epsilon_2}\,(z - b) \quad .$$

Again let us assume a simple realistic shape for $\epsilon(z)$ such as shown in Fig. 4(a). The approximate shapes for η, $d\eta/dx$ and $d^2\eta/dx^2$ are shown in Figs. 4(b), (c) and (d). q(x) will of course again have a shape similar to that of $d^2\eta/dx^2$. Therefore, in this case we shall be dealing with a q(x) having a shape like that shown in Fig. 5.

We solve the Schrödinger equation for this q(x) subject to the boundary conditions (38). For $x < A$ we have $z < a$; $\eta = 1$; $V(z) = u(x)$; and $x - A = z - a$ as before. The boundary conditions on V(z) for $z < a$ are also the same as above. Hence, again we obtain

$$B(k) = b(k)\, e^{2ik(a-A)} \quad .$$

For $x > B$ we have $z > b$; $\eta = \epsilon_2^{1/4}$; $V(z) = u(x)/\epsilon_2^{1/4}$; and $x - B = \sqrt{\epsilon_2}\,(z - b)$. Hence,

$$\begin{aligned} V(z) &= \frac{t(k)\, e^{ik\sqrt{\epsilon_2}\, z}}{\epsilon_2^{1/4}\sqrt{2\pi}}\, e^{ik(B-\sqrt{\epsilon_2}\, b)} \\ &= \frac{t(k)\, e^{ik\sqrt{\epsilon_2}\, z}}{\epsilon_2^{1/4}\sqrt{2\pi}}\, e^{-ik(a-A)}\, e^{ik(a-\sqrt{\epsilon_2}\, b+B-A)} \quad . \end{aligned}$$

In this case, the transmission coefficient t(k) of the associated Schrödinger equation and the T(k) of the transmission line equation are related in the following manner:

$$T(k) = \frac{t(k)}{\epsilon_2^{1/4}} e^{ik(a-\sqrt{\epsilon_2}b+B-A)} \quad ,$$

and again we are able to solve the transmission line problem by mapping it into the Schrödinger equation.

C. Reflection and Transmission, for Varying Frequencies, of a Plane Wave Normally Incident on a Dielectric Slab

Finally, we shall show that Eq. (32), the transmission line equation, also occurs in a simple electromagnetic problem. In this case, k and ϵ are defined differently, of course.

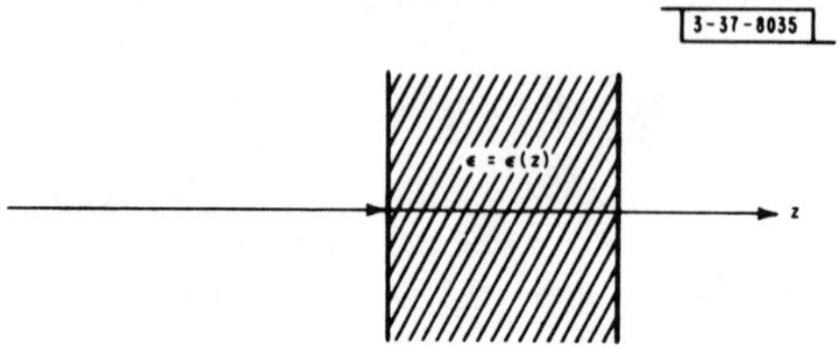

Fig. 6. Plane wave incident normally on a slab of dielectric constant, $\epsilon(z)$.

Consider normal incidence of a plane wave traveling in the z-direction on a slab of material whose dielectric constant $\epsilon = \epsilon(z)$ varies in the z-direction (Fig. 6). It was shown previously that, assuming a time dependence $e^{-i\omega t}$ and using Maxwell's equations and the constitutive equations, we find that the electric field $\underline{E}(\underline{x})$ must satisfy the equation

$$\underline{\nabla}^2 \underline{E}(\underline{x}) + \frac{\mu_0 \epsilon(z)\, \omega^2}{c^2} \underline{E}(\underline{x}) = -\underline{\nabla}\left[\frac{\underline{E}(\underline{x}) \cdot \underline{\nabla}\epsilon(z)}{\epsilon(z)}\right] \quad . \tag{41}$$

Since we are now assuming that $\epsilon = \epsilon(z)$ varies in the direction of propagation and since $\underline{E}(\underline{x})$ must be normal to this direction, we see at once that the right-hand side of (41) vanishes.

Let us assume further that

$$\underline{E}(\underline{x}) = [E_x(z), 0, 0] \quad .$$

Then we have at once

$$\frac{\partial^2 E_x(z)}{\partial z^2} + \frac{\mu_0 \omega^2}{c^2} \epsilon(z)\, E_x(z) = 0 \quad .$$

Similarly, if we had assumed that

$$\underline{E}(\underline{x}) = [0, E_y(z), 0] \quad ,$$

we would have obtained the same equation for $E_y(z)$, namely,

$$\frac{\partial^2 E_y(z)}{\partial z^2} + \frac{\mu_0 \omega^2}{c^2} \epsilon(z)\, E_y(z) = 0 \quad .$$

We can write either of these in the form of the transmission line equation (32):

$$\frac{\partial^2 E_v(z)}{\partial z^2} + k^2 \hat{\epsilon}(z)\, E_v(z) = 0 \quad , \tag{42}$$

where now v = x or y,

$$k^2 = \frac{\omega^2 \mu_o \epsilon_o}{c^2} \quad ,$$

$$\hat{\epsilon}(z) = \frac{\epsilon(z)}{\epsilon_o} \quad ,$$

and ϵ_o is the dielectric constant of free space.

IV. THE INVERSE PROBLEM FOR THE SCHRÖDINGER EQUATION – THE GEL'FAND-LEVITAN ALGORITHM AND ITS PROOF

The direct problem of solving the two time-independent equations arising from the Schrödinger equation has been treated earlier in this report.

To recapitulate briefly in this case specifying

(a) V(x),

(b) A_i the normalization of $\chi(x|E_i)$:

$$A_i = \int_{-\infty}^{\infty} |\chi(x|E_i)|^2\, dx \quad .$$

(c) The boundary conditions on $\chi(x|p)$:

$$\left.\begin{aligned} \lim_{x\to-\infty} \chi(x|p) &= \frac{e^{ipx}}{\sqrt{2\pi}} + \frac{b(p)\, e^{-ipx}}{\sqrt{2\pi}} \\ \lim_{x\to\infty} \chi(x|p) &= \frac{t(p)\, e^{ipx}}{\sqrt{2\pi}} \end{aligned}\right\} \quad \text{for } p > 0 \quad ,$$

$$\left.\begin{aligned} \lim_{x\to\infty} \chi(x|p) &= \frac{e^{ipx}}{\sqrt{2\pi}} + \frac{r(p)\, e^{-ipx}}{\sqrt{2\pi}} \\ \lim_{x\to-\infty} \chi(x|p) &= \frac{s(p)\, e^{ipx}}{\sqrt{2\pi}} \end{aligned}\right\} \quad \text{for } p < 0 \quad ,$$

we solve the two equations

$$\left[-\frac{d^2}{dx^2} + V(x)\right] \chi(x|p) = p^2 \chi(x|p) \qquad \text{for } p^2 > 0 \quad ,$$

$$\left[-\frac{d^2}{dx^2} + V(x)\right] \chi(x|E_i) = E_i \chi(x|E_i) \qquad \text{for } E_i < 0 \quad ,$$

and determine E_i, b(p), t(p), r(p), s(p), $\chi(x|p)$ and $\chi(x|E_i)$. It was also shown that the $\chi(x|p)$ and $\chi(x|E_i)$ thus determined satisfy the completeness and orthonormality conditions, and hence, that any function f(x) can be expanded in the following manner:

$$f(x) = \int_{-\infty}^{\infty} \chi(x|p)\, g(p)\, dp + \sum_i \chi(x|E_i)\, g_i \quad ,$$

where

$$g(p) = \int_{-\infty}^{\infty} \chi^*(x|p)\, f(x)\, dx \quad ,$$

$$A_i g_i = \int_{-\infty}^{\infty} \chi^*(x|E_i)\, f(x)\, dx \quad .$$

We shall next discuss the inverse problem, namely, the problem of determining V(x) from a knowledge of A_i, E_i, b(p) and the boundary conditions on $\chi(x|p)$. This problem was first treated by Gel'fand and Levitan. However, they dealt with the radial equation for $\ell = 0$, and we shall be dealing with the one-dimensional equation for the range $-\infty < x < \infty$.

We shall first give the Gel'fand-Levitan algorithm and then proceed to the proof. The proof will be given in two parts.

Gel'fand-Levitan Algorithm

If, given b(p), E_i, and A_i, we define

$$R(x) = \frac{1}{2\pi} \int_{-\infty}^{\infty} b(p)\, e^{-ipx}\, dp + \sum_i \frac{e^{\sqrt{-E_i}x}}{A_i} \quad , \tag{43}$$

and assume that the Gel'fand-Levitan equation for $x \geqslant y$,

$$K(x,y) = -R(x+y) - \int_{-\infty}^{x} K(x,z)\, R(z+y)\, dz \quad , \tag{44}$$

has a unique solution for K(x, y), then the potential V(x) is given by

$$V(x) = 2 \frac{d}{dx} [K(x,x)] \quad , \tag{45}$$

and for $p > 0$,

$$\left.\begin{aligned} \chi(x|p) &= \frac{e^{ipx}}{\sqrt{2\pi}} + \frac{b(p)\, e^{-ipx}}{\sqrt{2\pi}} + \frac{1}{\sqrt{2\pi}} \int_{-\infty}^{x} K(x,y)\, [e^{ipy} + b(p)\, e^{-ipy}]\, dy \\ \chi(x|E_i) &= e^{\sqrt{-E_i}x} + \int_{-\infty}^{x} K(x,y)\, e^{\sqrt{-E_i}y}\, dy \end{aligned}\right\} \tag{46}$$

In principle, the inverse problem which requires the solution of a linear integral equation – the Gel'fand-Levitan equation – is no more difficult than the direct problem which requires the solution of a linear differential equation. We note that for fixed x the Gel'fand-Levitan equation is a Fredholm equation and hence is of a well understood standard form.

Also, as $x \to -\infty$ we have $R(x) \to 0$. Each of the terms of the sum $\sum_i e^{\sqrt{-E_i}x}/A_1$ certainly tends to zero as $x \to -\infty$, and with the help of the Riemann-Lebesque lemma it can be shown that the integral in the expression for R(x) vanishes as $x \to -\infty$.† The given b(p) will of course satisfy the relations $b(-p) = b^*(p)$, $\lim_{|p| \to \infty} b(p) = 0$ in the upper half plane. We might point out that as a consequence of the first of these two relations R(x) is real and this in turn implies that K(x,y) is real.

In general, it is necessary that b(p), E_i, and A_i are specified before the Gel'fand-Levitan algorithm can be applied. However, for very short range potentials b(p) and E_i are sufficient and A_i need not be given.

Let us specify two functions $\psi(x|p)$ and $\psi(x|E_i)$ which are solutions of the equations

$$\left.\begin{aligned} \left[-\frac{d^2}{dx^2} + V(x)\right] \psi(x|p) &= p^2\psi(x|p) \qquad \text{for } p^2 > 0 \\ \left[-\frac{d^2}{dx^2} + V(x)\right] \psi(x|E_i) &= E_i\psi(x|E_i) \qquad \text{for } E_i < 0 \end{aligned}\right\} . \tag{47}$$

The $\psi(x|E_i)$ are to be identical with the $\chi(x|E_i)$ and hence are determined by imposing the same boundary condition on them as was imposed on the $\chi(x|E_i)$, namely,

$$\int_{-\infty}^{\infty} \psi^*(x|E_i)\, \psi(x|E_j)\, dx = A_i\delta_{ij} \quad .$$

However, the boundary condition

$$\lim_{x \to -\infty} \psi(x|p) = \frac{e^{ipx}}{\sqrt{2\pi}} \qquad \text{for all } p \tag{48}$$

which we impose on the $\psi(x|p)$ is different from the condition imposed on $\chi(x|p)$. One reason for the choice of this boundary condition is convenience. Also, with this condition the $\psi(x|p)$ are analytic functions for all p. The two independent solutions $\psi(x|p)$ and $\psi(x|-p)$ may now be superimposed and we write

$$\left.\begin{aligned} \chi(x|p) &= \psi(x|p) + b(p)\, \psi(x|-p) \qquad \text{for } p > 0 \\ \chi(x|p) &= \psi(x|p) + r(p)\, \psi(x|-p) \qquad \text{for } p < 0 \end{aligned}\right\} . \tag{49}$$

Next we proceed to Part I of the proof of the Gel'fand-Levitan algorithm. In this part of the proof we shall show that

$$\left.\begin{aligned} \psi(x|p) &= \frac{e^{ipx}}{\sqrt{2\pi}} + \frac{1}{\sqrt{2\pi}} \int_{-\infty}^{x} K(x,y)\, e^{ipy}\, dy \\ \psi(x|E_i) &= e^{\sqrt{-E_i}x} + \int_{-\infty}^{x} K(x,y)\, e^{\sqrt{-E_i}y}\, dy \end{aligned}\right\} \tag{50}$$

satisfy the equations (47) when (45) is satisfied. It is immediately apparent that $\psi(x|p)$ as given by (50) satisfies the boundary condition(48).

† The Riemann-Lebesque theorem (Lighthill, Introduction to Fourier Analysis and Generalized Functions, p.46) states that: If b(p) is an ordinary function absolutely integrable from $-\infty$ to ∞, then its Fourier transform $\to 0$ as $|x| \to \infty$. However, b(p) in this case consists of a given set of values and is therefore certainly absolutely integrable over $(-\infty,\infty)$.

Next we calculate $(d^2/dx^2)\,\psi(x|p) + p^2\psi(x|p)$.

$$\frac{d}{dx}\,\psi(x|p) = i\,\frac{pe^{ipx}}{\sqrt{2\pi}} + \frac{K(x,x)}{\sqrt{2\pi}}\,e^{ipx} + \frac{1}{\sqrt{2\pi}}\int_{-\infty}^{x} K_x(x,y)\,e^{ipy}\,dy \quad ,$$

$$\frac{d^2}{dx^2}\,\psi(x|p) = -p^2\,\frac{e^{ipx}}{\sqrt{2\pi}} + \frac{ipK(x,x)}{\sqrt{2\pi}}\,e^{ipx} + \frac{e^{ipx}}{\sqrt{2\pi}}\,\frac{d}{dx}\,K(x,x)$$

$$+ \frac{K_x(x,x)}{\sqrt{2\pi}}\,e^{ipx} + \frac{1}{\sqrt{2\pi}}\int_{-\infty}^{x} K_{xx}(x,y)\,e^{ipy}\,dy \quad ,$$

where

$$K_x(x,x) = \frac{d}{dx}\,[K(x,y)]\big|_{y=x} \quad ,$$

$$K_y(x,x) = \frac{d}{dy}\,[K(x,y)]\big|_{y=x} \quad . \tag{51}$$

Integrating by parts and making use of the fact that as $x \to -\infty$, $R(x) \to 0$, and therefore also $K(x,y) \to 0$, we have

$$p^2\psi(x|p) = \frac{p^2\,e^{ipx}}{\sqrt{2\pi}} + \frac{1}{\sqrt{2\pi}}\int_{-\infty}^{x} K(x,y)\,p^2\,e^{ipy}\,dy$$

$$= \frac{p^2\,e^{ipx}}{\sqrt{2\pi}} - \frac{ipK(x,x)}{\sqrt{2\pi}}\,e^{ipx} + \frac{1}{\sqrt{2\pi}}\int_{-\infty}^{x} K_y(x,y)\,ip\,e^{ipy}\,dy$$

$$= \frac{p^2\,e^{ipx}}{\sqrt{2\pi}} - \frac{ipK(x,x)}{\sqrt{2\pi}}\,e^{ipx} + \frac{K_y(x,x)}{\sqrt{2\pi}}\,e^{ipx} - \frac{1}{\sqrt{2\pi}}\int_{-\infty}^{x} K_{yy}(x,y)\,e^{ipy}\,dy \quad .$$

Hence,

$$\frac{d^2\psi(x|p)}{dx^2} + p^2\psi(x|p) = \frac{e^{ipx}}{\sqrt{2\pi}}\left\{\frac{d}{dx}\,[K(x,x)] + K_x(x,x) + K_y(x,x)\right\}$$

$$+ \frac{1}{\sqrt{2\pi}}\int_{-\infty}^{x} [K_{xx}(x,y) - K_{yy}(x,y)]\,e^{ipy}\,dy$$

$$= \frac{2\,e^{ipx}}{\sqrt{2\pi}}\,\frac{d}{dx}\,[K(x,x)] + \frac{1}{\sqrt{2\pi}}\int_{-\infty}^{x} [K_{xx}(x,y) - K_{yy}(x,y)]\,e^{ipy}\,dy \quad . \tag{52}$$

Since

$$K_x(x,x) + K_y(x,x) = \frac{d}{dx}\,[K(x,x)] \quad , \tag{53}$$

we proceed to find an expression for $K_{xx}(x,y) - K_{yy}(x,y)$. We have

$$K(x,y) = -R(x+y) - \int_{-\infty}^{x} K(x,z)\,R(z+y)\,dz \quad .$$

Hence,

$$K_x(x,y) = -R'(x+y) - K(x,x)\,R(x+y) - \int_{-\infty}^{x} K_x(x,z)\,R(z+y)\,dz \quad ,$$

$$K_{xx}(x,y) = -R''(x+y) - K(x,x)\,R'(x+y) - \frac{d}{dx}[K(x,x)]\,R(x+y)$$

$$-K_x(x,x)\,R(x+y) - \int_{-\infty}^{x} K_{xx}(x,z)\,R(z+y)\,dz \quad ,$$

$$K_y(x,y) = -R'(x+y) - \int_{-\infty}^{x} K(x,z)\,R'(z+y)\,dz$$

$$= -R'(x+y) - K(x,x)\,R(x+y) + \int_{-\infty}^{x} K_z(x,z)\,R(z+y)\,dz \quad ,$$

$$K_{yy}(x,y) = -R''(x+y) - K(x,x)\,R'(x+y) + \int_{-\infty}^{x} K_z(x,z)\,R'(z+y)\,dz$$

$$= -R''(x+y) - K(x,x)\,R'(x+y) + K_y(x,x)\,R(x+y) - \int_{-\infty}^{x} K_{zz}(x,z)\,R(z+y)\,dz \quad .$$

Then

$$K_{xx}(x,y) - K_{yy}(x,y) = -\left\{\frac{d}{dx}[K(x,x)] + K_x(x,x) + K_y(x,x)\right\} R(x+y)$$

$$-\int_{-\infty}^{x} [K_{xx}(x,z) - K_{zz}(x,z)]\,R(z+y)\,dz \quad ,$$

or using (53),

$$K_{xx}(x,y) - K_{yy}(x,y) = -2\,\frac{d}{dx}[K(x,x)]\,R(x+y)$$

$$-\int_{-\infty}^{x} [K_{xx}(x,z) - K_{zz}(x,z)]\,R(z+y)\,dz \quad . \tag{54}$$

This is an integral equation for $K_{xx}(x,y) - K_{yy}(x,y)$. Let us try to find a solution in the form

$$K_{xx}(x,y) - K_{yy}(x,y) = L(x,y)\,2\,\frac{d}{dx}[K(x,x)] \quad . \tag{55}$$

Substituting (55) in (54), we obtain

$$L(x,y) = -R(x+y) - \int_{-\infty}^{x} L(x,z)\,R(z+y)\,dz \quad . \tag{56}$$

Hence, $L(x,y)$ satisfies the same equation as $K(x,y)$. If we assume that this equation has a unique solution, then $L(x,y) = K(x,y)$ and

$$K_{xx}(x,y) - K_{yy}(x,y) = 2K(x,y)\,\frac{d}{dx}K[x,x] \quad . \tag{57}$$

Using (57) in expression (52), we obtain

$$\frac{d^2\psi(x|p)}{dx^2} + p^2\psi(x|p) = \frac{2\,e^{ipx}}{\sqrt{2\pi}}\frac{d}{dx}[K(x,x)]$$

$$+ \frac{1}{\sqrt{2\pi}}\int_{-\infty}^{x} 2K(x,y)\frac{d}{dx}[K(x,x)]\,e^{ipy}\,dy$$

$$= \frac{2d[K(x,x)]}{dx}\left[\frac{e^{ipx}}{\sqrt{2\pi}} + \frac{1}{\sqrt{2\pi}}\int_{-\infty}^{x} K(x,y)\,e^{ipy}\,dy\right]$$

$$= 2\frac{d}{dx}[K(x,x)]\,\psi(x|p) \quad .$$

Hence, if

$$V(x) = 2\frac{d}{dx}[K(x,x)] \quad ,$$

then $\psi(x|p)$ satisfies Eq. (47). The proof for $\psi(x|E_i)$ is identical. This completes Part I of the proof.

In Part I of the proof of the Gel'fand-Levitan algorithm we have shown that the $\psi(x|p)$ and $\psi(x|E_i)$ as given by (50) satisfy the equations (47). Also, $\psi(x|p)$ obviously obeys the boundary condition (48). In Part II of the proof we shall prove the following completeness relationship:

$$\int_{-\infty}^{\infty}\psi(x|p)\,\psi^*(x'|p)\,dp + \int_{-\infty}^{\infty}\psi(x|p)\,b(-p)\,\psi^*(x'|-p)\,dp$$

$$+ \sum_i \frac{\psi(x|E_i)\,\psi(x'|E_i)}{A_i} = \delta(x-x') \quad . \tag{58}$$

We shall then use this completeness relationship to show that the $\psi(x|p)$ are linearly independent, that the $\psi(x|E_i)$ are quadratically integrable, have the normalization A_i, and are therefore the point eigenvalues of the Schrödinger equation (47), and finally, that the b(p) as used in (43) are indeed the reflection coefficients of the Schrödinger equation.

Let us first prove (58). For $x > y$ we have

$$K(x,y) = -R(x+y) - \int_{-\infty}^{x} K(x,z)\,R(z+y)\,dz$$

$$= -R(x+y) - \frac{1}{2\pi}\int_{-\infty}^{x} dz\int_{-\infty}^{\infty} K(x,z)\,b(p)\,e^{-ip(z+y)}\,dp - \int_{-\infty}^{x}\sum_i \frac{K(x,z)\,e^{\sqrt{-E_i}(z+y)}}{A_i}\,dz$$

$$= -R(x+y) - \frac{1}{\sqrt{2\pi}}\int_{-\infty}^{\infty}\left[\psi(x|-p) - \frac{e^{-ipx}}{\sqrt{2\pi}}\right]b(p)\,e^{-ipy}\,dp$$

$$- \sum_i \frac{\left[\psi(x|E_i) - e^{\sqrt{-E_i}x}\right]e^{\sqrt{-E_i}y}}{A_i} \quad .$$

Therefore,

$$K(x,y) = -\frac{1}{\sqrt{2\pi}}\int_{-\infty}^{\infty} \psi(x|-p)\, b(p)\, e^{-ipy}\, dp - \sum_i \frac{\psi(x|E_i)\, e^{\sqrt{-E_i}y}}{A_i} \quad . \tag{59}$$

Also,

$$\begin{aligned} K(x,y) &= -\int_{-\infty}^{\infty} \psi(x|-p)\, b(p) \left[\psi(y|-p) - \frac{1}{\sqrt{2\pi}}\int_{-\infty}^{y} K(y,z)\, e^{-ipz}\, dz\right] dp \\ &\quad - \sum_i \frac{\psi(x|E_i)}{A_i}\left[\psi(y|E_i) - \int_{-\infty}^{y} K(y,z)\, e^{\sqrt{-E_i}z}\, dz\right] \\ &= -\int_{-\infty}^{\infty} \psi(x|-p)\, b(p)\, \psi(y|-p)\, dp - \sum_i \frac{\psi(x|E_i)\, \psi(y|E_i)}{A_i} \\ &\quad + \int_{-\infty}^{y} K(y,z)\left[\frac{1}{\sqrt{2\pi}}\int_{-\infty}^{\infty} \psi(x|-p)\, b(p)\, e^{-ipz}\, dp + \sum_i \frac{\psi(x|E_i)\, e^{\sqrt{-E_i}z}}{A_i}\right] dz \quad . \end{aligned}$$

But, according to (59), the expression in brackets in the last integral is just $-K(x,z)$. Hence, rearranging terms we now have

$$\begin{aligned} \int_{-\infty}^{\infty} \psi(x|-p)\, b(p)\, \psi(y|-p)\, dp + \sum_i \frac{\psi(x|E_i)\, \psi(y|E_i)}{A_i} &= -K(x,y) \\ &\quad - \int_{-\infty}^{y} K(y,z)\, K(x,z)\, dz \quad . \end{aligned} \tag{60}$$

This is true for $x > y$. However, the left-hand side of (60) is symmetric in x and y so that for $x < y$ we have

$$\begin{aligned} \int_{-\infty}^{\infty} \psi(y|-p)\, b(p)\, \psi(x|-p)\, dp + \sum_i \frac{\psi(y|E_i)\, \psi(x|E_i)}{A_i} &= -K(y,x) \\ &\quad - \int_{-\infty}^{x} K(x,z)\, K(y,z)\, dz \quad . \end{aligned}$$

Introducing the Heaviside step function,

$$\eta(x) = \begin{cases} 1 & \text{when } x > 0 \quad , \\ 0 & \text{when } x < 0 \quad , \end{cases} \tag{61}$$

and replacing p by $-p$ in the integral on the left-hand side, we finally have

$$\begin{aligned} \int_{-\infty}^{\infty} \psi(x|p)\, b(-p)\, \psi^*(y|-p)\, dp + \sum_i \frac{\psi(x|E_i)\, \psi(y|E_i)}{A_i} &= -\eta(x-y)\, K(x,y) - \eta(y-x)\, K(y,x) \\ - \eta(x-y)\int_{-\infty}^{y} K(y,z)\, K(x,z)\, dz - \eta(y-x)\int_{-\infty}^{x} K(x,z)\, K(y,z)\, dz \quad . \end{aligned} \tag{62}$$

Next, using (50), we write

$$\int_{-\infty}^{\infty} \psi(x|p)\, \psi^*(y|p)\, dp = \int_{-\infty}^{\infty} \left[\frac{e^{ipx}}{\sqrt{2\pi}} + \frac{1}{\sqrt{2\pi}} \int_{-\infty}^{x} K(x,z)\, e^{ipz}\, dz\right]$$

$$\times \left[\frac{e^{-ipy}}{\sqrt{2\pi}} + \frac{1}{\sqrt{2\pi}} \int_{-\infty}^{y} K(y,v)\, e^{-ipv}\, dv\right] dp$$

$$= \frac{1}{2\pi} \int_{-\infty}^{\infty} e^{ip(x-y)}\, dp + \frac{1}{2\pi} \int_{-\infty}^{y} K(y,v)\, dv \int_{-\infty}^{\infty} e^{ip(x-v)}\, dp$$

$$+ \frac{1}{2\pi} \int_{-\infty}^{x} K(x,z)\, dz \int_{-\infty}^{\infty} e^{ip(z-y)}\, dp$$

$$+ \frac{1}{2\pi} \int_{-\infty}^{x} K(x,z)\, dz \int_{-\infty}^{y} K(y,v)\, dv \int_{-\infty}^{\infty} e^{ip(z-v)}\, dp \quad .$$

But

$$\int_{-\infty}^{\infty} e^{ip(x-y)}\, dp = 2\pi\delta(x-y) \quad .$$

Also,

$$\frac{1}{2\pi} \int_{-\infty}^{y} K(y,v)\, dv \int_{-\infty}^{\infty} e^{ip(x-v)}\, dp = \int_{-\infty}^{y} K(y,v)\, \delta(x-v)\, dv$$

$$= \begin{cases} K(y,x) & \text{when } x < y \\ 0 & \text{when } x > y \end{cases}$$

$$= \eta(y-x)\, K(y,x) \quad ,$$

$$\frac{1}{2\pi} \int_{-\infty}^{x} K(x,z)\, dz \int_{-\infty}^{\infty} e^{ip(z-y)}\, dp = \int_{-\infty}^{x} K(x,z)\, \delta(z-y)\, dz$$

$$= \begin{cases} K(x,y) & \text{when } y < x \\ 0 & \text{when } y > x \end{cases}$$

$$= \eta(x-y)\, K(x,y) \quad ,$$

$$\frac{1}{2\pi} \int_{-\infty}^{x} K(x,z)\, dz \int_{-\infty}^{y} K(y,v)\, dv \int_{-\infty}^{\infty} e^{ip(z-v)}\, dp = \int_{-\infty}^{x} K(x,z)\, dz \int_{-\infty}^{y} K(y,v)\, dv\delta(v-z)$$

$$= \int_{-\infty}^{y} K(x,z)\, K(y,z)\, dz \quad \text{when } x > y$$

$$= \int_{-\infty}^{x} K(x,z)\, K(y,z)\, dz \quad \text{when } y > x$$

$$= \eta(x-y) \int_{-\infty}^{y} K(x,z)\, K(y,z)\, dz + \eta(y-x) \int_{-\infty}^{x} K(x,z)\, K(y,z)\, dz \quad .$$

Hence,

$$\int_{-\infty}^{\infty} \psi(x|p)\, \psi^*(y|p)\, dp = \delta(x-y) + \eta(y-x)\, K(y,x) + \eta(x-y)\, K(x,y)$$

$$+ \eta(x-y) \int_{-\infty}^{y} K(x,z)\, K(y,z)\, dz + \eta(y-x) \int_{-\infty}^{x} K(y,z)\, K(x,z)\, dz \quad . \tag{63}$$

Adding (62) and (63), we obtain

$$\int_{-\infty}^{\infty} \psi(x|p)\, \psi^*(y|p)\, dp + \int_{-\infty}^{\infty} \psi(x|p)\, b(-p)\, \psi^*(y|-p)\, dp$$
$$+ \sum_i \frac{\psi(x|E_i)\, \psi(y|E_i)}{A_i} = \delta(x-y) \quad .$$

This completes the proof of the completeness relationship (58).

A direct consequence of (58) is that any quadratically integrable function f(x) can be expanded in the following manner:

$$f(x) = \int_{-\infty}^{\infty} \psi(x|p)\, a(p)\, dp + \sum_i \psi(x|E_i)\, a_i \quad , \tag{64}$$

where

$$a(p) = \int_{-\infty}^{\infty} f(y)\, \psi^*(y|p)\, dy + \int_{-\infty}^{\infty} f(y)\, b(-p)\, \psi^*(y|-p)\, dy \quad , \tag{65}$$

$$a_i = \int_{-\infty}^{\infty} \frac{f(y)\, \psi(y|E_i)}{A_i}\, dy \quad . \tag{66}$$

Let us treat two special cases:

(a) $f(x) = \psi(x|p)$,

(b) $f(x) = \psi(x|E_i)$.

$\psi(x|p)$ is not quadratically integrable, but it is still symbolically possible to treat case (a).

(a) When $f(x) = \psi(x|p)$, we see at once that in this case we must have

$$a(p) = \delta(p - p') \quad ,$$

$$a_i = 0 \quad .$$

The first of these gives

$$\int_{-\infty}^{\infty} \psi(y|p')\, \psi^*(y|p)\, dy + \int_{-\infty}^{\infty} \psi(y|p')\, b(-p)\, \psi^*(y|-p)\, dy = \delta(p-p') \quad . \tag{67}$$

The second gives

$$\int_{-\infty}^{\infty} \frac{\psi(y|p')\, \psi(y|E_i)}{A_i}\, dy = 0 \quad . \tag{68}$$

Hence, the $\psi(x|p)$ and $\psi(x|E_i)$ are orthogonal.

(b) When $f(x) = \psi(x|E_i)$, we see at once that in this case we must have

$$a(p) = 0 \quad ,$$

$$a_i = \delta_{ij} \quad .$$

From the second of these we obtain

$$\int_{-\infty}^{\infty} \frac{\psi(y|E_j)\,\psi(y|E_i)}{A_i}\,dy = \delta_{ij} \quad ,$$

or

$$\int_{-\infty}^{\infty} \psi(y|E_j)\,\psi(y|E_i)\,dy = A_i\delta_{ij} \quad . \tag{69}$$

Hence, the $\psi(x|E_i)$ are quadratically integrable and have the proper normalization.

We shall now prove that the b(p) as used in the Gel'fand-Levitan algorithm are indeed the reflection coefficients.

Let us consider the outgoing wave, i.e., $\chi(x|p)$ for $p > 0$, and assume that the reflection coefficient is $\hat{b}(p)$ and not b(p). With the help of the completeness relationships for $\psi(x|p)$ and $\chi(x|p)$, we shall show that $\hat{b}(p) \equiv b(p)$.

Since $\chi(x|p)$ and $\psi(x|p)$ are solutions of the same Schrodinger equation for different boundary conditions, we may write $\chi(x|p)$ as a linear combination of the two independent solutions $\psi(x|p)$ and $\psi(x|-p)$. We write

$$\chi(x|p) = \psi(x|p) + \hat{b}(p)\,\psi(x|-p) \qquad \text{for } p > 0 \quad . \tag{70}$$

We have chosen $\psi(x|p)$ so that it satisfies the boundary condition (48), and as a result of this $\chi(x|p)$ as given by (70) will satisfy the correct boundary condition

$$\lim_{x\to-\infty} \chi(x|p) = \frac{e^{ipx}}{\sqrt{2\pi}} + \frac{\hat{b}(p)\,e^{-ipx}}{\sqrt{2\pi}} \qquad \text{for } p > 0 \quad .$$

However, the $\chi(x|p)$ satisfying this boundary condition is unique, and hence $\hat{b}(p)$ is the reflection coefficient as defined previously.

Similarly, uniqueness of a solution for a given boundary condition gives us

$$\chi(x|p) = \hat{t}(p)\,\psi(x|p) \qquad \text{for } p < 0 \quad . \tag{71}$$

$\hat{t}(p)$ is the transmission coefficient, and $|\hat{b}(p)|^2 + |\hat{t}(p)|^2 = 1$ for all p.

Again we choose the proper eigenfunctions $\chi(x|E_i)$ and $\psi(x|E_i)$ to be identical and to have the same normalization A_i. We might point out that since K(x,y) is real, $\psi(x|E_i)$ and $\chi(x|E_i)$ are real.

We now have the completeness relationship (11) for the $\chi(x|p)$ and $\chi(x|E_i)$, namely,

$$\int_{-\infty}^{\infty} \chi(x|p)\,\chi^*(x'|p)\,dp + \sum_i \frac{\chi(x|E_i)\,\chi(x'|E_i)}{A_i} = \delta(x-x') \quad ,$$

and the completeness relationship (58) for the $\psi(x|p)$ and $\psi(x|E_i) = \chi(x|E_i)$, namely,

$$\int_{-\infty}^{\infty} \psi(x|p)\,\psi^*(x'|p)\,dp + \int_{-\infty}^{\infty} \psi(x|p)\,b(-p)\,\psi^*(x'|-p)\,dp + \sum_i \frac{\psi(x|E_i)\,\psi(x'|E_i)}{A_i} = \delta(x-x') \quad .$$

Let us consider (11) and substitute $\chi(x|p)$ from (70) for the range 0 to ∞, $\chi(x|p)$ from (71) for the range $-\infty$ to 0 and $\chi(x|E_i) = \psi(x|E_i)$. Then we obtain

$$\int_0^\infty \psi(x|p)\, \psi^*(x'|p)\, dp + \int_0^\infty \psi(x|p)\, \hat{b}^*(p)\, \psi^*(x'|-p)\, dp$$

$$+ \int_0^\infty \hat{b}(p)\, \psi(x|-p)\, \psi^*(x'|p)\, dp + \int_0^\infty |\hat{b}(p)|^2\, \psi(x|-p)\, \psi^*(x'|-p)\, dp$$

$$+ \int_{-\infty}^0 |\hat{t}(p)|^2\, \psi(x|p)\, \psi^*(x'|p)\, dp + \sum_i \frac{\psi(x|E_i)\, \psi(x'|E_i)}{A_i} = \delta(x - x') \quad . \qquad (72)$$

We note at once that

$$\int_0^\infty |\hat{b}(p)|^2\, \psi(x|-p)\, \psi^*(x'|-p)\, dp = \int_{-\infty}^0 |\hat{b}(p)|^2\, \psi(x|p)\, \psi^*(x'|p)\, dp \quad .$$

Also, since $|\hat{b}(p)|^2 + |\hat{t}(p)|^2 = 1$, we can add the fourth and fifth integrals of (72) and obtain as their sum

$$\int_{-\infty}^0 \psi(x|p)\, \psi^*(x'|p)\, dp \quad .$$

This integral in turn, when added to the first integral of (72), gives for their sum

$$\int_{-\infty}^\infty \psi(x|p)\, \psi^*(x'|p)\, dp \quad .$$

Finally, we note that since $\hat{b}(-p) = \hat{b}^*(p)$,

$$\int_0^\infty \psi(x|p)\, \hat{b}^*(p)\, \psi^*(x'|-p)\, dp + \int_0^\infty \hat{b}(p)\, \psi(x|-p)\, \psi^*(x'|p)\, dp$$

$$= \int_0^\infty \psi(x|p)\, \hat{b}(-p)\, \psi^*(x'|-p)\, dp + \int_{-\infty}^0 \hat{b}(-p)\, \psi(x|p)\, \psi^*(x'|-p)\, dp$$

$$= \int_{-\infty}^\infty \psi(x|p)\, \hat{b}(-p)\, \psi^*(x'|-p)\, dp \quad .$$

With these simplifications, (72) now becomes

$$\int_{-\infty}^\infty \psi(x|p)\, \psi^*(x'|p)\, dp + \int_{-\infty}^\infty \psi(x|p)\, \hat{b}(-p)\, \psi^*(x'|-p)\, dp$$

$$+ \sum_i \frac{\psi(x|E_i)\, \psi(x'|E_i)}{A_i} = \delta(x - x') \quad . \qquad (73)$$

It now appears that, as a consequence of (73), we can expand any quadratically integrable function f(x) in the following manner:

$$f(x) = \int_{-\infty}^\infty \psi(x|p)\, \hat{a}(p)\, dp + \sum_i \psi(x|E_i)\, a_i \quad ,$$

where

$$\hat{a}(p) = \int_{-\infty}^{\infty} f(y)\, \psi^*(y|p)\, dy + \int_{-\infty}^{\infty} f(y)\, \hat{b}(-p)\, \psi^*(y|-p)\, dy \quad ,$$

$$a_i = \int_{-\infty}^{\infty} \frac{f(y)\, \psi(y|E_i)}{A_i}\, dy \quad .$$

However, it follows immediately that

$$\hat{a}(p) \equiv a(p) \quad ,$$

where a(p) is given by (65). Hence, for any quadratically integrable function f(x) we must have for all p

$$[\hat{b}(-p) - b(-p)] \int_{-\infty}^{\infty} f(x)\, \psi(x|p)\, dx \equiv 0 \quad .$$

This can be true only if $b(p) = \hat{b}(p)$ for all p.

V. SOME EXAMPLES OF THE SOLUTION OF INVERSE SCATTERING PROBLEMS

We shall now give some examples of the extreme potency of the Gel'fand-Levitan algorithm and show how it can be used to reconstruct the entire scattering problem from the scattering coefficients.

A. The Delta Function Potential

First let us consider the case for which there are no bound states and

$$b(p) = -i\, \frac{A}{2}\, \frac{e^{-2i\alpha p}}{p + \frac{iA}{2}} \quad . \tag{74}$$

This reflection coefficient is of the proper form, since it satisfies the relations

$$b(-p) = b^*(p) \quad ,$$

$$\lim_{|p|\to\infty} b(p) = 0 \quad \text{in the upper half plane} \quad .$$

In this case, we now have

$$R(x) = -\frac{iA}{4\pi} \int_{-\infty}^{\infty} \frac{e^{-ip(2\alpha + x)}}{p + \frac{iA}{2}}\, dp \quad . \tag{75}$$

Evaluating the integral in the usual fashion by contour integration, i.e., closing the contour in the upper half plane when $x < -2\alpha$, we obtain

$$R(x) = 0 \quad \text{for } x < -2\alpha \quad .$$

Similarly, closing the contour in the lower half plane when $x > -2\alpha$, we obtain

$$R(x) = -\frac{A}{2}\, e^{-\frac{A}{2}(x+2\alpha)} \quad \text{when } x > -2\alpha \quad .$$

Hence,

$$R(x) = -\frac{A}{2}\,\eta(x + 2\alpha)\,e^{-\frac{A}{2}(x+2\alpha)} \quad . \tag{76}$$

The Gel'fand-Levitan equation for $x > y$ is then, for the given example,

$$K(x,y) = \frac{A}{2}\,\eta(x + y + 2\alpha)\,e^{-\frac{A}{2}(x+y+2\alpha)} + \frac{A}{2}\int_{-\infty}^{x} K(x,z)\,\eta(z + y + 2\alpha)\,e^{-\frac{A}{2}(z+y+2\alpha)}\,dz \quad . \tag{77}$$

We note than when

$$x < -\alpha \quad ,$$

$$x + y < -2\alpha \quad ,$$

$$z + y < -2\alpha \quad ,$$

and

$$K(x,y) \equiv 0 \quad ,$$

and hence also

$$V(x) \equiv 0 \quad .$$

Equation (77) suggests that we write

$$K(x,y) = \eta(x + y + 2\alpha)\,g(x,y) \quad . \tag{78}$$

Substituting this expression for $K(x,y)$, we obtain

$$\eta(x + y + 2\alpha)\,g(x,y) = \frac{A}{2}\,\eta(x + y + 2\alpha)\,e^{-\frac{A}{2}(x+y+2\alpha)} + \frac{A}{2}\int_{-\infty}^{x} \eta(x + z + 2\alpha)\,\eta(y + z + 2\alpha)\,g(x,z)\,e^{-\frac{A}{2}(y+z+2\alpha)}\,dz \quad .$$

Consider the product of two step functions. Evidently,

$$\eta(x - a)\,\eta(x - b) = \begin{cases} \eta(x - a) & \text{when } a > b \\ \eta(x - b) & \text{when } b > a \end{cases}$$

$$= \eta(a - b)\,\eta(x - a) + \eta(b - a)\,\eta(x - b) \quad .$$

Hence,

$$\eta(x + z + 2\alpha)\,\eta(y + z + 2\alpha) = \eta(y - x)\,\eta(x + z + 2\alpha) + \eta(x - y)\,\eta(y + z + 2\alpha)$$

$$= \eta(y + z + 2\alpha) \quad ,$$

since $x > y$.

Using this expression for the product of the step functions, we now have

$$\eta(x + y + 2\alpha)\, g(x,y) = \frac{A}{2}\, \eta(x + y + 2\alpha)\, e^{-\frac{A}{2}(x+y+2\alpha)}$$

$$+ \frac{A}{2} \int_{-\infty}^{x} \eta(y + z + 2\alpha)\, g(x,z)\, e^{-\frac{A}{2}(y+z+2\alpha)}$$

$$= \frac{A}{2}\, \eta(x + y + 2\alpha)\, e^{-\frac{A}{2}(x+y+2\alpha)}$$

$$+ \frac{A}{2}\, \eta(x + y + 2\alpha) \int_{-(y+2\alpha)}^{x} g(x,z)\, e^{-\frac{A}{2}(y+z+2\alpha)}\, dz$$

We have already shown that, for $x + y + 2\alpha < 0$, $K(x,y) = 0$. Hence we now treat the case $x + y + 2\alpha > 0$ so that $\eta(x + y + 2\alpha) = 1$. Then

$$g(x,y) = \frac{A}{2}\, e^{-\frac{A}{2}(x+y+2\alpha)} + \frac{A}{2} \int_{-(y+2\alpha)}^{x} g(x,z)\, e^{-\frac{A}{2}(y+z+2\alpha)}\, dz$$

$$= e^{-\frac{A}{2}(y+2\alpha)}\, f(x,y) \quad , \tag{79}$$

where

$$f(x,y) = \frac{A}{2}\, e^{-\frac{A}{2}x} + \frac{A}{2}\, e^{-A\alpha} \int_{-(y+2\alpha)}^{x} e^{-Az}\, f(x,z)\, dz \quad , \tag{80}$$

and

$$\frac{df(x,y)}{dy} = \frac{A}{2}\, e^{A(y+\alpha)}\, f(x, -y - 2\alpha) \quad . \tag{81}$$

We can satisfy (81) by letting

$$f(x,y) = e^{\frac{A}{2}y}\, v(x) \quad ,$$

and substituting in (80) we obtain

$$e^{\frac{A}{2}y}\, v(x) = \frac{A}{2}\, e^{-\frac{A}{2}x} + \frac{A}{2}\, e^{-A\alpha}\, v(x) \int_{-(y+2\alpha)}^{x} e^{-\frac{A}{2}z}\, dz$$

$$= \frac{A}{2}\, e^{-\frac{A}{2}x} - e^{-A\alpha}\, v(x) \left[e^{-\frac{A}{2}x} - e^{\frac{A}{2}(y+2\alpha)} \right] \quad ,$$

which gives

$$v(x) = \frac{A}{2}\, e^{A\alpha} \quad .$$

Hence,

$$f(x,y) = \frac{A}{2}\, e^{\frac{A}{2}(y+2\alpha)} \quad ,$$

and

$$g(x,y) = \frac{A}{2} \quad .$$

Then

$$K(x,y) = \eta(x + y + 2\alpha)\,\frac{A}{2} \quad , \tag{82}$$

and

$$V(x) = 2\,\frac{d}{dx}\,[K(x,x)]$$
$$= A\delta(x + \alpha) \quad . \tag{83}$$

Having determined K(x, y), we can now also find the eigenfunctions $\psi(x|p)$:

$$\psi(x|p) = \frac{e^{ipx}}{\sqrt{2\pi}} + \frac{A}{2\sqrt{2\pi}} \int_{-\infty}^{x} \eta(y + x + 2\alpha)\, e^{ipy}\, dy$$

$$= \frac{e^{ipx}}{\sqrt{2\pi}} + \frac{A}{2\sqrt{2\pi}}\,\eta(x + \alpha) \int_{-(x+2\alpha)}^{x} e^{ipy}\, dy$$

$$= \frac{e^{ipx}}{\sqrt{2\pi}}\left[1 - \frac{Ai}{2p}\,\eta(x + \alpha)\right] + \frac{Ai}{2p}\,\eta(x + \alpha)\, e^{-2ip\alpha}\,\frac{e^{-ipx}}{\sqrt{2\pi}} \quad . \tag{84}$$

From this we see at once that $\psi(x|p)$ satisfies (48), namely,

$$\lim_{x\to-\infty} \psi(x|p) = \frac{e^{ipx}}{\sqrt{2\pi}} \quad \text{for all } p \quad .$$

Also,

$$\frac{d\psi(x|p)}{dx} = \frac{ip\, e^{ipx}}{\sqrt{2\pi}}\left[1 - \frac{Ai}{2p}\,\eta(x + \alpha)\right] - \frac{Ai}{2p}\,\delta(x + \alpha)\,\frac{e^{ipx}}{\sqrt{2\pi}}$$
$$+ \frac{Ai}{2p}\,\delta(x + \alpha)\, e^{-2ip\alpha}\,\frac{e^{-ipx}}{\sqrt{2\pi}} + \frac{A}{2}\,\eta(x + \alpha)\, e^{-2ip\alpha}\,\frac{e^{-ipx}}{\sqrt{2\pi}}$$

$$= \frac{ip\, e^{ipx}}{\sqrt{2\pi}} + \frac{Ai}{2p}\,\frac{\delta(x + \alpha)}{\sqrt{2\pi}}\left(e^{-2ip\alpha}\, e^{-ipx} - e^{ipx}\right)$$
$$+ \frac{A}{2}\,\frac{\eta(x + \alpha)}{\sqrt{2\pi}}\left(e^{ipx} + e^{-2ip\alpha}\, e^{-ipx}\right)$$

$$= \frac{ip\, e^{ipx}}{\sqrt{2\pi}} + \frac{A}{2}\,\frac{\eta(x + \alpha)}{\sqrt{2\pi}}\left(e^{ipx} + e^{-2ip\alpha}\, e^{-ipx}\right) \quad ,$$

$$\frac{d^2\psi(x|p)}{dx^2} = -p^2\,\frac{e^{ipx}}{\sqrt{2\pi}} + \frac{A}{2}\,\frac{\delta(x + \alpha)}{\sqrt{2\pi}}\left(e^{ipx} + e^{-2ip\alpha}\, e^{-ipx}\right)$$
$$+ \frac{ipA}{2}\,\frac{\eta(x + \alpha)}{\sqrt{2\pi}}\left(e^{ipx} - e^{-2ip\alpha}\, e^{-ipx}\right)$$

$$\frac{d^2\psi(x|p)}{dx^2} = -p^2 \frac{e^{ipx}}{\sqrt{2\pi}} + \frac{ipA}{2} \frac{\eta(x+\alpha)}{\sqrt{2\pi}} (e^{ipx} - e^{-2ip\alpha} e^{-ipx}) + \frac{A}{\sqrt{2\pi}} e^{-ip\alpha} .$$

Hence,

$$-\frac{d^2\psi(x|p)}{dx^2} + V(x)\,\psi(x|p) = p^2 \left\{ \frac{e^{ipx}}{\sqrt{2\pi}} [1 - \frac{Ai}{2p} \eta(x+\alpha)] + \frac{Ai}{2p} \eta(x+\alpha)\, e^{-2ip\alpha} \frac{e^{-ipx}}{\sqrt{2\pi}} \right\} - \frac{A}{\sqrt{2\pi}} e^{-ip\alpha}$$

$$+ A\delta(x+\alpha) \left\{ \frac{e^{ipx}}{\sqrt{2\pi}} [1 - \frac{Ai}{2p} \eta(x+\alpha)] + \frac{Ai}{2p} \eta(x+\alpha)\, e^{-2ip\alpha} \frac{e^{-ipx}}{\sqrt{2\pi}} \right\} = p^2\psi(x|p) .$$

$\psi(x|p)$ satisfies the Schrödinger equation:

$$\left[-\frac{d^2}{dx^2} + A\delta(x+\alpha) \right] \psi(x|p) = p^2\psi(x|p) , \tag{85}$$

subject to the boundary condition

$$\lim_{x\to-\infty} \psi(x|p) = \frac{e^{ipx}}{\sqrt{2\pi}} \quad \text{for all } p ,$$

and hence for $p > 0$,

$$\chi(x|p) = \psi(x|p) + b(p)\,\psi(x|-p) \tag{86}$$

is the unique solition which satisfies the same Schrödinger equation subject to the different boundary condition

$$\lim_{x\to-\infty} \chi(x|p) = \frac{e^{ipx}}{\sqrt{2\pi}} + \frac{b(p)\, e^{-ipx}}{\sqrt{2\pi}} .$$

Substituting in (86) from (74) and (84), we have

$$\chi(x|p) = \psi(x|p) + b(p)\,\psi(x|-p)$$

$$= \frac{e^{ipx}}{\sqrt{2\pi}} [1 - \frac{Ai}{2p} \eta(x+\alpha)] + \frac{Ai}{2p} \eta(x+\alpha)\, e^{-2ip\alpha} \frac{e^{-ipx}}{\sqrt{2\pi}}$$

$$- \frac{Ai}{2} \frac{e^{-2ip\alpha}}{(p + \frac{iA}{2})} \frac{e^{-ipx}}{\sqrt{2\pi}} [1 + \frac{Ai}{2p} \eta(x+\alpha)] - \frac{A^2\eta(x+\alpha)\, e^{ipx}}{4p(p + i\frac{A}{2})\sqrt{2\pi}}$$

$$= \frac{e^{ipx}}{\sqrt{2\pi}} \left[1 - \frac{Ai\eta(x+\alpha)}{2p} - \frac{A^2\eta(x+\alpha)}{4p(p + i\frac{A}{2})} \right]$$

$$+ \frac{e^{-ipx}}{\sqrt{2\pi}} \left\{ \frac{Ai}{2p} \eta(x+\alpha) - \frac{Ai}{2} \frac{[1 + \frac{Ai}{2p} \eta(x+\alpha)]}{(p + \frac{iA}{2})} \right\} e^{-2i\alpha p} .$$

Letting $x \to \infty$, we can now also find the transmission coefficient $t(p)$:

$$\lim_{x\to\infty} \chi(x|p) = \frac{e^{ipx}}{\sqrt{2\pi}} \left[1 - \frac{Ai}{2p} - \frac{A^2}{4p(p + \frac{iA}{2})}\right] + \frac{e^{-ipx}}{\sqrt{2\pi}} \left[\frac{Ai}{2p} - \frac{Ai}{2p} \frac{(1 + \frac{Ai}{2p})}{(1 + \frac{Ai}{2p})}\right] e^{-2i\alpha p} ,$$

or

$$\lim_{x\to\infty} \chi(x|p) = t(p) \frac{e^{ipx}}{\sqrt{2\pi}} ,$$

where

$$t(p) = 1 - \frac{Ai}{2p} - \frac{A^2}{2p(2p + iA)} ,$$

or

$$t(p) = \frac{2p}{2p + iA} . \tag{87}$$

Also, as has been shown to be generally true,

$$t(-p) = \frac{-2p}{-2p + iA} = \frac{2p}{2p - iA} = t^*(p) ,$$

and

$$|b(p)|^2 + |t(p)|^2 = \frac{A^2}{4p^2 + A^2} + \frac{4p^2}{4p^2 + A^2} = 1 .$$

We can also find r(p), the reflection coefficient, from the other side by proceeding in the following manner.

For $p < 0$,

$$\lim_{x\to-\infty} \chi(x|p) = s(p) \frac{e^{ipx}}{\sqrt{2\pi}} .$$

It was shown earlier in this report that

$$s(p) = t(p) \quad \text{for all } p .$$

Hence, we write for $p < 0$:

$$\chi(x|p) = s(p)\, \psi(x|p)$$

$$= \frac{2p}{2p - iA} \left\{\frac{e^{ipx}}{\sqrt{2\pi}} [1 - \frac{Ai}{2p} \eta(x + \alpha)] + \frac{Ai}{2p} \eta(x + \alpha)\, e^{-2ip\alpha} \frac{e^{-ipx}}{\sqrt{2\pi}}\right\} .$$

In order to find r(p), we now let $x \to +\infty$ and find

$$\lim_{x\to\infty} \chi(x|p) = \frac{2p}{2p - iA} \left[\frac{e^{ipx}}{\sqrt{2\pi}} (1 - \frac{Ai}{2p}) + \frac{Ai}{2p} e^{-2ip\alpha} \frac{e^{ipx}}{\sqrt{2\pi}}\right]$$

$$= \frac{e^{ipx}}{\sqrt{2\pi}} + \frac{Ai}{2p - Ai} e^{-2ip\alpha} \frac{e^{-ipx}}{\sqrt{2\pi}} .$$

Hence,

$$r(p) = \frac{Ai}{2p - Ai} e^{-2ip\alpha} .$$

We might note that only for $\alpha = 0$ do we have $r(p) = b(p)$ for all p.

Before proceeding to further examples, we emphasize again the extreme potency of the Gel'fand-Levitan algorithm. In the previous example, we have reconstructed the entire scattering problem from a knowledge of the reflection coefficient on one side. We have found the potential function producing this reflection coefficient, the corresponding eigenfunctions of the Schrödinger equation, the transmission coefficient, and the reflection coefficient on the other side.

B. Reflectionless Potential

In this example, we shall use the Gel'fand-Levitan algorithm to construct a reflectionless potential.

We put $b(p) = 0$, and let us also assume that there is but one bound state, E_1, having a normalization A_1. Then

$$R(x) = \frac{e^{\sqrt{-E_1}x}}{A_1} , \tag{88}$$

$$K(x,y) = -\frac{e^{\sqrt{-E_1}(x+y)}}{A_1} - \int_{-\infty}^{x} K(x,z) \frac{e^{\sqrt{-E_1}(z+y)}}{A_1} dz$$

$$= -\frac{e^{\sqrt{-E_1}y}}{A_1} \left\{ e^{\sqrt{-E_1}x} + \int_{-\infty}^{x} K(x,z)\, e^{\sqrt{-E_1}z} \right\} dz . \tag{89}$$

Equation (89) suggests that we write

$$K(x,y) = -\frac{e^{\sqrt{-E_1}y}}{A_1} f(x) , \tag{90}$$

where

$$f(x) = e^{\sqrt{-E_1}x} - \frac{f(x)}{A_1} \int_{-\infty}^{x} e^{2\sqrt{-E_1}z} dz$$

$$= e^{\sqrt{-E_1}x} - \frac{f(x)\, e^{2\sqrt{-E_1}x}}{2A_1\sqrt{-E_1}} , \tag{91}$$

or

$$f(x) = \frac{2A_1\sqrt{-E_1}}{2A\sqrt{-E_1}\, e^{-\sqrt{-E_1}x} + e^{\sqrt{-E_1}x}} , \tag{92}$$

and

$$K(x,y) = \frac{-2\sqrt{-E_1}\, e^{\sqrt{-E_1}y}}{e^{\sqrt{-E_1}x} + 2A\sqrt{-E_1}\, e^{-\sqrt{-E_1}x}}$$

$$= -\sqrt{-E_1}\, e^{\sqrt{-E_1}(y-x_o)} \operatorname{sech}\sqrt{-E_1}\,(x - x_o) \quad , \tag{93}$$

where

$$x_o = \frac{1}{2\sqrt{-E_1}} \ln 2A_1\sqrt{-E_1} \quad . \tag{94}$$

Hence, we obtain a reflectionless potential V(x) given by

$$V(x) = 2\frac{d}{dx}[K(x,x)] = 2E_1 \operatorname{sech}^2\sqrt{-E_1}\,(x - x_o) \quad . \tag{95}$$

Also,

$$\psi(x|p) = \frac{e^{ipx}}{\sqrt{2\pi}} - \frac{\sqrt{-E_1}}{\sqrt{2\pi}} \int_{-\infty}^{x} e^{\sqrt{-E_1}(y-x_o)} \operatorname{sech}\sqrt{-E_1}\,(x - x_o)\, e^{ipy}\, dy$$

$$= \frac{e^{ipx}}{\sqrt{2\pi}} - \sqrt{-E_1}\, \frac{\operatorname{sech}\sqrt{-E_1}\,(x - x_o)\, e^{\sqrt{-E_1}(x-x_o)}\, e^{ipx}}{(\sqrt{-E_1} + ip)\sqrt{2\pi}}$$

$$= \frac{e^{ipx}}{\sqrt{2\pi}} \left[1 - \frac{\sqrt{-E_1}\operatorname{sech}\sqrt{-E_1}\,(x - x_o)\, e^{\sqrt{-E_1}(x-x_o)}}{(\sqrt{-E_1} + ip)}\right] \quad . \tag{96}$$

Since b(p) = 0,

$$\chi(x|p) = \frac{e^{ipx}}{\sqrt{2\pi}} \left[1 - \frac{\sqrt{-E_1}\operatorname{sech}\sqrt{-E_1}\,(x - x_o)\, e^{\sqrt{-E_1}(x-x_o)}}{\sqrt{-E_1} + ip}\right] \quad . \tag{97}$$

In order to determine t(p), we consider

$$\lim_{x\to\infty} \chi(x|p) = \frac{e^{ipx}}{\sqrt{2\pi}} \left[1 - \frac{2\sqrt{-E_1}}{\sqrt{-E_1} + ip}\right] \quad .$$

Hence,

$$t(p) = \frac{p + i\sqrt{-E_1}}{p - i\sqrt{-E_1}} \quad . \tag{98}$$

We may now write for $p < 0$,

$$\chi(x|p) = \frac{e^{ipx}}{\sqrt{2\pi}} \frac{(p - i\sqrt{-E_1})}{(p + i\sqrt{-E_1})} \left[1 - \frac{\sqrt{-E_1}\operatorname{sech}\sqrt{-E_1}\,(x - x_o)\, e^{\sqrt{-E_1}(x-x_o)}}{\sqrt{-E_1} + ip}\right] \quad . \tag{99}$$

Then

$$\lim_{x\to\infty} \chi(x|p) = \frac{e^{ipx}}{\sqrt{2\pi}} \quad ,$$

and hence r(p), the reflection coefficient on the other side, is also zero. That this must be so is apparent at once from the relations

$$|b(p)|^2 + |t(p)|^2 = 1 \qquad \text{for } p > 0 \quad ,$$

$$|r(p)|^2 + |s(p)|^2 = 1 \qquad \text{for } p < 0 \quad ,$$

$$s(p) = t(p) \qquad \text{for all } p \quad .$$

Finally, we might mention that in this case

$$\chi(x|E_1) = \psi(x|E_1) = e^{\sqrt{-E_1}x} - \sqrt{-E_1}\int_{-\infty}^{x} e^{\sqrt{-E_1}(2y-x_o)} \operatorname{sech}\sqrt{-E_1}\,(x - x_o)\,dy$$

$$= e^{\sqrt{-E_1}x} - \frac{\sqrt{-E_1}\operatorname{sech}\sqrt{-E_1}\,(x - x_o)\, e^{\sqrt{-E_1}(2x-x_o)}}{2\sqrt{-E_1}}$$

$$= \frac{e^{\sqrt{-E_1}x}}{2} \operatorname{sech}\sqrt{-E_1}\,(x - x_o) \quad . \tag{100}$$

Again the entire scattering problem has been reconstructed with the help of the Gel'fand-Levitan algorithm.

VI. REFLECTION COEFFICIENTS FOR POTENTIALS WHICH VANISH IDENTICALLY FOR $x < -\alpha$

Unfortunately, the potentials we have been obtaining so far are not of practical interest, since they do extend to infinity and are therefore impossible to construct in actuality.

We shall next give conditions on the reflection coefficient b(p) which will result in potentials which will be identically zero on one side at least.

We first prove the following theorem:

Theorem

$$\text{If } R(x) \equiv 0 \qquad \text{for } x < -2\alpha \quad ,$$

$$\text{Then } V(x) \equiv 0 \qquad \text{for } x < -\alpha \quad .$$

We hope to be able to choose b(p) and the point eigenvalues E_i so that $R(x) \equiv 0$ when $x < -2\alpha$ and thus obtain a corresponding potential V(x) which is identically zero for $x < -\alpha$. To prove the theorem, we use the Gel'fand-Levitan equation for $x > y$:

$$K(x,y) = -R(x + y) - \int_{-\infty}^{x} K(x,z)\,R(y + z)\,dz \quad .$$

If

$$x < -\alpha \quad ,$$

then

$$y < -\alpha \quad ,$$

and

$$x + y < -2\alpha \quad ,$$

hence,

$$R(x + y) \equiv 0 \quad .$$

Also,

$$z < x < -\alpha \quad ,$$

and hence

$$y + z < -2\alpha \quad ,$$

and

$$R(z + y) = 0 \quad .$$

Therefore,

$$K(x,y) \equiv 0 \quad \text{when } x < -\alpha \quad ,$$

and

$$V(x) = 2 \frac{d}{dx} [K(x,x)] \equiv 0 \quad \text{when } x < -\alpha \quad .$$

This completes the proof of the theorem.

Let us now consider b(p) in the complex plane. Suppose b(p) has poles at $p = i\tau_j$ where $\tau_j > 0$. We shall show that, if $E_j = -\tau_j^2$ and A_j is chosen properly, we will be able to make $R(x) \equiv 0$ for $x < -2\alpha$; hence, b(p) will be the reflection coefficient for a potential V(x) which is identically zero for $x < -\alpha$.

Assume that b(p) has the form

$$b(p) = g(p)\, e^{-2i\alpha p} \quad , \tag{101}$$

where

$$\left.\begin{aligned} g(-p) &= g^*(p) && \text{on the real axis} \\ g(p) &= 0(p^{-K+1}) && \text{where } K \geqslant 1 \end{aligned}\right\} \quad , \tag{102}$$

and g(p) has poles of residue r_j at the points

$$p = i\tau_j = i\sqrt{-E_j} > 0 \quad . \tag{103}$$

Then

$$R(x) = \frac{1}{2\pi} \int_{-\infty}^{\infty} g(p)\, e^{-ip(x+2\alpha)}\, dp + \sum_j \frac{e^{\sqrt{-E_j}x}}{A_j} \quad . \tag{104}$$

When $x < -2\alpha$,

$$\frac{1}{2\pi}\int_{-\infty}^{\infty} g(p)\, e^{-ip(x+2\alpha)}\, dp = i \sum_j r_j\, e^{2\alpha\tau_j + \tau_j x}$$

$$= \sum_j i r_j\, e^{2\alpha\tau_j}\, e^{\sqrt{-E_j}\,x} \quad .$$

Hence, if we choose A_j so that

$$\frac{1}{A_j} = -i r_j\, e^{2\alpha\tau_j} \quad , \tag{105}$$

then $R(x) \equiv 0$ for $x < -2\alpha$, and hence $V(x) \equiv 0$ for $x < -\alpha$.

As an example, consider the case where

$$\left.\begin{aligned} b(p) &= \frac{iB}{2}\,\frac{e^{-2i\alpha p}}{p - \frac{iB}{2}} \\ E_1 &= -\frac{B^2}{4} \end{aligned}\right\} \tag{106}$$

In this case,

$$\frac{1}{2\pi}\int_{-\infty}^{\infty} b(p)\, e^{-ipx}\, dp = \frac{iB}{4\pi}\int_{-\infty}^{\infty} \frac{e^{-ip(x+2\alpha)}}{p - \frac{iB}{2}}\, dp$$

$$= -\frac{B}{2}\, e^{B\alpha}\, e^{\frac{B}{2}x} \qquad \text{for } x < -2\alpha \quad .$$

Hence,

$$R(x) = -\frac{B}{2}\, e^{B\alpha}\, e^{\frac{B}{2}x} + \frac{e^{\frac{B}{2}x}}{A_1} \qquad \text{for } x < -2\alpha \quad . \tag{107}$$

If we choose

$$\frac{1}{A_1} = \frac{B}{2}\, e^{B\alpha} \quad , \tag{108}$$

then

$$R(x) \equiv 0 \qquad \text{for } x < -2\alpha \quad .$$

When $x > -2\alpha$, we close the contour used to evaluate the integral in the expression for $R(x)$ in the lower half plane. This contour contains no poles of the integrand, and hence in this case the value of the integral is zero. Then

$$R(x) = \frac{e^{\frac{B}{2}x}}{A_1} = \frac{B}{2}\, e^{\frac{B}{2}(x+2\alpha)} \qquad \text{for } x > -2\alpha \quad . \tag{109}$$

We now have

$$R(x) = \frac{B}{2}\,\eta(x + 2\alpha)\, e^{\frac{B}{2}(x+2\alpha)} \quad , \tag{110}$$

and for $x > y$,

$$K(x,y) = -\frac{B}{2}\,\eta(x + y + 2\alpha)\, e^{\frac{B}{2}(x+y+2\alpha)}$$

$$-\frac{B}{2}\int_{-\infty}^{x} K(x,z)\,\eta(y + z + 2\alpha)\, e^{\frac{B}{2}(y+z+2\alpha)}\, dz \quad , \tag{111}$$

when

$$x < -\alpha \quad ,$$

$$x + y < -2\alpha \quad ,$$

$$z + y < -2\alpha \quad ,$$

and

$$K(x,y) \equiv 0 \quad ,$$

and hence,

$$V(x) \equiv 0 \quad .$$

Equation (111) suggests that we write

$$K(x,y) = \eta(x + y + 2\alpha)\, g(x,y) \quad , \tag{112}$$

and, proceeding as in a previous example given in this report, we obtain

$$g(x,y) = -\frac{B}{2} \quad , \tag{113}$$

$$K(x,y) = -\frac{B}{2}\,\eta(x + y + 2\alpha) \quad , \tag{114}$$

$$V(x) = -B\delta(x + \alpha) \quad . \tag{115}$$

As we expected, this potential is identically zero for $x < -\alpha$. Suppose now that

$$b(p) = \frac{iB}{2}\,\frac{1}{p - \frac{iB}{2}} \tag{116}$$

and that there are no point eigenvalues. We will now not be able to make $R(x) \equiv 0$ for $x < -2\alpha$. Let us see what the corresponding potential will be in this case.

We have

$$R(x) = \begin{cases} -\frac{B}{2}\, e^{\frac{B}{2}x} & \text{for } x < 0 \quad , \\ 0 & \text{for } x > 0 \quad , \end{cases}$$

or

$$R(x) = -\frac{B}{2}\,\eta(-x)\, e^{\frac{B}{2}x} \quad . \tag{117}$$

Hence, for $x > y$,

$$K(x,y) = \frac{B}{2}\,\eta(-x-y)\,e^{\frac{B}{2}(x+y)} + \frac{B}{2}\int_{-\infty}^{x} K(x,z)\,\eta(-y-z)\,e^{\frac{B}{2}(y+z)}\,dz \quad . \tag{118}$$

The solution of this equation may be shown to be

$$K(x,y) = \frac{B\eta(-x)\,e^{\frac{B}{2}(x+y)}}{2-e^{Bx}} + \frac{B}{2}\,\eta(x)\,\eta(x+y) + B\eta(x)\,\eta(-x-y)\,e^{\frac{B}{2}(x+y)} \quad . \tag{119}$$

Then

$$K(x,x) = \frac{B\eta(-x)\,e^{Bx}}{2-e^{Bx}} + \frac{B}{2}\,\eta(x) \quad , \tag{120}$$

and

$$V(x) = 2\,\frac{d}{dx}\,[K(x,x)] = \frac{4B^2\eta(-x)\,e^{Bx}}{(2-e^{Bx})^2} + B\delta(x)\left(\frac{2-3\,e^{Bx}}{2-e^{Bx}}\right)$$

$$= \frac{4B^2\eta(-x)\,e^{Bx}}{(2-e^{Bx})^2} - B\delta(x) \quad . \tag{121}$$

Hence, in this case the potential has an exponential tail for $x < 0$.

Finally, let us consider as an example the case for which

$$\left.\begin{aligned} b(p) &= \frac{-e^{-2i\alpha p}}{(p+i)\,(p-i)} \\ E_1 &= -1 \end{aligned}\right\} \quad . \tag{122}$$

In this case,

$$R(x) = -\frac{1}{2\pi}\int_{-\infty}^{\infty} \frac{e^{-ip(x+2\alpha)}}{(p+i)\,(p-i)}\,dp + \frac{e^{x}}{A_1} \quad . \tag{123}$$

When $x < -2\alpha$, the exponent in the integral is positive, and we close the contour in the upper half plane. Evaluating the integral thus, we obtain for $x < -2\alpha$:

$$-\frac{1}{2\pi}\int_{-\infty}^{\infty} \frac{e^{-ip(x+2\alpha)}}{(p+i)\,(p-i)}\,dp = -i\,\frac{e^{x+2\alpha}}{2i} = -e^{x}\,\frac{e^{2\alpha}}{2} \quad .$$

If we choose

$$\frac{1}{A_1} = \frac{e^{2\alpha}}{2} \quad , \tag{124}$$

then $R(x) \equiv 0$ for $x < -2\alpha$, and from the theorem we expect $V(x) \equiv 0$ for $x < -\alpha$. When $x > -2\alpha$, the exponent in the integral is negative, and we close the contour in the lower half plane. Thus we obtain for $x > -2\alpha$:

$$-\frac{1}{2\pi}\int_{-\infty}^{\infty} \frac{e^{-ip(x+2\alpha)}}{(p+i)\,(p-i)}\,dp = \frac{2\pi i}{2\pi}\,\frac{e^{-x-2\alpha}}{(-2i)} = -e^{-x}\,\frac{e^{-2\alpha}}{2} \quad .$$

Hence,

$$R(x) = -e^{-x}\frac{e^{-2\alpha}}{2} + e^{x}\frac{e^{2\alpha}}{2} = \sinh(x + 2\alpha) \qquad \text{for } x < -2\alpha \quad ,$$

or

$$R(x) = \eta(x + 2\alpha)\sinh(x + 2\alpha) \quad . \tag{125}$$

For $x > y$,

$$K(x,y) = -\eta(x + y + 2\alpha)\sinh(x + y + 2\alpha) - \int_{-\infty}^{x} K(x,z)\,\eta(y + z + 2\alpha)\sinh(y + z + 2\alpha)\,dz \tag{126}$$

We note that when

$$x < -\alpha \quad ,$$

$$x + y < -2\alpha \quad ,$$

$$z + y < -2\alpha \quad ,$$

hence

$$K(x,y) \equiv 0 \quad ,$$

and

$$V(x) \equiv 0 \qquad \text{when } x < -\alpha$$

as we expected. Let us now try to solve the integral equation (126) and find $V(x)$ for $x > -\alpha$. If we write

$$K(x,y) = \eta(x + y + 2\alpha)\,g(x,y) \quad , \tag{127}$$

then

$$\begin{aligned}\eta(x + y + 2\alpha)\,g(x,y) &= -\eta(x + y + 2\alpha)\sinh(x + y + 2\alpha)\\ &\quad - \int_{-\infty}^{x} g(x,z)\,\eta(x + z + 2\alpha)\,\eta(y + z + 2\alpha)\sinh(y + z + 2\alpha)\,dz\\ &= -\eta(x + y + 2\alpha)\sinh(x + y + 2\alpha)\\ &\quad - \int_{-\infty}^{x} g(x,z)\,\eta(y + z + 2\alpha)\sinh(y + z + 2\alpha)\,dz\\ &= -\eta(x + y + 2\alpha)\sinh(x + y + 2\alpha)\\ &\quad - \eta(x + y + 2\alpha)\int_{-(y+2\alpha)}^{x} g(x,z)\sinh(y + z + 2\alpha)\,dz \quad .\end{aligned}$$

Hence,

$$g(x,y) = -\sinh(x + y + 2\alpha) - \int_{-(y+2\alpha)}^{x} g(x,z)\sinh(y + z + 2\alpha)\,dz \quad . \tag{128}$$

Also,

$$\frac{dg(x,y)}{dy} = -\cosh(x+y+2\alpha) - \int_{-(y+2\alpha)}^{x} g(x,z)\cosh(x+y+2\alpha)\,dz \quad ,$$

$$\frac{d^2g(x,y)}{dy^2} = -\sinh(x+y+2\alpha) - \int_{-(y+2\alpha)}^{x} g(x,z)\sinh(x+y+2\alpha)\,dz - g(x,-y-2\alpha) \quad ,$$

or

$$\frac{d^2g(x,y)}{dy^2} = g(x,y) - g(x,-y-2\alpha) \quad . \tag{129}$$

This is satisfied by

$$g(x,y) = v_1(x)\,u(y) + v_2(x) \quad , \tag{130}$$

where u(y) satisfied

$$\frac{d^2u(y)}{dy^2} = u(y) - u(-y-2\alpha) \quad . \tag{131}$$

From (131),

$$u(y) = \sinh\sqrt{2}\,(y+\alpha) \quad , \tag{132}$$

and

$$g(x,y) = v_1(x)\sinh\sqrt{2}\,(y+\alpha) + v_2(x) \quad . \tag{133}$$

Substituting this expression for g(x, y) in the integral equation (128), we obtain

$$\begin{aligned}
v_1(x)\sinh\sqrt{2}\,(y+\alpha) + v_2(x) &= -\sinh(x+y+2\alpha) \\
&\quad - \int_{-(y+2\alpha)}^{x} [v_1(x)\sinh\sqrt{2}\,(z+\alpha) + v_2(x)]\,[\sinh(y+z+2\alpha)]\,dz \\
&= -\sinh(x+y+2\alpha) - v_2(x)\,[\cosh(y+z+2\alpha)]_{-(y+2\alpha)}^{x} \\
&\quad - v_1(x)\left\{\frac{1}{2(1+\sqrt{2})}\sinh[(1+\sqrt{2})\,z + y + (2+\sqrt{2})\,\alpha]\right. \\
&\quad \left. - \frac{1}{2(1-\sqrt{2})}\sinh[(1-\sqrt{2})\,z + y + (2-\sqrt{2})\,\alpha]\right\}_{-(y+2\alpha)}^{x} \\
&= -\sinh(x+y+2\alpha) - v_2(x)\,[\cosh(y+x+2\alpha) - 1] \\
&\quad - v_1(x)\left\{\frac{1}{2(1+\sqrt{2})}\sinh[x+y+2\alpha+\sqrt{2}\,(x+\alpha)]\right. \\
&\quad + \frac{1}{2(1+\sqrt{2})}\sinh\sqrt{2}\,(y+\alpha) \\
&\quad - \frac{1}{2(1-\sqrt{2})}\sinh[x+y+2\alpha-\sqrt{2}\,(x+\alpha)] \\
&\quad \left. + \frac{1}{2(1-\sqrt{2})}\sinh\sqrt{2}\,(y+\alpha)\right\} \quad ,
\end{aligned}$$

or

$$0 = -\sinh(x + y + 2\alpha) - v_2(x)\cosh(x + y + 2\alpha)$$

$$-v_1(x)\left\{\frac{1}{2(1+\sqrt{2})}\sinh[(x + y + 2\alpha) + \sqrt{2}\,(x + \alpha)]\right.$$

$$\left.-\frac{1}{2(1-\sqrt{2})}\sinh[x + y + 2\alpha - \sqrt{2}\,(x + \alpha)]\right\} \quad ,$$

or

$$0 = -e^{x+y+2\alpha} + e^{-(x+y+2\alpha)} - v_2(x)\,[e^{x+y+2\alpha} + e^{-(x+y+2\alpha)}]$$

$$-v_1(x)\left\{\frac{1}{2(1+\sqrt{2})}[e^{x+y+2\alpha+\sqrt{2}(x+\alpha)} - e^{-(x+y+2\alpha)-\sqrt{2}(x+\alpha)}]\right\}$$

$$-\frac{1}{2(1-\sqrt{2})}[e^{x+y+2\alpha-\sqrt{2}(x+\alpha)} - e^{-(x+y+2\alpha)+\sqrt{2}(x+\alpha)}]$$

$$= e^{x+y+2\alpha}\left[-1 - v_2(x) - \frac{v_1(x)\,e^{\sqrt{2}(x+\alpha)}}{2(1+\sqrt{2})} + \frac{v_1(x)\,e^{-\sqrt{2}(x+\alpha)}}{2(1-\sqrt{2})}\right]$$

$$+ \; e^{-(x+y+2\alpha)}\left[1 - v_2(x) + \frac{v_1(x)\,e^{-\sqrt{2}(x+\alpha)}}{2(1+\sqrt{2})} - \frac{v_1(x)\,e^{\sqrt{2}(x+\alpha)}}{2(1-\sqrt{2})}\right] \quad .$$

Hence, $v_1(x)$ and $v_2(x)$ must satisfy

$$\left.\begin{aligned} 1 + v_2(x) + \frac{v_1(x)}{2}\left[\frac{e^{\sqrt{2}(x+\alpha)}}{1+\sqrt{2}} - \frac{e^{-\sqrt{2}(x+\alpha)}}{1-\sqrt{2}}\right] &= 0 \\ 1 - v_2(x) - \frac{v_1(x)}{2}\left[\frac{e^{\sqrt{2}(x+\alpha)}}{1-\sqrt{2}} - \frac{e^{-\sqrt{2}(x+\alpha)}}{1+\sqrt{2}}\right] &= 0 \end{aligned}\right\} \quad . \qquad (134)$$

Therefore,

$$2 + \frac{v_1(x)}{2}\left\{e^{\sqrt{2}(x+\alpha)}\left[\frac{1}{1+\sqrt{2}} - \frac{1}{1-\sqrt{2}}\right] + e^{-\sqrt{2}(x+\alpha)}\left[\frac{1}{1+\sqrt{2}} - \frac{1}{1-\sqrt{2}}\right]\right.$$

$$= 2 + v_1(x)\sqrt{2}\,[e^{\sqrt{2}(x+\alpha)} + e^{-\sqrt{2}(x+\alpha)}] = 0 \quad ,$$

or

$$v_1(x) = -\frac{1}{\sqrt{2}}\,\frac{2}{[e^{\sqrt{2}(x+\alpha)} + e^{-\sqrt{2}(x+\alpha)}]} = -\frac{1}{\sqrt{2}\cosh\sqrt{2}(x + \alpha)} \quad . \qquad (135)$$

Also,

$$1 + v_2(x) + \frac{[e^{\sqrt{2}(x+\alpha)} \quad e^{-\sqrt{2}(x+\alpha)}] \quad \sqrt{2}\,[e^{\sqrt{2}(x+\alpha)} + e^{-\sqrt{2}(x+\alpha)}]}{\sqrt{2}\,[e^{\sqrt{2}(x+\alpha)} + e^{-\sqrt{2}(x+\alpha)}} = 0 \quad ,$$

and

$$v_2(x) = -\frac{1}{\sqrt{2}} \cdot \frac{[e^{\sqrt{2}(x+\alpha)} - e^{-\sqrt{2}(x+\alpha)}]}{[e^{\sqrt{2}(x+\alpha)} + e^{-\sqrt{2}(x+\alpha)}]}$$

$$= -\frac{1}{\sqrt{2}} \frac{\sinh\sqrt{2}\,(x+\alpha)}{\cosh\sqrt{2}\,(x+\alpha)} \quad . \tag{136}$$

Hence,

$$g(x,y) = -\frac{1}{\sqrt{2}} \frac{\sinh\sqrt{2}\,(x+\alpha) + \sinh\sqrt{2}\,(y+\alpha)}{\cosh\sqrt{2}\,(x+\alpha)} \quad , \tag{137}$$

and

$$K(x,y) = -\frac{\eta(x+y+2\alpha)}{\sqrt{2}} \frac{[\sinh\sqrt{2}(x+\alpha) + \sinh\sqrt{2}(y+\alpha)]}{\cosh\sqrt{2}(x+\alpha)} \quad , \tag{138}$$

$$K(x,x) = -\sqrt{2}\,\eta(x+\alpha)\tanh\sqrt{2}\,(x+\alpha) \quad , \tag{139}$$

and

$$V(x) = 2\frac{d}{dx}[K(x,x)]$$

$$= -4\eta(x+\alpha)\,\mathrm{sech}^2\sqrt{2}\,(x+\alpha) \quad . \tag{140}$$

Hence, again we have obtained a potential which is identically zero for $x < -\alpha$.

BIBLIOGRAPHY

A complete bibliography, to the date of publication, of the inverse scattering problem is given by:

L. D. Faddeyev (translated from the Russian by B. Seckler), "The Inverse Problem in the Quantum Theory of Scattering," Math. Phys. 4, 72 (1963).

Papers which we have particularly drawn upon for inverse scattering and the one-dimensional Schrödinger equation are:

1. I. Kay, "On the Determination of a Linear System from the Reflection Coefficients," New York University, Institute of Mathematical Sciences, Division of Electromagnetic Research Report No. EM-74.

2. K. O. Friedrichs, "Spectral Representation of Linear Operators," Lecture Notes, New York University, Institute of Mathematical Sciences (Summer 1948).

3. I. Kay and H. E. Moses, "The Determination of the Scattering Potential from the Spectral Measure Function":

   Part III: "Calculation of the Scattering Potential from the Scattering Operator for the One-Dimensional Schrödinger Equation," Il Nuovo cimento, Series X, 3, 276 (1956).

   Part IV: "Pathological Scattering Problems in One Dimension," Il Nuovo cimento, No. 1 del Supplemento al Vol. 5, Series X, 230 (1957).

4. I. Kay and H. E. Moses, "Reflectionless Transmission Through Dielectrics and Scattering Potentials," J. Appl. Phys. 27, 1503 (1956).

As a reference for electromagnetic propagation, and in particular the mapping of the transmission line equations given in Sec. II, see, e.g., S. A. Schelkunoff, "Remarks Concerning Wave Propagation in Stratified Media," The Theory of Electromagnetic Waves, A Symposium (Interscience, 1951), p. 185.

For the transmission line equations themselves, see Simon Ramo and John R. Whinnery, Fields and Waves in Modern Radio (Wiley, New York; Chapman and Hall, London, 1953), p. 23.